Undergraduate Texts in Mathematics

Editors

S. Axler
F.W. Gehring
K.A. Ribet

T0215031

Springer
New York
Berlin
Heidelberg
Hong Kong
London
Milan
Paris
Tokyo

Undergraduate Texts in Mathematics

(continued after index)

Ronald S. Irving

Integers, Polynomials, and Rings

A Course in Algebra

 Springer

Ronald S. Irving
College of Arts and Sciences
University of Washington
Seattle, WA 98195
USA

Mathematics Subject Classification (2000): 12-01, 13-01

Library of Congress Cataloging-in-Publication Data
Irving, Ronald S., 1952–
    Integers, polynomials, and rings : a course in algebra / Ronald S. Irving.
      p. cm. — (Undergraduate texts in mathematics)
    Includes bibliographical references and index.
    ISBN 0-387-40397-3 (hard cover : acid-free paper) — ISBN 0-387-20172-6 (softcover :
      acid-free paper)
      1. Algebra, Abstract.   I. Title.   II. Series.
    QA162.I58 2003   512—dc22   2003059135

ISBN 0-387-40397-3 (hardcover)   Printed on acid-free paper.
ISBN 0-387-20172-6 (softcover)

Printed in the United States of America.

9 8 7 6 5 4 3 2 1   SPIN 10792243 (hardcover)   SPIN 10947241 (softcover)

Springer-Verlag is a part of *Springer Science+Business Media*

springeronline.com

For Gail

Preface

This book began life as a set of notes that I developed for a course at the University of Washington entitled *Introduction to Modern Algebra for Teachers*. Originally conceived as a text for future secondary-school mathematics teachers, it has developed into a book that could serve well as a text in an undergraduate course in abstract algebra or a course designed as an introduction to higher mathematics.

This book differs from many undergraduate algebra texts in fundamental ways; the reasons lie in the book's origin and the goals I set for the course. The course is a two-quarter sequence required of students intending to fulfill the requirements of the teacher preparation option for our B.A. degree in mathematics, or of the teacher preparation minor. It is required as well of those intending to matriculate in our university's Master's in Teaching program for secondary mathematics teachers. This is the principal course they take involving abstraction and proof, and they come to it with perhaps as little background as a year of calculus and a quarter of linear algebra. The mathematical ability of the students varies widely, as does their level of mathematical interest.

With such an audience, I have chosen to focus less on content and more on the doing of mathematics. Content matters, of course, but as much as a vehicle for mathematical insight as an end in itself. I wish for students to leave the course with an in-depth experience, perhaps their only one, in reading mathematics, speaking about mathematics, listening to others speak about mathematics, and writing mathematics. Surely, we hope that secondary mathematics teachers develop these skills in their own students; I want to ensure that they have the opportunity themselves. The course content becomes the raw material through which the students develop the ability to understand and communicate mathematics. I love algebra. I want my students to love algebra. But I also want them to master such skills as learning what a mathematical statement is, what a mathematical argument or proof is, how to present an argument orally, how to present an argument in writing, how to recognize a correct proof written or spoken by someone else, and how to converse effectively about mathematics.

In order to help my students achieve these goals, I strive to keep my lecturing to a minimum, and this has forced me to write notes that students can rely on as the primary source of the mathematical material. I expect my students to learn in large measure by reading the material in the book, working on the exercises, and, in groups of four or five students, discussing their findings with their colleagues. Through a combination of group discussion, individual attempts at discovering and writing solutions, further reading, and discussions with me and the course teaching assistant, the students ultimately master—to a greater or lesser extent—the material. Their understanding is expressed through their written proofs and oral explanations.

Learning by working in groups is natural in this course for two reasons. First, the intellectual processes of proof and mathematical communication are best learned by practice; no one practices when I lecture. Second, the course is part of the students' preparation to become mathematics teachers; as teachers, they will communicate mathematical ideas to others and listen as others do the same to them. By practicing in class, the students gain an appreciation of the difficulty and the importance of expressing mathematical ideas effectively.

An important facet of the course is writing. I insist that the students write their arguments in well-structured English prose. The teaching assistant and I provide detailed comments to help them learn how to do this. Some of the students are good writers, but they may not have realized previously that they can apply their writing skills to mathematics.

For this approach to work, the material has to be handed to the students in manageable portions, with small gaps between the portions for the student to fill in by doing the exercises, which, in turn, are themselves structured as sequences of questions with even smaller gaps between them. Whenever textbooks that I have used, or my own written materials, leave too large a gap, the students fall into the resulting chasm. For learning to occur, the steps must be the right size. The steps taken by standard undergraduate texts may suit other audiences (though not as many, in my experience, as one might suppose), but they are too large for this one, as I discovered when I first taught the course in the fall of 1996. Within a month, I began to supplement the chosen text with commentary that attempted to fill gaps in certain proofs. Then I began to write my own assignments, interweaving my material with passages in the text. Ultimately, I left the text behind, writing and rewriting assignments repeatedly as I taught the course until they evolved into this book.

One nonstandard feature of the book is that I prove only a few of the theorems in full. Most proofs are left as exercises, and these exercises form the heart of the course. Sometimes, the treatment of a single result is stretched out over several pages, as I ask the student to prove it in a sequence of cases, building up to the general case. An example of this is the treatment of Eisenstein's criterion in Section 11.3. Almost every exercise in which a student is asked to prove a theorem contains a detailed hint or outline of the proof. Indeed, to a mathematically experienced reader, some of my outlines may appear to be complete proofs themselves. Yet, for almost all students, the outlines are far from complete. Unwinding their meaning can be a significant challenge, and the unwinding process serves as the catalyst for learning. Students try to understand what is written; discuss their understandings with each other, the teaching assistant, and me; write drafts of proofs; use class time to show the drafts to each other and us; turn in the proofs; receive comments; and try again.

I have found that the material in this book consistently challenges all the students in the course. The few students at the top of each class ultimately succeed in meeting the challenge, while the large majority of students complete the course with some gaps in understanding. They all leave with a much firmer appreciation of the mathematical enterprise. A few students may fall by the wayside. Perhaps this is just as well. Not everyone is intended to be a secondary mathematics teacher, and if this course provides a few with the opportunity to discover this, it has served a useful purpose.

As for the content of the course, the book has three parts: "Integers," "Polynomials," and "All Together Now." In Part I, some fundamental ideas of algebra are introduced in the concrete context of integers, with rings brought in only in Chapter 6 as a way of organizing some of the ideas. The high point of Part I is Chapter 7, in which Fermat's and Euler's theorems on congruence are proved and RSA encryption is discussed. Part II treats polynomials with coefficients chosen from the integers or various fields. Again, the treatment is concrete at first, but ultimately makes contact with abstract ideas of ring theory. In Part III, the parallels we have seen between rings of integers and rings of polynomials are placed in the broader setting of Euclidean rings, for which some general theorems are proved and applied to the ring of Gaussian integers. The irreducible Gaussian integers are determined, and simultaneously we determine, as Fermat first did, which prime numbers are sums of two integer squares.

The choice of content—rings of integers and rings of polynomials—is a natural one for a course intended for future secondary teachers, who will go on to teach these topics in some form themselves. In the course, the topics are studied more deeply and more abstractly than at the high-school level, especially with the introduction of rings and fields. This provides the students with the opportunity to acquire a more advanced viewpoint on material that is at the core of secondary mathematics. The material can work equally well for a much wider range of mathematics students, and may be well suited for self-study. An important theme in the book is a familiar one in undergraduate algebra courses: Ideas introduced initially for the ring of integers make sense as well for rings of polynomials over fields and more generally still for Euclidean rings. The reader is taken through these settings, then returns from the abstraction of Euclidean rings to the concrete example of the Gaussian integers. There is nothing novel about this choice of topics. Almost every standard junior–senior algebra text includes many of them, perhaps even from the same perspective. Moreover, most standard texts include material on many other topics as well. This one does not. I have not aimed for comprehensiveness. Rather, I have aimed to provide a limited amount of material, with the goal of having students come to grips with, or even master, a small number of serious mathematical ideas.

The most significant omission from this book is group theory. There are a few arguments over the course of the book that might lend themselves to a group-theoretic perspective, in particular, those having to do with the group of units of a ring, but my sense is that introducing groups in this setting would do an injustice to the subject. I prefer to introduce groups in a geometric context, as symmetry groups of regular polygons and regular polyhedra, for example. Such a treatment did not seem to fit comfortably into this book, so I have chosen to omit groups altogether. The result is a book lacking the flexibility to be a comprehensive introduction to abstract algebra, but such a book is not what I have aimed to write. On this point, I wish to acknowledge my debt to Lindsay Childs, whose *A Concrete Introduction to Higher Algebra* provided me with the model of an algebra text that has integers and polynomials as its focus and yet is full of beautiful mathematics suitable for an undergraduate course.

The book contains more material than I can cover in our two-quarter course. The best of mathematics majors might be able to move quickly, but they may also be better served by a more traditional groups–rings–fields-style undergraduate algebra course. I do not aim to move quickly, and moreover, since I intend for much of the learning to be done by the students without lecture, I must allow them extra time. Let me describe what I envision as the core of the course, and what I might choose to cover myself within the constraints of two quarters.

In the first quarter, I would do all of Chapters 1 to 7. This is essentially a self-contained course on integers, induction, and some topics in elementary number theory, culminating in a description of RSA encryption. I would not have time for Chapter 8, and would leave this as recommended reading. I introduce the inclusion–exclusion principle in Chapter 7 and use it to provide a means of calculating the Euler phi function. The treatment of binomial coefficients in Chapter 8 allows us to give a proof of inclusion–exclusion, but the ideas are not used again later in the book, and this topic can be safely omitted. In a two-semester course, I would happily include Chapter 8 in the first semester.

The core of the second quarter is Chapters 9 through 16, with Chapter 10 done lightly and with Chapter 17 left as supplementary reading. Chapter 10 contains an extended treatment of cubic and quartic polynomials with real coefficients. This represents a bit of a detour from the main flow of the book. I would certainly cover Sections 10.1 and 10.2, since this allows the student to put the quadratic formula in a broader context. For the same reason, I might wish to cover Sections 10.4 and 10.7 as well. In a one-year course, I would cover the entire chapter. The analysis of cubic and quartic polynomials is one of the great stories in the history of algebra, and represents the most important mathematics done in the sixteenth century. Every student of algebra should study this, especially future secondary teachers. Indeed, I have twice taught a course for preservice and in-service teachers on this material, and I have incorporated some of my notes for that course in Chapter 10. But, again, the chapter in its entirety takes us away from the flow of the book, since the main goal of Part II is for the student to see polynomials with coefficients in a fixed field as a natural analogue of integers. By Chapter 12, this idea should be apparent. Part II concludes, in Chapter 14, with the construction of the polynomial analogue of the ring of integers modulo a prime number, and this construction allows us to obtain the roots of a nonconstant polynomial f with coefficients in a field F through the construction of an extension field K of F that contains those roots. In Chapter 15, our work with rings of integers and rings of polynomials is placed in a broader and more abstract setting, applicable to other rings, and in Chapter 16, we focus on one such ring, the ring of Gaussian integers. Chapter 17 treats topics that have been considered throughout the book, but in less than the usual detail and without complete proofs. In a one-year course, I would include this material, and might even wish to say more about the missing proofs.

I do not have any experience with paring the material down to an amount suitable for a one-semester course. If I were to do so, I might cover Chapters 1 through 6, the first four sections of Chapter 9, the first two sections of Chapter 10, and Chapter 12.

Acknowledgments

I learned algebra from Michael Artin, my Ph.D. advisor, a great educator and great mathematician. In his undergraduate algebra course at M.I.T. and the book he lovingly developed from the course, he has introduced thousands of students to the beauty of algebra.

The Department of Mathematics at the University of Washington has provided me with a wonderful home for mathematics and the teaching of mathematics. I am grateful to many colleagues. I will mention just a few. I had the great privilege to work with and learn from Robert B. Warfield, Jr., in the years before his untimely death. He did everything well and with passion: research, teaching, department and university service, and above all, his role as husband and father. From him I learned that it is possible to teach all one's courses, from business algebra for five hundred to graduate algebra for five, with an eye toward active engagement of the students. My colleague Steve Monk offered our algebra course for teaching majors for many years before I did, setting the example I took as my starting point. He used a standard undergraduate algebra text, but he emphasized student understanding of what it means to do mathematics. David Collingwood, James Morrow, and John Sullivan have inspired me as well, with their passion for teaching and for students. I have been fortunate to have colleagues willing to use portions of my notes in their own teaching: Ed Curtis, Julianne Harris, John Sullivan, Monty McGovern, and John Palmieri. I have also been fortunate to work with three extraordinary teaching assistants: Adam Nyman, Rebekah Hahn, and Joan Lind. And I have been the continuing beneficiary of Brooke Miller's wise counsel on students and much else.

As I have taught my students, I have learned from them how to teach mathematics and how to write mathematics. I am enormously in their debt. Among the many students who have inspired me are Christine Fraser, Melissa Johnson, Joy Scott, Leonard Barnett, Maria Cason, Leslie Chen, Cutts Peaslee, Lisa Behmer, and of course, Susan Sturms, my friend and constant prod.

That my course notes, idiosyncratic as they were, might be publishable as a book is an idea for which I must thank Ina Lindemann, of Springer-Verlag. Her appreciation of my vision, her encouragement, and her patience made this book a reality. But her greatest gift to me was David Kramer. Two years ago, shortly after she and I agreed to go ahead with publication of this book, I assumed administrative duties that took me away from teaching and from the book. I had hoped to give it a line-by-line reading in order to improve it both locally and globally and to add an index. However, this work was not getting done, and moreover, it might have been better done with the help of another reader. A year ago, I proposed to Ina that we might find such a reader. Ina granted my wish, and David appeared on my (electronic) doorstep. He is the reader of my dreams. We have collaborated off and on over the past eight months, finding no mathematical, grammatical, or stylistic detail too small for lively and stimulating debate. We have had fun, and we have made a better book. I could not list the ways, large and small, in which David has improved it, from sharpening an argument here to adding a problem there, but let me note at least that his amplification of certain passages has brought into the text his special warmth, wit, and humor. Working with David has been an unexpected and joyous privilege.

My first teachers of mathematics, and of much more, were my parents. They teach me still, about how to live a mentally and emotionally vibrant life well into one's ninth decade. I have also benefited from decades of support given to me by Annie McNab and my siblings, Jeff and Gail.

From my wife, Gail, I have learned much of what I know about grace, generosity, and love. She is a gifted teacher. If only I were a better student. Jessica and Joel have been the longest-suffering of my many students. Joel, you can't tease me anymore about never being able to finish a book. Here it is.

Seattle, Washington
June 30, 2003

Contents

Part III All Together Now

1

Introduction: The McNugget Problem

Mathematics is often regarded as the study of calculation, but in fact, mathematics is much more. It combines creativity and logic in order to arrive at abstract truths. This introductory chapter is intended to illustrate how calculation, creativity, and logic can be combined to solve a particular problem. We will see that calculation alone is not enough. Needed in addition are an idea and an argument. The idea helps us decide what calculations are needed, and the argument demonstrates that the proposed solution to the problem is correct. The problem we will study is a practical one, but it can be translated into a question about numbers and arithmetic. The ideas that underlie its solution will play an important role throughout this book.

Problem 1.1. McDonald's sells Chicken McNuggets in boxes of 6, 9, and 20. By buying various quantities of each box size, you can buy many different amounts of Chicken McNuggets. However, you cannot buy every possible amount. For instance, you cannot buy 5 McNuggets, or 8 McNuggets. What other amounts of Chicken McNuggets is it not possible for you to buy? Is there a largest such amount, and if so, what is it?

Let us restate the problem in a little more detail. Recall that the *positive integers* are the numbers of the form 1, 2, 3, and so on; the *negative integers* are the numbers -1, -2, -3, and so on; and the *integers* are the positive integers, the negative integers, and 0. We will also use the phrase *nonnegative integers* to refer to the collection of positive integers and 0. Here is our restatement:

Problem 1.2. McDonald's sells Chicken McNuggets in boxes of 6, 9, and 20. Find all the positive integers that represent amounts of Chicken McNuggets that you cannot buy. Furthermore, if there is a largest such amount, find a positive integer N with the following two properties:

1. You cannot buy N Chicken McNuggets by ordering a suitable number of boxes of 6, 9, or 20 nuggets.
2. For every integer n with $n > N$, you can buy n Chicken McNuggets by ordering a suitable number of boxes of 6, 9, or 20 nuggets.

To solve this problem, we might try working our way through the positive integers one by one, in increasing order. Clearly, we cannot buy 1, 2, 3, 4, or 5 Chicken McNuggets, but we can buy 6. Continuing, we see that we cannot buy 7 or 8, we can buy 9, we cannot buy 10 or 11, and we can buy 12. This is easy enough, and we can continue on for a while.

Exercise 1.1. Do continue on, up to $n = 49$, determining for each n whether you can buy n Chicken McNuggets.

Does this help? The answer depends on our goal. If our plan is to keep doing this until we get an answer, without applying any thought, then we will never succeed, since calculating up to a particular positive integer does not tell us what the results of the calculations above that integer will be, and we certainly cannot work our way through every possible positive integer. As valuable as calculation is, it can produce only a finite amount of data, and what we seem to need here is an infinite amount of data.

In order to get by with a finite amount of data, we need to complement the data with two items. First, we need an idea. Second, armed with this idea, we need an argument to show that we can buy n Chicken McNuggets for all n above a particular integer N, if, in fact, such an integer N exists. Conceivably, there may be infinitely many integers n with the property that you cannot buy n nuggets by ordering a suitable number of boxes of 6, 9, or 20. As interesting as the problem of finding the desired number N is the problem of showing that there is such an N, that is, that we can buy every possible amount of nuggets beyond a certain point. With a good idea, we may be able to get away with making only a finite number of calculations.

Exercise 1.2. Think of an idea that lets you solve the Chicken McNugget problem, using your data from 1 to 49. Armed with your idea, solve the problem.

Have you succeeded? If so, great. If not, let us analyze the problem more closely. Let us work with easier versions of the Chicken McNugget problem, returning to the original problem after we get some insights. We can simplify the problem in two ways: (a) by using nugget boxes with fewer nuggets, and (b) by allowing only two different box sizes. For example, suppose the Chicken McNugget boxes come in two sizes, the smaller size having 2 nuggets and the larger size having 3. What are the possibilities for the number of nuggets we can buy?

This simpler problem is easy to solve. Since we can buy as many boxes of 2 as we wish, we can buy 2 or 4 or 6 or 8 or 10 or *any* even number of nuggets. That leaves only odd numbers to worry about. We can buy 3 nuggets by choosing a box of 3. We can add to this a box of 2 in order to buy 5, and then another box of 2 in order to buy 7, and then additional boxes of 2 in order to buy 9 or 11 or 13 or any odd number n of nuggets, with $n \geq 3$. We see that we can buy any number of nuggets we please except 1. Thus N in this case is 1: We cannot buy 1 nugget, but we can buy n nuggets for $n > 1$.

This solution did not depend on calculation alone; there was an idea. The solution is so simple that one may not realize that an idea is actually involved. The idea is to divide the positive integers into two families, the even integers and the odd integers, and to treat each separately. Since successive integers in each family differ by 2, and since we can always buy 2 nuggets, once we can buy a certain number of nuggets in the family, we can buy all the larger sizes of nuggets in that family.

Suppose next that Chicken McNuggets come in boxes of sizes 2 and 4. How many nuggets can we buy? This problem is easier still, but its solution has a different form. The boxes of 4 are not needed, for we can use two boxes of 2 to buy 4. Thus whatever we can do with boxes of 2 and 4 we can do with boxes of 2 alone. What can we do with boxes of 2? We can buy 2, 4, 6, 8, or any other even number of nuggets, but we cannot buy any odd number of nuggets. Thus, in contrast to the previous example, there is no N such that for every $n > N$ we can buy n nuggets. Instead, there are infinitely many positive integers n for which it is impossible to buy n nuggets, and those are the odd n's.

Before proceeding with more examples, let us introduce some terminology that will free us from the discussion of nuggets and turn this into a purely mathematical problem. For positive integers a and b, let us say that another positive integer n is (a, b)-*accessible* if n can be obtained by adding some combination of a's and b's together. More precisely, n is (a, b)-accessible if there exist nonnegative integers r and s such that $n = ra + sb$. We allow r or s to be 0 as well as positive; this is why we say that r and s can be nonnegative integers. We will say that n is (a, b)-*inaccessible* if in contrast, n is not (a, b)-accessible. Similarly, if we take a third positive integer c, the positive integer n is (a, b, c)-accessible if there exist nonnegative integers r, s, and t such that $n = ra + sb + tc$, and n is (a, b, c)-inaccessible if it is not (a, b, c)-accessible.

Using the new terminology, we have shown that the $(2, 3)$-accessible positive integers are the integers greater than 1 and that the $(2, 4)$-accessible positive integers are the even integers. This terminology allows us to restate the Chicken McNugget problem:

Problem 1.3. Find the positive integers that are $(6, 9, 20)$-accessible. Determine whether there is a largest $(6, 9, 20)$-inaccessible positive integer N, above which all the positive integers are $(6, 9, 20)$-accessible, and if so, determine N.

Here are some problems to solve, expressed using the terminology of accessibility.

Exercise 1.3. Use the idea of dividing positive integers into the even and the odd families to solve the following problems:

1. Find the $(2, 5)$-accessible and the $(2, 7)$-accessible positive integers.
2. For m an arbitrary positive odd integer, find the $(2, m)$-accessible positive integers.

3. For m an arbitrary positive even integer, find the $(2, m)$-accessible positive integers.

The next-simplest accessibility problem to analyze is that of finding the $(3, 4)$-accessible positive integers. There are several ways to proceed. One way is to review what we have done for $(2, m)$-accessibility, breaking the integers up into two families, even and odd. What analogue exists here? We can let 3 play the role that 2 played before, leading to the family of positive integers divisible by 3 and the family of positive integers not divisible by 3. The positive integers divisible by 3 are certainly $(3, 4)$-accessible. To study those that are not $(3, 4)$-accessible, we can refine our analysis further.

Think of the odd integers as those with a remainder of 1 when we divide by 2 and the even integers as those with remainder 0 when we divide by 2. Similarly, the positive integers fall into three families with respect to 3: those with a remainder of 1 when we divide by 3, those with a remainder of 2 when we divide by 3, and those with a remainder of 0 when we divide by 3. This produces three lists of positive integers:

$$1, 4, 7, 10, 13, \ldots \ ;$$
$$2, 5, 8, 11, 14, \ldots \ ;$$
$$3, 6, 9, 12, 15, \ldots \ .$$

Let us examine $(3, 4)$-accessibility for the numbers in each of these lists. Every number in the third list is $(3, 4)$-accessible, of course. In the first list, 1 is not $(3, 4)$-accessible, but 4 is. Every number above 4 on the first list is obtained by adding some number of 3's to 4, so every number above 4 in the first list is also $(3, 4)$-accessible. In the second list, it is easy to see that 2 and 5 are not $(3, 4)$-accessible, but 8 is. Every number above 8 in the second list is obtained by adding some number of 3's to 8, so every number above 8 in the second list is also $(3, 4)$-accessible. We have taken care of all the positive integers, finding that the only ones that are not $(3, 4)$-accessible are 1, 2, and 5. Again, the key to the solution is an idea, in this case the idea of breaking the positive integers into three families according to their remainders when divided by 3.

Exercise 1.4. Use the idea of breaking the positive integers into three families according to their remainders when divided by 3 to solve the following problems:

1. If m is divisible by 3, what are the $(3, m)$-accessible positive integers?
2. Find the $(3, 5)$-accessible positive integers, the $(3, 7)$-accessible positive integers, and the $(3, 8)$-accessible positive integers.
3. Suppose m is a positive integer not divisible by 3. Using the preceding examples, guess what the largest N might be such that N is $(3, m)$-inaccessible but every integer $n > N$ is $(3, m)$-accessible.
4. Make an argument showing that your guess is correct.

We are ready to return to the Chicken McNugget problem. Below is an outline of an approach to solving the problem. You may wish to try to come up with your own solution before reading the outline.

Exercise 1.5. Analyze $(6, 9, 20)$-accessibility.

1. We have broken the positive integers into two families, based on remainders when divided by 2, and three families, based on remainders when divided by 3. Observe that the positive integers can also be broken into six families, depending on what their remainders are when divided by 6.
2. Write down the first few positive integers in each of the six families. For instance, one family begins with

$$1, 7, 13, 19, 25, 31, 37, 43, 49.$$

3. For each of the six lists, find the first number in the list that is $(6, 9, 20)$-accessible.
4. Explain why every number after that number in the list is also $(6, 9, 20)$-accessible.
5. Produce a complete list of the $(6, 9, 20)$-inaccessible positive integers.
6. Using the list, deduce that there is a largest integer that is $(6, 9, 20)$-inaccessible and find the positive integers that are $(6, 9, 20)$-accessible.
7. What ideas went into this solution? How did these ideas allow you to reduce the calculations necessary to a rather small amount?

There is a way to buy only 4 Chicken McNuggets. To do so, you must buy a Chicken McNugget Happy Meal. In addition to a container of four nuggets, you will get fries, a drink, and a toy. If all you want is 4 nuggets, you pay a premium in buying the meal. It would be cheaper to buy six nuggets and throw two away. Nonetheless, you can buy 4 nuggets, as an alternative to 6, 9, and 20. Let us redo the Chicken McNugget problem with this new information.

Exercise 1.6. Determine the positive integers that are $(4, 6, 9, 20)$-accessible and the positive integers that are $(4, 6, 9, 20)$-inaccessible.

1. Notice first that 20 is irrelevant (why?), so that you are really determining the $(4, 6, 9)$-accessible and $(4, 6, 9)$-inaccessible positive integers.
2. Determine these integers.
3. Assuming that you are willing to buy Happy Meals, what is the largest number of Chicken McNuggets that it is not possible for you to buy at McDonald's?

Part I

Integers

Induction and the Division Theorem

2.1 The Method of Induction

In the Introduction we discussed a mathematical problem whose solution required the verification of an infinite family of statements. We needed to show, for each integer $n > 43$, that by ordering a suitable number of Chicken McNugget boxes of size 6, 9, or 20 we can buy n Chicken McNuggets. Equivalently, we needed to show that each integer $n > 43$ is $(6, 9, 20)$-accessible. This is an infinite family of statements: 44 is $(6, 9, 20)$-accessible, 45 is $(6, 9, 20)$-accessible, 46 is $(6, 9, 20)$-accessible, 47 is $(6, 9, 20)$-accessible, and so on.

The Chicken McNugget problem is typical of a number of mathematical problems in that the solution consists of an infinite family of statements, indexed by the integers n larger than some fixed integer N. An important means of handling such statements is the *principle of mathematical induction*. We did not discuss this principle explicitly in our treatment of the Chicken McNugget problem, but it was used implicitly, as will become evident. Let us introduce additional problems whose solution can be regarded as an infinite family of statements, and see by example how induction is used.

We begin with the problem of adding all the integers from 1 to a given positive integer n. This problem figures in a familiar story about the childhood of Carl Friedrich Gauss, one of the greatest mathematicians in history. Gauss lived from 1777 to 1855. If you have studied physics, you may be familiar with his name as the unit of measurement of magnetic field strength. In the story, which may be apocryphal, a schoolteacher asked Gauss's class one day to add all the integers from 1 to 100. As the story goes, the teacher did this not because of the mathematical interest of the problem but in order to keep the students busy. To the teacher's astonishment, Gauss produced the correct answer almost instantly.

Gauss may have answered the question by performing addition quickly, but more likely he knew a formula for the sum of all the numbers from 1 to a positive integer n. Let us obtain such a formula ourselves, so we too can answer the teacher's question instantly.

Exercise 2.1. We wish to obtain, for each positive integer n, a formula for the sum of all the numbers from 1 to n. Let us begin by considering a different question that turns out to be simpler. Recall that an integer is *even* if it is divisible by 2 and *odd* otherwise. Thus an even integer can be written in the form $2n$ for some integer n, but an odd integer cannot.

1. Calculate the sums of the first few positive odd integers: 1, $1 + 3$, $1 + 3 + 5$, $1 + 3 + 5 + 7$. Do you recognize the results as familiar numbers?
2. Guess a formula for the sum $1 + 3 + 5 + \cdots + (2n - 1)$ of the first n positive odd integers.
3. Now draw the following sequence of pictures: a single dot, a 2×2 array of four dots, a 3×3 array of 9 dots, and a 4×4 array of 16 dots. How many dots must you add to each array to obtain the next larger array?
4. Using these arrays, explain how each square is built by a sequence of odd numbers, and how this explains the formula that you have guessed.

Exercise 2.2. Suppose n is a positive integer.

1. What can you do to each of the numbers in the sum

$$1 + 3 + 5 + \cdots + (2n - 1)$$

in order to obtain the sum

$$2 + 4 + 6 + \cdots + 2n$$

of the first n positive even integers.
2. Using this idea, and your formula for the sum of the odd integers from 1 to $2n - 1$, obtain a formula for the sum $2 + 4 + 6 + \cdots + 2n$ of the even integers from 2 to $2n$.
3. Using the formula you just obtained for the sum of the first n even integers, perform a simple division to obtain a formula for the sum

$$1 + 2 + 3 + \cdots + n$$

of all the positive integers from 1 to n.
4. Using this formula, calculate Gauss's sum $1 + 2 + \cdots + 100$.

We have succeeded in finding a formula for the sum of the first n positive integers for every positive integer n. However, our approach required a bit of cleverness, and the argument in the last part of Exercise 2.1 is not entirely satisfactory.

Let us develop the notion of mathematical induction and then return to the sum formula. The technique of induction comes in handy when we have an infinite family of statements we wish to prove, one for each positive integer n. For instance, in the example above, the nth statement would say that the sum of the integers from 1 to n is $(n^2 + n)/2$; that is, the nth statement is the equality

$$1 + 2 + 3 + \cdots + n = \frac{n^2 + n}{2}.$$

The principle of mathematical induction gives us a way to proceed. We can think of it as a technique for climbing a *stairway to heaven*. Suppose there is such a stairway, starting on the ground. Let us label the ground level 0, and let us give each step above the ground a label as well, the first step being level 1, the second level 2, and so on. By climbing all the way to the top, if we can, we will reach heaven. To do so, we need a technique for *getting off the ground* and a technique for *continuing*:

1. First we want a technique that allows us to climb from the ground onto step 1. This is what is meant by "getting off the ground." Getting from the ground onto the first step may be easy, or it may not, depending on how high the step is.
2. Second, we want a technique that allows us to go from each step k to the next step $k + 1$. This is what is meant by "continuing."

Suppose we have these two techniques. Employing the first one, we can get onto step 1. Employing the second, we can get from step 1 to step 2. Employing the second technique again, we can get from step 2 to step 3. Employing it yet again, we can get from step 3 to step 4, and so on. Repeated use of the second technique should allow us to climb all the way to the top. The principle of mathematical induction states that this is true: If we have a technique for getting off the ground and a technique for continuing, we can climb all the way up the stairway and reach heaven.

Let us state this a bit more formally. Suppose we have a family of statements that we wish to prove, one for every positive integer n. Let us call the nth statement Statement(n). The principle of induction says that we can prove all these statements at once if we do the following:

1. First prove Statement(1).
2. Second, show that for every positive integer k, if Statement(k) holds, then so does Statement($k+1$). In doing this, we do not assume that we actually know that Statement(k) holds. We only *assume* it as a hypothesis in our attempt to prove Statement($k + 1$). Continuing the staircase metaphor, we do not assume that we can actually ever reach any of the particular steps $1, 2, 3, \ldots$. We only show that *if* we ever find ourselves on step k, *then* we can get to step $k + 1$.

Suppose we have performed both of these actions. From the first action, we know that Statement(1) is true. From the second action, it follows that Statement(2) is true. Since Statement(2) is true, it follows using the second action again that Statement(3) is true. Since Statement(3) is true, it follows using the second action yet again that Statement(4) is true. By continuing in this way, we can conclude that Statement(n) is true for every positive integer n. This is the method of mathematical induction.

It is important to realize that action one is as important as action two. We may be able to climb from any step of the stairway to the next, but if we cannot get onto the first step, our ability to climb the remaining steps is useless.

Let us use induction in some examples. First, we return to the problem of showing that

$$1 + 3 + 5 + \cdots + (2n - 1) = n^2.$$

We had a geometric argument earlier, based on counting dots in square arrays, that was somewhat vague. In particular, it depended too much on intuition. Using induction, one could instead proceed as follows.

Theorem 2.1. *Let n be a positive integer. Then*

$$1 + 3 + 5 + \cdots + (2n - 1) = n^2.$$

Proof. We have an infinite family of statements we wish to prove, one for each positive integer n. Statement 1 says that $1 = 1^2$, Statement 2 says that $1 + 3 = 2^2$, and so on. We wish to climb the stairway that says that these infinitely many statements are all true. To get off the ground, we must climb the first step, which means we must show that $1 = 1^2$. This is certainly true, and this obvious equality gets us on the first step successfully, with very little effort.

Suppose we find ourselves on step k. We wish to show that we can get to step $k + 1$. To be standing on step k means that the equality of the theorem is true for $n = k$:

$$1 + 3 + 5 + \cdots + (2k - 1) = k^2.$$

In saying that we assume ourselves to be standing on step k we are saying that we can assume that statement k is true. What we must show next is that *under this assumption*, we can get to step $k + 1$. This means that we must show that the equality of the theorem holds for $n = k + 1$, *on the assumption that it holds for $n = k$.* The equality for $n = k + 1$ takes the form

$$1 + 3 + 5 + \cdots + (2(k + 1) - 1) = (k + 1)^2,$$

which we obtain by substituting $k + 1$ for n.

Since $2(k + 1) - 1 = 2k + 1$, we can rewrite the last equality as

$$1 + 3 + 5 + \cdots + (2k + 1) = (k + 1)^2.$$

To review, we are *assuming* that the equality

$$1 + 3 + 5 + \cdots + (2k - 1) = k^2$$

is true, and we wish to show that the equality

$$1 + 3 + 5 + \cdots + (2k + 1) = (k + 1)^2$$

is true on that assumption.

One way to proceed is to take the equality that we are assuming to be true and add the odd number $2k + 1$ to both sides. We then get

$$1 + 3 + 5 + \cdots + (2k - 1) + (2k + 1) = k^2 + (2k + 1).$$

But $k^2 + 2k + 1 = (k + 1)^2$. Therefore,

$$1 + 3 + 5 + \cdots + (2k - 1) + (2k + 1) = (k + 1)^2,$$

which is exactly what we needed to prove. Thus we have shown that if the equality of the theorem holds when $n = k$, then it holds when $n = k + 1$; if we can get to step k, we can get to step $k + 1$. The principle of induction allows us to conclude that we have verified our desired statement for all values of n. We have climbed the stairway to heaven, and the proof is complete.

Exercise 2.3. Using the principle of mathematical induction, as above, prove that the following equalities hold for every positive integer n.

1. The sum of the first n integers:

$$1 + 2 + 3 + \cdots + n = \frac{n(n + 1)}{2}$$

2. The sum of the squares of the first n integers:

$$1^2 + 2^2 + 3^2 + \cdots + n^2 = \frac{n(n + 1)(2n + 1)}{6}.$$

Sometimes a family of statements we wish to prove true is indexed by a sequence of integers that starts not with 1 but with another integer, such as 0 or 2 or 3. The principle of induction still works, with a slight modification. We first show that the lowest-numbered statement holds, and then show that if Statement(k) holds, then Statement($k + 1$) holds.

Here is an example. Recall that for a positive integer n the symbol $n!$ (pronounced n *factorial*) represents the product

$$n \times (n - 1) \times (n - 2) \times \cdots \times 3 \times 2 \times 1.$$

For instance, 4!, or "four factorial," is the product $4 \times 3 \times 2 \times 1$, which equals 24. Let us prove the following result by induction. Notice that the inequality of the theorem is false if n equals 1, 2, or 3.

Theorem 2.2. *For every integer* $n \geq 4$, *the inequality* $n! > 2^n$ *holds.*

Proof. We will proceed in two stages, by induction. First we show that the desired inequality $n! > 2^n$ holds for $n = 4$. This is clear, since it is simply the statement that $24 > 16$, which is obviously true. We have gotten off the ground onto step number 4. For the second stage, suppose that for some $k \geq 4$, the inequality $k! > 2^k$ holds. We wish to show that $(k+1)! > 2^{k+1}$. Since we are assuming that $k! > 2^k$ and since it is evident that $(k+1)! = (k+1) \times k!$, we may conclude that

$$(k+1)! = (k+1) \times k! > (k+1) \times 2^k.$$

But since k is at least 4, the number $k+1$ is certainly greater than 2. Hence

$$(k+1) \times 2^k > 2 \times 2^k = 2^{k+1}.$$

Stringing the two sequences of equalities and inequalities together, we conclude that $(k+1)! > 2^{k+1}$, as desired. By the principle of induction, we can conclude that for every integer $n \geq 4$ we have the inequality $n! > 2^n$.

Exercise 2.4. Prove the two statements below:

1. For every integer $n \geq 3$, the inequality $n^2 > 2n+1$ holds. (Hint: You can prove this by induction if you wish, but alternatively, you can prove it directly, without induction.)
2. For every integer $n \geq 5$, the inequality $2^n > n^2$ holds. (Hint: Use induction and the inequality in the previous part of the exercise.)

One must be careful in using induction that errors do not creep into the argument. Below is a statement and an attempted proof of the statement by induction.

Theorem 2.3. *All cats have the same color.*

Proof. We will show that for every collection of k cats there is a color such that all the cats in the collection have that color. Certainly this is true for $k = 1$: There is a color such that the single cat in the collection has that color. Suppose the statement is true for every collection of k cats and suppose you come upon a collection of $k+1$ cats. You can line them up in a row and number them, so that cat 1 is on the left, cat 2 is next, and so on, with cat $k+1$ on the right end. Consider the k cats to the left, cats 1 to k. By assumption, they all have the same color. Consider the k cats to the right, from 2 to $k+1$. By assumption again, they all have the same color. But then cat 1 has the same color as the cats in the middle, but cat $k+1$ also has the same color as the cats in the middle, and so all $k+1$ cats have the same color. We have thus proved, by the principle of mathematical induction, that for every positive integer n, all the cats in a collection of n cats have the same color. Thus, all cats have the same color.

Exercise 2.5. Theorem 2.3 cannot be correct, so there must be an error in the proof. Study the proof, determine where the error lies, and explain what is wrong.

2.2 The Tower of Hanoi

Another illustration of the use of induction arises in analyzing the game called the Tower of Hanoi. There are infinitely many versions of the game, depending on the choice of a positive integer n. The game has three vertical poles, with n disks of different diameters initially stacked on one pole and the other two poles empty. The initial pile of n disks has the largest one at the bottom, the smallest at the top, and the disks in between arranged so that every disk is smaller than the disks below and larger than the disks above. The object of the game is to move the disks one by one from pole to pole, following the rule that no disk may be placed on top of a smaller disk, so that at the end one succeeds in transferring the pile from its initial pole to one of the initially empty poles. Doing so wins the game, or solves the n-disk puzzle.

A little experimentation with small values of n shows that you can solve the puzzle in one move if $n = 1$, in three moves if $n = 2$, and in seven moves if $n = 3$. For instance, with $n = 1$, there is only one disk, and you can simply move it in one step to another pole; with $n = 2$, you can move the smaller disk to pole 2, the larger disk to pole 3, and then the smaller disk to pole 3. You should work out solutions for $n = 3$, 4, and 5 at least, continuing further if you wish, going far enough so that you can make a guess about what is true in general. Do not read on until you have done so.

Did your evidence suggest anything? You might have guessed that for larger n, the puzzle can be solved in $2^n - 1$ moves. Let us verify this by induction, or by climbing the stairway to heaven.

Theorem 2.4. *For each positive integer n, the Tower of Hanoi puzzle with n disks can be solved in $2^n - 1$ moves.*

Proof. We have again an infinite family of statements we wish to prove, one for each positive integer n. Statement 1 says that the puzzle with 1 disk can be solved in $2^1 - 1$ $(= 1)$ move. Statement 2 says that the puzzle with 2 disks can be solved in $2^2 - 1$ moves, or 3 moves. Statement n says that the puzzle with n disks can be solved in $2^n - 1$ moves. The stairway we wish to climb this time is the stairway that says that these infinitely many statements are all true. To get off the ground, we must climb the first step, which means we must show that the puzzle with 1 disk can be solved in 1 move. We can show this just by doing it, as already discussed. This gets us on the first step successfully.

Suppose we have reached step k. We wish to show that we can get to step $k + 1$. To be on step k means that we are assuming that the puzzle with k disks can be solved in $2^k - 1$ moves. In saying that we suppose that we have reached step k, we are saying that we are assuming that statement k is true. What we must show next is that the puzzle with $k + 1$ disks can be solved in $2^{k+1} - 1$ steps *on the assumption that the puzzle with k disks can be solved in* $2^k - 1$ *steps.*

To proceed, imagine a three-stage approach to solving the $(k+1)$-disk puzzle. In stage one, we move the k smallest disks from the initial pole to another of the poles, which we will call the second pole, leaving the largest disk on the initial pole. Of course, we do this following the rules of the game, never placing a disk upon a smaller disk. In stage two, we move the largest disk to the remaining empty pole, the third pole. In stage three, we complete the solution by moving the k smallest disks, which currently lie on pole two, onto pole three atop the largest disk.

Let us count how many moves it will take for us to do each stage. The first stage, moving k disks from pole 1 to pole 2, is really a version of the k-disk puzzle. We are assuming that we can solve the k-disk puzzle in $2^k - 1$ moves. Hence we can complete the first stage in $2^k - 1$ moves. The second stage, moving the largest disk to pole three, takes one move. The third stage, moving k disks from pole 2 to pole 3, is another version of the k-disk puzzle. Therefore, like the first stage, it can be done in $2^k - 1$ moves. Adding together the $2^k - 1$ moves used for the first stage, the 1 move used for the second, and the $2^k - 1$ moves used for the third, we get a total of

$$\left(2^k - 1\right) + 1 + \left(2^k - 1\right)$$

moves, and this equals $2^{k+1} - 1$.

We have shown that *if* we can solve the k-puzzle in $2^k - 1$ moves, *then* we can solve the $(k+1)$-puzzle in $2^{k+1} - 1$ moves. Thus we have shown that if we can get to step k, we can get to step $k+1$. Since we have also shown that we can obviously get to step 1, the principle of induction allows us to conclude that we have verified our desired statement for all positive integer values of n, and the proof is complete. We have again climbed the stairway to heaven.

To illustrate induction again, let us prove another fact about the Tower of Hanoi puzzle. We have shown that the puzzle with n disks can be solved in $2^n - 1$ moves. Can it be solved in fewer moves? Again, experimentation with small values of n will suggest an answer: It cannot. This is clear for $n = 1$ and $n = 2$. Let us prove it in general.

Theorem 2.5. *For each positive integer n, the Tower of Hanoi puzzle with n disks requires at least $2^n - 1$ moves to be solved.*

Proof. Let Theorem(n) be the statement that the n-disk puzzle requires at least $2^n - 1$ moves to be solved. We will use induction to prove Theorem(n) for each positive integer n. To begin, we must show that Theorem(1) is true. This is the statement that at least one move is required to solve the 1-disk puzzle. Obviously, we cannot solve it with 0 moves; at least 1 move is needed. Thus Theorem(1) is true.

Next we will assume that for some positive integer k, Theorem(k) is true, and show that on that assumption Theorem$(k+1)$ holds. Theorem(k) states that at least $2^k - 1$ moves are required to solve the k-disk puzzle. This is what we are assuming.

To solve the $(k + 1)$-disk puzzle, we must at some point move the largest disk from the initial pole to another pole. In order to do this, we must have already moved the k smallest disks off the largest disk onto the other two poles. If in doing so we split the k disks among the other two poles so each pole has some disk on it, we will not be allowed to move the largest disk, for it cannot be placed on a smaller disk. Thus we must in fact move all k smallest disks from the initial pole to a single other pole before we are free to move the largest disk to another pole. Moving these k disks is a version of the k-puzzle, which by the assumption of Theorem(k) requires at least $2^k - 1$ moves. After performing the necessary moves, we are free to move the largest disk to the vacant pole. This takes 1 move. To complete the puzzle in the fewest possible moves, we want to move onto the largest disk the k smallest disks, which are at this point on a single pole. This again requires at least $2^k - 1$ moves, by Theorem(k). Adding up, we see that we must make at least

$$\left(2^k - 1\right) + 1 + \left(2^k - 1\right)$$

moves to solve the $(k + 1)$-puzzle. Since

$$2^k - 1 + 1 + 2^k - 1 = 2^{k+1} - 1,$$

we have proved that if we must make at least $2^k - 1$ moves to solve the k-disk puzzle, then we must make at least $2^{k+1} - 1$ moves to solve the $(k + 1)$-disk puzzle. Thus, we have proved Theorem($k + 1$) under the assumption of Theorem(k). The principle of induction allows us to conclude that we have verified Theorem(n) for all values of n.

Exercise 2.6. Let us introduce a modified version of the Tower of Hanoi game. We place the three poles in a straight line and make a new rule: A disk can be moved only from one pole to an adjacent pole. Suppose the goal of the modified game is to move the usual stack of n disks from a pole at one end to the pole at the other end.

1. Solve the modified puzzle for small values of n, and determine how many moves are required.
2. From these examples, guess a general formula for the number of moves needed to solve the puzzle.
3. Use induction to prove that the puzzle can be solved in the guessed number of moves.

2.3 The Division Theorem

With induction available as a method of proof, we can move on to the study of integers. Recall that an integer a is *divisible* by an integer n if there is another integer m such that $a = mn$. If an integer a is divisible by an integer n, we also say that n *divides* a. Let us begin with some simple facts about divisibility.

Theorem 2.6. *The following divisibility facts hold:*

1. *The integer 0 is divisible by every integer.*
2. *Suppose n, a, and r are integers, and n divides a. Then n divides ra.*
3. *Suppose n, a, and b are integers, and n divides both a and b. Then n divides $a + b$ and $a - b$.*
4. *Suppose r divides s and s divides t. Then r divides t.*

Proof. For the first part, for every integer r, we have $0 = r \cdot 0$. Thus 0 is divisible by r. For the second part, we can proceed as follows. Since n divides a, by definition there is an integer m such that $a = mn$. Therefore $ra = rmn$. Using the definition of divisibility again, we see that n divides ra.

Exercise 2.7. Prove the third and fourth parts of Theorem 2.6 above. Then use Theorem 2.6 to prove Theorem 2.7 below.

Theorem 2.7. *Suppose a and b are integers divisible by an integer n. Then the integer $ra + sb$ is also divisible by n for every pair of integers r and s.*

The key to solving the Chicken McNugget problem in the Introduction was the fact that if a positive integer is divided by 6, a remainder occurs of 0, 1, 2, 3, 4, or 5. This is a special case of a result that may seem so obvious it hardly requires discussion. Nonetheless, discuss it we will, for it underlies much of what we will do in the coming chapters. Let us lead up to it with a question.

Suppose you ask a class of students to do the division below and to state the quotient and the remainder:

$$397 \div 14.$$

If you do this yourself, you will quickly get an answer (what is it?), and you probably expect everyone in the class, assuming that no calculational errors are made, to obtain the same answer as yours. Suppose, however, that one student comes up with a quotient of 27 and a remainder of 19, another a quotient of 28 and a remainder of 5, and yet another a quotient of 26 and a remainder of 33. Are they all correct? It is true, after all, that

$$397 = (14 \times 27) + 19,$$

that

$$397 = (14 \times 28) + 5,$$

and that

$$397 = (14 \times 26) + 33.$$

After some thought, you realize that what is wrong with the answers of the first and third students is that their remainders are too big. The remainder should be smaller than the divisor. This is part of what we mean when we speak of a remainder.

Once everyone agrees on what a remainder is, is it true that the problem of obtaining a quotient and a remainder upon dividing one positive integer by another always has a solution, and is this solution unique? These are really two different questions. Let us begin with the first one. We want the following statement to be true, a statement called the division theorem.

Theorem 2.8 (Division Theorem). *For every two positive integers a and b, there exist nonnegative integers q and r, with $r < a$, such that*

$$b = aq + r.$$

The division theorem asserts the familiar fact that when we divide b by a, we get a nonnegative integer q as the quotient and a nonnegative integer r less than the divisor a as a remainder.

If we fix a to be a specific positive integer, then the division theorem becomes an infinite sequence of statements, one for each positive integer b. It guarantees that for each such b, when we divide b by the given a, a quotient q exists and a remainder r between 0 and $a - 1$ exists. For example, for $a = 6$, the division theorem states that for every positive integer b there exist nonnegative integers q and r, with $r < 6$, such that $b = 6q + r$. Thus when we divide b by 6, we get a quotient q and a remainder r, with r between 0 and 5. This is exactly the result we needed to solve the Chicken McNugget problem. Similarly, the division theorem for $a = 2$ and $a = 3$ underlay our solutions in the Introduction to the simpler nugget problems.

Exercise 2.8. Let us consider the division theorem in the special case $a = 2$.

1. Explain why the division theorem can be restated in this case as follows: For a positive integer b, there exists a nonnegative integer q such that either $b = 2q$ or $b = 2q + 1$.
2. As trivial as this result may seem, let us prove it, using induction. The statement we wish to prove is that every positive integer b can be written either as $2q$ or as $2q + 1$ for some nonnegative integer q. Follow the outline below, using induction:
   (a) First show that 1 can be written in the desired form.
   (b) Now suppose an integer b that is greater than or equal to 1 can be written in the desired form. Show that $b + 1$ can also be so written. (Hint: There are two cases here, and they must be dealt with separately. First assume that b has the form $2q$ for some nonnegative integer q and show that $b+1$ can be written in the desired form. Then assume that b has the form $2q + 1$ for some nonnegative integer q and show that $b + 1$ can be written in the desired form.)
   (c) Conclude by the principle of induction that every positive integer b can be written as $2q$ or as $2q + 1$ for some nonnegative integer q.

Next let us see what the division theorem says in the special case that $a = 3$. This is more complicated than the $a = 2$ case.

Exercise 2.9. Explain why the division theorem for $a = 3$ can be restated as follows: For a positive integer b, there exists a nonnegative integer q such that b equals either $3q$, $3q + 1$, or $3q + 2$. Prove this result by induction, following the outline below:

1. First show that 1 can be written in the desired form.
2. Now suppose an integer b that is greater than or equal to 1 can be written in the desired form. Show that $b + 1$ can also be so written. (Hint: There would appear to be three cases here, depending on whether b has the form $3q$, $3q + 1$, or $3q + 2$, but really the first two can be combined into a single case. First assume that b has the form $3q$ or $3q + 1$ for some nonnegative integer q and show that $b + 1$ can be written in one of the three forms. Then assume that b has the form $3q + 2$ for some nonnegative integer q and show that $b + 1$ can be written in one of the three forms.)
3. Conclude by the principle of induction that every positive integer b can be written as $3q$, $3q + 1$, or $3q + 2$ for some nonnegative integer q.

These two examples serve as models for a proof of the division theorem in general.

Exercise 2.10. Prove the division theorem by induction. (Hint: Take a to be a fixed positive integer and let b vary. Prove the theorem for varying b by induction. First treat the case $b = 1$. Then assume that the theorem is true for a given b and show that it holds for $b + 1$.)

The division theorem can be strengthened by adding a statement about the uniqueness of q and r:

Theorem 2.9. *For positive integers a and b, there exists a unique choice of nonnegative integers q and r, with $r < a$, such that*

$$b = aq + r.$$

In Theorem 2.8 we merely asserted that q and r exist; in Theorem 2.9, we are asserting in addition that q and r are unique. What we mean by this is that only one choice of q and one choice of r will work, with the restrictions that q and r are nonnegative integers and that $r < a$. To prove this stronger version, we need only prove the uniqueness statement, since we already know that suitable q and r exist.

Let us discuss further what is meant by "unique." Suppose Ilya divides b by a and comes up with a quotient of q and a remainder of r, with q and r nonnegative and with $r < a$. Suppose Anya divides b by a and comes up with a quotient of s and a remainder of t, with s and t nonnegative integers and $t < a$. We hope that the answers of Ilya and Anya agree. In other words, we hope that Ilya's quotient q equals Anya's quotient s and Ilya's remainder r equals Anya's remainder t. That this must be the case is what uniqueness means. We can formulate this in the following more explicit form:

Theorem 2.10. *For positive integers a and b, suppose q and r are nonnegative integers with $r < a$ such that*

$$b = aq + r$$

and suppose also that s and t are nonnegative integers with $t < a$ such that

$$b = as + t.$$

Then $q = s$ and $r = t$.

Theorem 2.9 should be regarded as two statements, which we can call the existence statement and the uniqueness statement. Theorem 2.8 is the existence statement alone, Theorem 2.10 is the uniqueness statement alone, and Theorem 2.9 is the combination of Theorem 2.8 and Theorem 2.10.

Exercise 2.11. Prove Theorem 2.10. You may find the following outline useful.

1. Assume first that $r \leq t$, so that $t - r \geq 0$.
2. Observe that in this case $t - r < a$ and use the given equalities to show that a divides $t - r$.
3. Conclude, using the fact that $0 \leq t - r < a$, that $t - r = 0$ and $t = r$.
4. Deduce that $q = s$.
5. Assume next that $r \geq t$ and make a similar argument.
6. Conclude that since at least one of $r \leq t$ and $r \geq t$ is true, we have obtained the desired equalities.

3

The Euclidean Algorithm

3.1 Greatest Common Divisors

Suppose a and b are two positive integers. A *common divisor* of a and b is an integer r that divides both of them, and the *greatest common divisor* of a and b is the largest integer that divides both of them. For example, the common divisors of 330 and 420 are the integers 1, 2, 3, 5, 6, 10, 15, 30, and their negatives, from which we see that 30 is the greatest common divisor of 330 and 420. We will write (a, b) for the greatest common divisor of a and b. Thus $(330, 420) = 30$. Notice that $(a, b) = (b, a)$, so it does not matter in what order we write the pair of numbers. Notice also that 1 divides every integer, so the greatest common divisor of two integers is at least 1. Two integers whose greatest common divisor is 1 are said to be *relatively prime*.

A positive integer n that factors as the product rs also factors as the product $(-r)(-s)$, and $-n$ factors as $(r)(-s)$ or $(-r)(s)$. The data of these four factorizations are redundant. In particular, the divisors of a positive integer a are the same as the divisors of $-a$, so that

$$(a, b) = (-a, b) = (a, -b) = (-a, -b).$$

Therefore, when studying divisors, we can assume that a and b are positive integers, and we will usually do so.

Exercise 3.1. To get a feeling for the notion of greatest common divisor, make the following calculations:

1. Find the greatest common divisor of
   (a) 10 and 100;
   (b) 6 and 24;
   (c) 2 and 234 786 991 302.
2. More generally, show that if a and b are positive integers and a divides b, then the common divisors of a and b are precisely the divisors of a. Conclude in this case that a is the greatest common divisor of a and b.

3. Find the greatest common divisor of
   (a) 6 and 9;
   (b) 25 and 40;
   (c) 14 and 85;
   (d) 30 and 84;
   (e) 66 and 561;
   (f) 70 and 1869;
   (g) 227 761 and 661 643.

Common divisors of two positive integers a and b arose in our study of the Chicken McNugget problem. Suppose packages of nuggets come in sizes a and b. Then the possible quantities of nuggets we can buy are what we called the (a, b)-accessible positive integers. These are the integers n such that n can be written as $ar + bs$ for some choice of nonnegative integers r and s. The greatest common divisor of a and b plays a role in the analysis. To see why, recall Theorem 2.7, which states that for integers a and b divisible by an integer d and for every choice of integers r and s, the integer $ar + bs$ is also divisible by d.

Theorem 2.7 implies that if n is (a, b)-accessible and d is a common divisor of a and b, then d divides n. Thus any quantity of nuggets that can be purchased by buying boxes of sizes a and b must be divisible by d. Suppose in particular that d is the greatest common divisor of a and b. If $d > 1$, then there are infinitely many positive integers m not divisible by d, and for each such m, it is impossible to buy m nuggets. Only if the greatest common divisor d is 1 can we hope for an affirmative answer to the question raised in the Introduction: Is there a positive integer N such that for every integer $n > N$, we can buy n nuggets? We will see later that if the greatest common divisor of a and b is 1, then the answer is yes for the (a, b)-nugget problem. The point to observe now is that in analyzing the (a, b)-nugget problem, we must begin by studying the greatest common divisor of two positive integers a and b.

What method did you use to calculate greatest common divisors of the integer pairs above? One method arises naturally from the definition of greatest common divisor. It is guaranteed to work, but it can take a long time. Suppose a and b are positive integers, with $a < b$. You could try dividing a and b by every positive integer from 1 to a, one by one. In doing so, you will obtain a list of the common divisors of a and b. The largest common divisor in your list is their greatest common divisor. (Since we are looking for the *greatest* common divisor, it would generally be faster to start with a and work your way backwards to 1.) Another approach is to factor each of a and b into smaller integers and then stare at the factorizations to figure out what the greatest common divisor is. For instance, $66 = 2 \times 3 \times 11$ and $561 = 3 \times 11 \times 17$, so you may see from this that 33 is the greatest common divisor.

One weakness of both these approaches is that factoring an integer can take a long time. The example of the pair 227 761 and 661 643 already suggests how laborious this can be. For integers with sufficiently many digits, even the fastest computers cannot obtain factorizations in our lifetimes.

Fortunately, there is another way to proceed, one that has been known for 2500 years and that is unexpectedly fast. It produces the greatest common divisor without performing a long sequence of trial divisions. Let us work through the basic ideas of the method.

Exercise 3.2. Use Theorem 2.6 to show that for positive integers a and b with $a < b$, the list of common divisors of a and b coincides with the list of common divisors of a and $b - a$. Showing that the two lists coincide requires two steps. First you must show that every integer on the first list, the list of common divisors of a and b, is on the second list, the list of common divisors of a and $b - a$. Then you must go the other way, showing that every integer on the second list is on the first list. Once you have shown that the two lists coincide, you may conclude that

$$(a, b) = (a, b - a).$$

Using this result, let us go back to some earlier calculations. Suppose we want to find the greatest common divisor d of 30 and 84. The result implies that d is also the greatest common divisor of 30 and 54. (Why?) We can repeat the process to see that d is also the greatest common divisor of 30 and 24, and finally that d is the greatest common divisor of 24 and 6, which is 6.

Let us try the calculation of $(70, 1869)$ with this approach. We start by repeatedly subtracting 70 from 1869. All the subtractions of 70 get a bit tedious. How many subtractions would we actually make? We can predict this simply by dividing 1869 by 70. We find that the quotient is 26, with a remainder of 49. Thus after 26 subtractions of 70 from 1869, we will be left with 49:

$$(1869, 70) = (1869 - 70, 70) = (1869 - (2 \times 70), 70)$$
$$= (1860 - (3 \times 70), 70) = \cdots$$
$$= (1869 - (26 \times 70), 70) = (49, 70).$$

In other words, we can replace 1869 in this calculation with the remainder, 49, that we obtain when we divide 1869 by 70. The calculation of $(49, 70)$ is certain to be easier than the calculation of $(1869, 70)$.

This idea can be repeated (or *iterated*, as mathematicians say). Iterating yields

$$(70, 49) = (21, 49) = (49, 21).$$

Iterating again yields

$$(49, 21) = (49 - (2 \times 21), 21) = (7, 21).$$

Our sequence of reductions has brought us to the point where we can read off the answer directly: $(1869, 70) = (21, 7) = 7$.

The first set of calculations above showed that $(1869, 70) = (70, 49)$. We could obtain this either by repeatedly subtracting 70 from 1869 or by simplifying our work through division. This simplification procedure can be applied in general. For positive integers a and b with $a < b$, the division theorem ensures the existence of nonnegative integers q and r with $r < a$ such that $b = aq + r$. By subtracting a from b a total of q times, we find that

$$(b, a) = (b - a, a) = (b - 2a, a) = (b - 3a, a) = \cdots$$
$$= (b - qa, a) = (r, a).$$

This allows us to conclude that $(b, a) = (a, r)$. However, hidden by the dots in the sequence of equalities above is an argument that needs to be made more carefully.

Exercise 3.3. Using Theorem 2.7, prove that if a and b are positive integers with $b = aq + r$, for nonnegative integers q and r, then

$$(b, a) = (a, r).$$

In other words, the greatest common divisor of b and a is the same as the greatest common divisor of a and r.

In trying to calculate (b, a), the idea of replacing (b, a) with (a, r) can be iterated. Let us see what happens. Continue to assume that a and b are positive integers with $a < b$. We begin with the equality $(b, a) = (a, r)$, where r is the remainder obtained on dividing b by a. Then we start over again with the new pair r and a. The benefit of this is that r is likely to be a lot smaller than b. We continue by subtracting as many r's from a as we can:

$$(a, r) = (a - r, r) = (a - 2r, r) = \cdots .$$

How far can we go? We use the division theorem again, to divide a by r and get a quotient and a remainder. There is a problem with notation here, since we cannot use q and r again as the names of our quotient and remainder. One way to deal with this is to use indices.

Start again, regarding the numbers q and r in the equation $b = aq + r$ as just the first of possibly a long sequence of quotients and remainders. With this possibility in mind, let us rewrite the equation

$$b = aq + r$$

as

$$b = aq_1 + r_1.$$

The 1's are attached to q and r to indicate that q_1 is the *first quotient* and r_1 is the *first remainder*. From the preceding discussion, we know that $(b, a) = (a, r_1)$.

The next step is to use the division theorem on a and r_1. We obtain $a = r_1 q_2 + r_2$ for some nonnegative integers q_2 and r_2 with $r_2 < r_1$. The 2's attached to q and r in this equation signify that q_2 is the *second quotient* and r_2 is the *second remainder*. We obtain

$$(a, r_1) = (a - r_1 q_2, r_1) = (r_2, r_1).$$

Let us keep going. Since $r_2 < r_1$, we can keep subtracting r_2 from r_1 to get

$$(r_1, r_2) = (r_1 - r_2, r_2) = (r_1 - 2r_2, r_2) = \cdots .$$

Again, the division theorem tells us how far to go. Say $r_1 = r_2 q_3 + r_3$ for nonnegative integers q_3 and r_3 with $r_3 < r_2$. Then

$$(r_1, r_2) = (r_1 - r_2 q_3, r_2) = (r_3, r_2).$$

We can keep going, continuing to use the division theorem to obtain new quotients and remainders, and rewriting the pair of integers whose greatest common divisor we are calculating in terms of smaller integers:

$$(b, a) = (a, r_1) = (r_1, r_2) = (r_2, r_3) = (r_3, r_4) = \cdots .$$

This process eventually stops. The reason is that the remainders r_i are nonnegative integers that keep getting smaller. As we shall see, it stops at the point at which one of the remainders r_n divides the previous remainder r_{n-1}. The result is that we obtain the sequence

$$(b, a) = (a, r_1) = (r_1, r_2) = \cdots = (r_{n-1}, r_n) = r_n.$$

This is the Euclidean algorithm.

3.2 The Euclidean Algorithm

Let us use the Euclidean algorithm on some examples. After that we will find a way to state the algorithm formally and prove that it always works.

As a first example, what is the greatest common divisor of 330 and 420? We proceed by making a sequence of calculations of quotients and remainders. First we divide 420 by 330, obtaining

$$420 = (330 \times 1) + 90.$$

From this and the result of Exercise 3.3, we can conclude that $(420, 330)$, the greatest common divisor of 330 and 420, is the same as $(330, 90)$, the greatest common divisor of 90 and 330. Now we continue:

$$330 = (90 \times 3) + 60.$$

This implies that $(330, 90) = (90, 60)$. Again,

$$90 = (60 \times 1) + 30,$$

and this implies that $(90, 60) = (60, 30)$. Finally,

$$60 = (30 \times 2) + 0,$$

and this implies that $(60, 30) = (30, 0) = 30$. The algorithm stops, and we can conclude that $(330, 420) = 30$. The sequence of calculations we have made can be summarized thus:

$$330 = (90 \times 3) + 60,$$
$$90 = (60 \times 1) + 30,$$
$$60 = (30 \times 2) + 0.$$

Let us do another example, again displaying the calculations. Suppose we want to find the greatest common divisor $(3145, 23001)$ of 3145 and 23 001. Certainly we cannot tell at a glance. We use the division theorem to calculate the following sequence of remainders:

$$23\,001 = (3145 \times 7) + 986,$$
$$3145 = (986 \times 3) + 187,$$
$$986 = (187 \times 5) + 51,$$
$$187 = (51 \times 3) + 34,$$
$$51 = (34 \times 1) + 17,$$
$$34 = (17 \times 2) + 0.$$

We can conclude from this that

$$(23\,001, 3145) = (3145, 986) = (986, 187) = (187, 51)$$
$$= (51, 34) = (34, 17) = 17.$$

Thus 17 is the greatest common divisor of 3145 and 23 001.

Let us do one more example. Suppose we want to find the greatest common divisor $(627, 2015)$ of 627 and 2015. We use the division theorem again to calculate the following sequence of remainders:

$$2015 = (627 \times 3) + 134,$$
$$627 = (134 \times 4) + 91,$$
$$134 = (91 \times 1) + 43,$$
$$91 = (43 \times 2) + 5,$$
$$43 = (5 \times 8) + 3,$$
$$5 = (3 \times 1) + 2,$$
$$3 = (2 \times 1) + 1,$$
$$2 = (1 \times 2) + 0.$$

We can conclude from this that

$$(2015, 627) = (627, 134) = (134, 91) = (91, 43) = (43, 5)$$
$$= (5, 3) = (3, 2) = (2, 1) = 1.$$

Thus the greatest common divisor of 627 and 2015 is 1.

Exercise 3.4. Use the Euclidean algorithm to calculate the greatest common divisor of:

1. 6 and 9;
2. 25 and 40;
3. 14 and 85
4. 66 and 561;
5. 70 and 1869;
6. 568 and 4292;
7. 17 017 and 18 900;
8. 227 761 and 661 643.

We have used the Euclidean algorithm, but we have not proved that it is guaranteed to work. To do so, we first need a formal statement of the algorithm:

Algorithm 3.1 *Let a and b be positive integers with $a < b$, and suppose a does not divide b. Call a and b the* inputs *of the algorithm. Perform the division theorem iteratively, starting with a and b, to obtain the sequence of equalities*

$$b = aq_1 + r_1,$$
$$a = r_1 q_2 + r_2,$$
$$r_1 = r_2 q_3 + r_3,$$
$$r_2 = r_3 q_4 + r_4,$$
$$\cdots$$
$$r_{n-1} = r_n q_{n+1} + r_{n+1}.$$

Here $0 \leq r_1 < a$, and each successive r_{i+1} satisfies $0 \leq r_{i+1} < r_i$. Stop when a remainder of 0 is reached. In this case, we may suppose that $r_n \neq 0$ and $r_{n+1} = 0$, so that the last two equalities produced by the algorithm before it terminates have the form

$$r_{n-2} = r_{n-1} q_n + r_n,$$
$$r_{n-1} = r_n q_{n+1} + 0.$$

Call the last nonzero remainder, r_n, the output *of the algorithm.*

For positive integers a and b, we wish prove two statements:

1. The algorithm just described, with a and b as the input, terminates; that is, a remainder of zero is eventually obtained.
2. The output r_n of the algorithm is the greatest common divisor of a and b.

The following theorem combines these two statements. Read its proof closely.

Theorem 3.2. *Let a and b be positive integers with $a < b$. If a divides b, then the greatest common divisor of a and b is a. If a does not divide b, then the Euclidean algorithm applied to a and b terminates after a finite number n of steps. The output of the algorithm, r_n, is the greatest common divisor of a and b.*

Proof. We have already seen that if a divides b, then $(a, b) = a$. Thus we may assume that a does not divide b. We must show first that the Euclidean algorithm terminates. The remainders satisfy the sequence of strict inequalities

$$a > r_1 > r_2 > r_3 > \cdots .$$

Each time we perform another iteration of the division theorem in carrying out the algorithm, the new remainder is a nonnegative integer smaller than the previous one. After at most a iterations, the remainder must be 0, and the algorithm terminates.

To show that the output is the greatest common divisor of a and b, we proceed by induction. What should we do an induction on? We do it on the number of steps n taken before the Euclidean algorithm terminates. Suppose first that a and b are positive integers for which the algorithm terminates after one step. This means that the algorithm takes the form

$$b = aq_1 + r_1,$$
$$a = r_1 q_2 + 0.$$

The output of the algorithm is r_1. We know that $(b, a) = (a, r_1)$. But the second equality shows that r_1 divides a. Hence $(a, r_1) = r_1$, and we conclude that $(b, a) = (a, r_1) = r_1$. Thus the output is indeed the desired greatest common divisor.

Now take k to be a positive integer and assume that whenever an input of c and d is given for which the Euclidean algorithm terminates after k steps, then the algorithm's output is the greatest common divisor (c, d). Suppose a and b are a pair of positive integers such that the algorithm, when fed a and b as input, takes $k + 1$ steps to terminate. We wish to show that the output of the algorithm is (a, b).

The algorithm with input a and b starts with

$$b = aq_1 + r_1,$$
$$a = r_1 q_2 + r_2.$$

Notice that what we do after this in running the algorithm is exactly what we would do if we started the algorithm running with the pair r_1 and a as input rather than the pair a and b. Thus, since the algorithm takes $k+1$ steps to terminate with a and b as input, it takes only k steps to terminate with r_1 and a as input. Moreover, the output is the same, whether we start with input a and b or with input r_1 and a.

The inductive assumption can be applied to the input pair of r_1 and a. Therefore, the output, r_{k+1}, is the greatest common divisor (a, r_1) of a and r_1. However, we already know from the equality $b = aq_1 + r_1$ that $(b, a) = (a, r_1)$. Therefore, the output r_{k+1} is also the greatest common divisor (a, b) of a and b, as we wished to show. We have proved by induction on the number of steps in the algorithm that the output is the greatest common divisor of the input.

3.3 Bézout's Theorem

The Euclidean algorithm can be used to do more than determine the greatest common divisor (a, b) of a pair of positive integers a and b. It also allows us to express (a, b) in terms of a and b. At first, this may not appear to be important, but it turns out to have many significant consequences.

For arbitrary integers a and b, we say that another integer n is an *integer linear combination* of a and b if n can be written as $ar + bs$ for some integers r and s. The use of the word "integer" comes from the fact that the coefficients r and s are taken to be integers, and "linear" comes from the fact that a and b appear to the first power, which is the power to which the variables x and y appear in the equation $y = mx + b$ of a straight *line*. In contrast, an expression of the form $a^2r + b^2s$ might be called an integer quadratic combination of a and b. What we will show in the following theorem is that the greatest common divisor of two integers can be written as an integer linear combination of those integers.

Theorem 3.3 (Bézout). *Let a and b be integers with greatest common divisor d. Then there exist integers r and s such that*

$$d = ar + bs.$$

Thus, the greatest common divisor of a and b is an integer linear combination of a and b. Moreover, the data of the Euclidean algorithm can be used to determine an explicit pair of integers r and s such that $d = ar + bs$.

Bézout's theorem has both computational and theoretical uses, as we will see. Before proving Bézout's theorem and seeing what consequences it has, let us look at some examples.

Consider the integer pair 70 and 1869. We calculated their greatest common divisor earlier, using the following sequence of equalities:

$$1869 = (70 \times 26) + 49,$$
$$70 = (49 \times 1) + 21,$$
$$49 = (21 \times 2) + 7,$$
$$21 = (7 \times 3) + 0.$$

The output of the algorithm—the last nonzero remainder—is 7. Thus 7 is the greatest common divisor of 70 and 1869. Bézout's theorem asserts in this case that 7 is an integer linear combination of 70 and 1869. Indeed, you can check that this is true, since

$$7 = (1869 \times 3) + (70 \times (-80)) = (1869 \times 3) - (70 \times 80).$$

How do we find these coefficients, 3 and -80? We do not need to guess. We can use the data of the Euclidean algorithm in reverse. Take each row in the sequence of equalities above, ignoring the last one, and rewrite it as an equality with the remainder on one side and the rest on the other side:

$$7 = 49 - (21 \times 2),$$
$$21 = 70 - (49 \times 1),$$
$$49 = 1869 - (70 \times 26)$$

We wish to express 7 as an integer linear combination of 70 and 1869. These equalities let us do so.

The first equality allows us to write 7 as an integer linear combination of 49 and 21. The second lets us write 21 as an integer linear combination of 70 and 49. Substituting the value of 21 from the second equality into the first equality, we obtain

$$7 = 49 - (\mathbf{21} \times 2) = 49 - (\mathbf{70} - (\mathbf{49} \times 1)) \times 2 = (49 \times 3) - (70 \times 2),$$

where we have used boldface numerals to indicate where the substitution has been made. This expresses 7 as an integer linear combination of 49 and 70. The third equality expresses 49 as an integer linear combination of 70 and 1869. Substituting this relationship into our last equality, we obtain

$$7 = (\mathbf{49} \times 3) - (70 \times 2) = (\mathbf{1869} - (\mathbf{70} \times 26)) \times 3 - (70 \times 2).$$

After collecting terms, we obtain

$$7 = (1869 \times 3) - (70 \times 80),$$

as desired.

Let us try this procedure on the pair 3145 and 23 001. Their greatest common divisor is 17. We would like to express 17 as an integer linear combination of 3145 and 23001. We take the sequence of equalities constructed earlier that summarizes the data of the Euclidean algorithm applied to 3145 and 23 001, and we rewrite it from bottom to top, with the remainders placed on the left side. Doing so produces a new sequence of equalities:

$$17 = 51 - (34 \times 1),$$
$$34 = 187 - (51 \times 3),$$
$$51 = 986 - (187 \times 5),$$
$$187 = 3145 - (986 \times 3),$$
$$986 = 23\,001 - (3145 \times 7).$$

This is all the data we need. We now work our way down, line by line, expressing 17 as an integer linear combination of 51 and 34, as an integer linear combination of 187 and 51, as an integer linear combination of 986 and 187, as an integer linear combination of 3145 and 986, and finally, as an integer linear combination of 23001 and 3145. Here are the calculations:

$$17 = 51 - (\mathbf{34} \times 1)$$
$$= 51 - (\mathbf{187} - (\mathbf{51} \times \mathbf{3})) \times 1 = -187 + (\mathbf{51} \times 4)$$
$$= -187 + (\mathbf{986} - (\mathbf{187} \times \mathbf{5})) \times 4 = (986 \times 4) - (\mathbf{187} \times 21)$$
$$= (986 \times 4) - (\mathbf{3145} - (\mathbf{986} \times \mathbf{3})) \times 21$$
$$= -(3145 \times 21) + (\mathbf{986} \times 67),$$
$$= -(3145 \times 21) + (\mathbf{23\,001} - (\mathbf{3145} \times \mathbf{7})) \times 67$$
$$= (23\,001 \times 67) - (3145 \times 490).$$

We find in the end that

$$17 = (23\,001 \times 67) - (3145 \times 490).$$

With these two examples as guides, you should be able to express the greatest common divisor d of a pair of integers a and b as an integer linear combination of a and b.

Exercise 3.5. For each pair of integers a and b below, use the Euclidean algorithm to find integers r and s such that the greatest common divisor (a, b) of a and b can be expressed as the integer linear combination $ar + bs$.

1. 6 and 9;
2. 25 and 40;
3. 14 and 85
4. 66 and 561;
5. 70 and 1869;
6. 568 and 4292;
7. 17,017 and 18 900;
8. 227 761 and 661 643.

Exercise 3.6. Prove Bézout's theorem. (Hint: As in the proof that the Euclidean algorithm yields a greatest common divisor, use induction on the number of steps before the Euclidean algorithm terminates for a given input pair.)

We now take a look at some of the consequences of Bézout's theorem. For positive integers a and b, their greatest common divisor d is defined as the largest of all the common divisors of a and b. Examples suggest that every other common divisor of a and b divides the greatest common divisor. For instance, we have seen that the common divisors of 330 and 420 are 1, 2, 3, 5, 6, 10, 15, 30, and their negatives. All of these integers divide the greatest common divisor, 30. Bézout's theorem can be used to show that in general, each common divisor of a and b divides their greatest common divisor (a, b).

Exercise 3.7. Use Bézout's theorem to prove that for positive integers a and b, if an integer e is a common divisor of both a and b, then it divides their greatest common divisor d. (Hint: Write down what Bézout's theorem says about the relationship among a, b, and d, and then use the fact that e divides both a and b.)

3.4 An Application of Bézout's Theorem

We can use Bézout's theorem to prove a famous result of René Descartes, the seventeenth-century mathematician and philosopher after whom *Cartesian* coordinates are named. What is needed for Descartes's result is not Bézout's theorem itself but rather the following consequence. Recall that two integers are *relatively prime* if their greatest common divisor is 1.

Theorem 3.4. *Suppose that a and b are two relatively prime integers, and suppose that c is an integer such that a divides the product bc. Then a divides c.*

Exercise 3.8. Prove Theorem 3.4. (Hint: To do so, you must make use of the relative primality of a and b. Use Bézout's theorem to write down an equation involving a and b; then multiply both sides of the equation by c.) Show that if the assumption in Theorem 3.4 that a and b are relatively prime is dropped, then the conclusion of the theorem may fail. Specifically, give an example of three integers a, b, and c such that a divides the product bc but a divides neither b nor c.

Exercise 3.9. Use induction and Theorem 3.4 to prove Corollary 3.5 below.

Corollary 3.5 *Let a and b be relatively prime integers, and suppose that c is an integer and n a positive integer such that a divides the product $b^n c$. Then a divides c.*

We are ready for Descartes's result. Recall that a *rational number* is a number of the form a/b, where a and b are integers and b is nonzero. The expression a/b is *reduced* if a and b are relatively prime. This means that since a and b have no common factor greater than 1, we cannot cancel any factors from a and b.

An important problem in algebra is the solution of polynomial equations

$$a_n x^n + a_{n-1} x^{n-1} + \cdots + a_1 x + a_0 = 0.$$

Depending on the situation in which the polynomial arises, the coefficients a_i may be arbitrary real numbers, rational numbers, integers, or some other restricted set of numbers. If the coefficients are rational, we can multiply both sides of the polynomial equation by a suitable integer to clear denominators, obtaining in this way an equivalent equation with integer coefficients. Thus, if we are studying polynomial equations with rational coefficients, we may as well restrict the coefficients further to the set of integers. Here is Descartes's theorem.

Theorem 3.6 (Descartes). *Suppose $a_0, a_1, \ldots, a_n$ are integers, and suppose further that r and s are relatively prime integers such that the rational number r/s is a solution to the equation*

$$a_n x^n + a_{n-1} x^{n-1} + \cdots + a_1 x + a_0 = 0.$$

Then s divides a_n and r divides a_0.

Descartes's theorem takes a stronger form if the coefficient of highest degree is 1:

Theorem 3.7. *Suppose the equation*

$$x^n + a_{n-1} x^{n-1} + \cdots + a_1 x + a_0 = 0$$

has integer coefficients. Every rational number solution to this equation is an integer that divides a_0.

Exercise 3.10. Prove Descartes's theorem in its first form, Theorem 3.6. Then use Theorem 3.6 to deduce Theorem 3.7. In proving Theorem 3.6, you can follow the outline below:

1. Substitute r/s into the equation and multiply through by s^n to clear denominators, obtaining the equation

$$a_n r^n + a_{n-1} r^{n-1} s + \cdots + a_1 r s^{n-1} + a_0 s^n = 0.$$

(Check that this is true.)

2. Rewrite the left side as

$$a_n r^n + \left(a_{n-1} r^{n-1} + \cdots + a_0 s^{n-1} \right) s$$

and deduce that s divides $a_n r^n$.

3. Use Corollary 3.5 to deduce that s divides a_n.
4. Use a similar argument, regrouping the terms differently, to show that r divides a_0.

Exercise 3.11. Use Theorem 3.7 to find all rational number solutions of the equations below:

1. $x^3 + x - 1 = 0$;
2. $x^3 - 2x + 1 = 0$;
3. $x^3 + x^2 + x + 1 = 0$;
4. $x^3 - 24x + 5 = 0$;
5. $x^5 - 3x^4 - x^3 + 3x^2 + 9x - 6 = 0$.

3.5 Diophantine Equations

Bézout's theorem can be interpreted as saying that for integers a and b with greatest common divisor d, there are integer solutions to the equation

$$ax + by = d.$$

Here x and y are the unknowns, or variables, while a, b, and d are known constant numbers. For instance, suppose we want to solve the equation

$$70x + 1869y = 7.$$

Bézout's theorem guarantees that 7 is an integer linear combination of 70 and 1869, since 7 is their greatest common divisor. Thus Bézout's theorem guarantees the existence of integer solutions x and y to the equation $70x + 1869y = 7$. Moreover, Bézout's theorem asserts that we can find a solution explicitly by using the Euclidean algorithm, and we did so. We found that $x = -80$ and $y = 3$ is a solution.

Equations with integer coefficients and with variables that take integer values are called *Diophantine* equations, after the Greek mathematician Diophantus, who lived around 250 C.E. Diophantine equations can be hard to solve; some Diophantine equations have no solutions at all. For example, consider the equation

$$2x + 4y = 17.$$

Notice that regardless of the integer values that x and y are allowed to assume, the integers $2x$ and $4y$ are even, so their sum $2x + 4y$ is also even. But 17 is odd. Therefore, there can be no solution. Similarly,

$$6x + 21y = 133$$

has no solution, since the left side must be divisible by 3 and the right side is not.

Another example of a Diophantine equation is the equation

$$x^n + y^n = z^n,$$

where x, y, and z are variables and n is a fixed positive integer. If $n = 2$, you know many solutions, such as $x = 3$, $y = 4$, $z = 5$, and $x = 5$, $y = 12$, $z = 13$. These are *Pythagorean triples*, which form the sides of a right triangle. For $n > 2$, there are trivial solutions arising when we let one of the variables take on the value 0, but it is hard to find any others. Indeed, the brilliant French mathematician Pierre de Fermat (1601–1665) made the famous conjecture that there are no solutions other than the trivial ones. He came up with this conjecture while reading Diophantus's book *Arithmetica*, and wrote in the margin of that book that he had a marvelous proof for the statement but that the margin, alas, did not allow him enough room to write it down.

In those days much of mathematics was carried on by private correspondence, with mathematicians posing challenges to their colleagues rather than publishing proofs of their results. Consequently, Fermat never produced complete proofs of a number of his assertions. In the century after his death all of his unproved assertions were proved by other mathematicians, except for that marginal note about the lack of nontrivial solutions to the equation $x^n + y^n = z^n$ for $n > 2$. As Fermat's last unproved statement, it came to be known as "Fermat's last theorem." A better name would have been "Fermat's conjecture," since a theorem is a true mathematical statement, and without a proof, we cannot *know* whether a mathematical statement is true. We can only conjecture that it *might* be true. Fermat's conjecture remained just that, a conjecture, for over three centuries, and a proof of this conjecture became the most sought-after prize in mathematics, tantalizing generations of the world's most accomplished mathematicians, as well as hosts of mathematical amateurs.

Over the years Fermat's conjecture was proved for many values of n, including all values up to some huge number. But no one had been able prove that the statement was true for *every* integer n greater than 2 until 1994, when Andrew Wiles finished years of work by finding a proof. His achievement received worldwide media coverage, including a story on the front page of the *New York Times*. It was the greatest mathematical accomplishment of the past century.

Whether Fermat himself had found a proof remains unknown, although it is most unlikely that he did, since Wiles's proof uses an impressive battery of the heavy machinery developed by twentieth-century mathematics. It far more likely that Fermat had a proof only for the special cases $n = 3$ and $n = 4$ and had assumed, incorrectly, that the property of unique factorization enjoyed by the integers (we will study this property in Chapter 5) holds in all *number rings* of the type that we will study in Chapter 6.

Now that we know that some Diophantine equations can be very difficult to solve, let us return to some simpler Diophantine equations, such as $2x + 4y =$

17. The line of reasoning we used above to show that this equation has no solutions can be used more generally to prove the following statement:

Theorem 3.8. *For integers a, b, and e, if (a, b) does not divide e, then the equation*

$$ax + by = e$$

has no integer solution; that is, there are no integer values of x and y that solve the equation.

Many statements in mathematics take the form "If P is true, then Q is true." In logic, the alternative statement "If Q is not true, then P is not true" is called the *contrapositive* of the original one. The two statements are *logically equivalent*: They say the same thing, in a different form, and they are both either simultaneously true or simultaneously false. For instance, the contrapositive of Theorem 3.8 is the following:

Theorem 3.9. *For integers a, b, and e, if the equation*

$$ax + by = e$$

has an integer solution, then (a, b) divides e.

Exercise 3.12. Prove Theorem 3.9, thereby proving the logically equivalent Theorem 3.8. (Hint: Use Theorem 2.7.)

Theorem 3.8 is a negative result, telling us that certain Diophantine equations cannot be solved. Bézout's theorem can be regarded as a positive result, telling us that for integers a and b with greatest common divisor d, the Diophantine equation $ax + by = d$ has a solution, and that the solution can be found explicitly using the Euclidean algorithm. More generally, in contrast to the setup of Theorem 3.8, let us consider integers a, b, and e for which the greatest common divisor (a, b) *does* divide e. Can we solve the equation $ax + by = e$?

Theorem 3.10. *For integers a, b, and e, if (a, b) divides e, then the equation*

$$ax + by = e$$

has an integer solution.

Proof. Let us write d for (a, b). We are assuming that d divides e, which means that there is another integer t with $e = dt$. By Bézout's theorem, there are solutions $x = r$ and $y = s$ to the equation $ax + by = d$; that is, there are integers r and s such that

$$ar + bs = d.$$

Multiplying both sides of this last equation by t, we obtain

$$a(rt) + b(st) = dt = e.$$

Thus, the equation $ax + by = e$ has the solution $x = rt$ and $y = st$.

Not only have we proved Theorem 3.10, but also we have shown how to obtain a solution. First we use the Euclidean algorithm to solve $ax+by = (a, b)$ for x and y, then we multiply these solutions by the integer $e/(a, b)$.

We can put Theorems 3.9 and 3.10 together into one statement. Let us again consider a statement of the form "If P is true, then Q is true." The statement "If Q is true, then P is true" is called its *converse*. These two statements are *not* logically equivalent. Both could be true, one could be true while the other is false, or both could be false. They are independent of each other. If both are true, this can be summarized by saying "P is true if and only if Q is true." The phrase "if and only if" is a way of combining two separate statements, each the converse of the other. For instance, combining Theorems 3.9 and 3.10, we obtain the following result:

Theorem 3.11. *For integers a, b, and e, the equation*

$$ax + by = e$$

has an integer solution if and only if (a, b) divides e.

Exercise 3.13. For each of the following equations, decide whether it can be solved in integers. If so, find a solution:

1. $6x + 9y = 2$;
2. $6x + 9y = -33$;
3. $25x + 40y = 345$;
4. $14x + 85y = 3$;
5. $66x + 561y = 22$;
6. $66x + 561y = 99$;
7. $70x + 1869y = 35$;
8. $3145x + 23\,001y = 4$;
9. $3145x + 23\,001y = -85$.

We have just discussed how to use the Euclidean algorithm to obtain a solution to the equation $ax + by = e$ for integers a, b, and e such that (a, b) divides e. However, there will be many more solutions. To see why, let us consider an example.

Suppose we wish to solve $6x+15y = 21$. Since $(6, 15) = 3$ and 3 divides 21, Theorem 3.10 ensures that there is a solution. Our procedure for finding the solution is first to use the Euclidean algorithm to solve $6r + 15s = 3$. Doing so yields the solution $r = -2$ and $s = 1$, with

$$(6 \times (-2)) + (15 \times 1) = 3.$$

If we multiply both sides of this equality by 7, we obtain

$$(6 \times (-14)) + (15 \times 7) = 21. \qquad\qquad (\star)$$

Thus $x = -14$ and $y = 7$ is another solution to $6x + 15y = 21$.

These two solutions are not the only ones. For instance, $6 + 15 = 21$, so $x = y = 1$ is another solution. We can explain why more solutions exist by writing down the obvious equality

$$6 \times 15 = 15 \times 6.$$

Let us divide the right factors of both sides of this equation by 3, the greatest common divisor of 6 and 15. We get

$$6 \times 5 = 15 \times 2.$$

Next multiply both sides of this equation by a generic integer u (that is, u represents an arbitrary integer) to get

$$6 \times 5u = 15 \times 2u$$

and rewrite this as

$$\big(6 \times (5u)\big) + \big(15 \times (-2u)\big) = 0. \tag{$\star\star$}$$

Equation $(\star\star)$ is just a fancy way of saying that zero equals zero, and if we add equation $(\star\star)$ to equation $(\star)$ above, we obtain

$$6(-14 + 5u) + 15(7 - 2u) = 21.$$

We have thus found a recipe for writing down infinitely many integer solutions to the equation $6x + 15y = 21$ once we have a single solution. For instance, taking $u = 3$, we get the solution $x = y = 1$, and taking $u = -29$, we get the solution $x = -159$, $y = 65$.

Exercise 3.14. Using the idea above, describe infinitely many integer solutions to each of the equations below:

1. $6x + 9y = -33$;
2. $25x + 40y = 345$;
3. $14x + 85y = 3$.

Let us finish the discussion of Bézout's theorem and equations of the form $ax + by = e$ with a puzzle.

Exercise 3.15. Suppose you have two hourglasses, the first measuring a time period of a minutes and the second measuring a time period of b minutes. How can you use them to measure a time period of c minutes, where a, b, and c are as follows?

1. $a = 3$, $b = 7$, and $c = 8$;
2. $a = 5$, $b = 7$, and $c = 11$;
3. $a = 6$, $b = 11$, and $c = 13$.

(Hint: Here is the solution of a simpler version. Suppose $a = 2$, $b = 3$, and $c = 1$. You can use the two-minute and three-minute hourglasses to measure a period of 1 minute by starting them at the same time. The time between the moment when the two-minute hourglass runs out of sand and the three-minute hourglass does is one minute.)

4

Congruences

4.1 Congruences

In this chapter we will begin our study of congruence, a notion that makes it possible to formulate certain statements about divisibility in a concise way. Initially, congruences may appear to be no more than a notational device for stating arithmetic facts. However, once we start using the notation, a new and important way of thinking will emerge.

For a positive integer m and integers a and b, we follow Gauss and write

$$a \equiv b \pmod{m}$$

to mean that a and b have the same remainder upon division by m. In words, the notation is read as

a is congruent to b modulo m.

The expression $a \equiv b \pmod{m}$ is called a *congruence*, and m is the *modulus* of the congruence. For example,

$$12 \equiv 26 \pmod 7$$

is read as "12 is congruent to 26 modulo 7" and means that 12 and 26 have the same remainder, namely 5, upon division by 7.

Every integer is congruent to every other integer modulo 1, since every integer has the same remainder, namely 0, upon division by 1, so congruences modulo 1 are not interesting. For this reason, we generally assume in working with congruences that the modulus m is an integer greater than 1.

Exercise 4.1. To develop your understanding of the definitions, do the following:

1. Determine whether

$$3 \equiv 123 \pmod 5$$

and whether

$$42 \equiv 88 \pmod{17}.$$

2. Decide whether an even integer can be congruent to an odd integer modulo 2.

Exercise 4.2. An equivalent definition of congruence is that $a \equiv b \pmod{m}$ if m divides $a - b$. Show that our two definitions of congruence are equivalent; that is, show that for integers a and b, a positive integer m divides $a - b$ if and only if the remainder obtained when a is divided by m equals the remainder obtained when b is divided by m.

The following result is essentially a restatement of the division theorem.

Theorem 4.1. *Let m be an integer greater than 1. Every integer a is congruent modulo m to exactly one of the integers $0, 1, \ldots, m - 1$.*

Exercise 4.3. Theorem 4.1 is really two statements. First, it asserts that an integer a is congruent modulo m to at least one of the integers from 0 to $m - 1$. Second, it asserts that there is a unique such integer; that is, if a is congruent to an integer r and to an integer s, both between 0 and $m - 1$, then r and s must be the same integer; that is, $r = s$. Prove both statements.

For an integer $m > 1$ and another integer a, the unique integer between 0 and $m - 1$ to which a is congruent modulo m is called the *least nonnegative residue* of a modulo m. For example, the least nonnegative residue of 79 modulo 9 is 7. After all,

$$79 \equiv 7 \pmod{9},$$

and no nonnegative integer smaller than 7 is congruent to 79 modulo 9. (We know this by Theorem 4.1, which tells us that 79 is congruent to exactly one integer in the set $0, 1, \ldots, 8$, and once we know that 79 is congruent to 7 modulo 9, we know that it cannot be congruent to 0 or 1 or 2 or 3 or 4 or 5 or 6 modulo 9.

As another example, the least nonnegative residue of the integer

$$123\,332\,334\,567$$

modulo 2 is 1, and the least nonnegative residue of the integer

$$124\,333\,222\,555\,666\,777\,222\,119$$

modulo 100 is 19. Make sure that you understand why these two statements are true. Recall that the least nonnegative residue of a modulo m is the remainder obtained when a is divided by m.

What is the least nonnegative residue of 3^{99} modulo 5? The answer this time is not apparent. We can multiply 3 times itself 99 times to find out, but this will take a long time. We will see soon that the answer can be found quickly by using some basic facts about congruences. These facts are the analogues for congruences of some familiar facts for equalities.

The reason that such analogues exist is that congruences behave in certain ways like equalities. Two numbers are equal if they are exactly the same, while two numbers are congruent (modulo some integer m) if they are "sort of" the same in the sense that their remainders upon division by m are the same. This "sort of" equality, or *equivalence relation*, gives congruences some important properties shared by true equalities.

For integers a, b, and c, if $a = b$ and $b = c$, then of course, $a = c$. Our first basic congruence fact is analogous.

Proposition 4.2 *Fix an integer $m > 1$. Suppose a, b, and c are integers such that*

$$a \equiv b \pmod{m} \quad and \quad b \equiv c \pmod{m}.$$

Then

$$a \equiv c \pmod{m}.$$

Proof. The congruence $a \equiv b \pmod{m}$ implies that m divides $a - b$, and $b \equiv c \pmod{m}$ implies that m divides $b - c$. We have already proved that if m divides two integers, then it divides their sum. Therefore, m divides $(a - b) + (b - c)$, which is $a - c$. This proves that $a \equiv c \pmod{m}$, as desired.

Let us review another familiar fact about equalities. For integers a, b, e, and f, if

$$a = e \quad and \quad b = f,$$

then

$$a + b = e + f \quad and \quad ab = ef.$$

Informally, we say that we can add and multiply equalities. There is a parallel result for congruences; that is, they can be added and multiplied:

Proposition 4.3 *Fix an integer $m > 1$. Suppose a, b, e, and f are integers satisfying*

$$a \equiv e \pmod{m} \quad and \quad b \equiv f \pmod{m}.$$

Then

$$a + b \equiv e + f \pmod{m} \quad and \quad ab \equiv ef \pmod{m}.$$

Proof. Let us prove the statement about products. When we translate it into a statement about division, it takes the following form. If m divides $a - e$ and m divides $b - f$, then m divides $ab - ef$. This is by no means obvious. Let us rephrase it yet again. To say that m divides $a - e$ is to say that there is an integer r such that $a - e = rm$. Similarly, there is an integer s such that $b - f = sm$. In other words, $a = e + rm$ and $b = f + sm$. Therefore,

$$ab = (e + rm)(f + sm) = ef + esm + frm + rsm^2.$$

We can rewrite this as

$$ab - ef = (es + fr + rsm)m,$$

from which we see that m divides $ab - ef$.

We can use Proposition 4.3 to deal with the problem of finding the least nonnegative residue of 3^{99} modulo 5. Raising 3 to a high power might take considerable effort, but raising 1 to a high power is easy: We always get 1. We therefore seek a (preferably low) power of 3 that is congruent to 1 modulo 5. Let us do a bit of calculation:

$$3^1 = 3 \equiv 3 \pmod 5,$$
$$3^2 = 9 \equiv 4 \pmod 5,$$
$$3^3 = 27 \equiv 2 \pmod 5,$$
$$3^4 = 81 \equiv 1 \pmod 5.$$

With the fact $81 \equiv 1 \pmod 5$ in hand, in Proposition 4.3 we take 81 as both a and b, and we take 1 as both e and f, thereby obtaining

$$81 \times 81 \equiv 1 \times 1 \pmod 5,$$

or

$$81^2 \equiv 1 \pmod 5.$$

Repeating this, we find that

$$81 \times 81 \times 81 \equiv 1 \times 1 \times 1 \pmod 5,$$

or

$$81^3 \equiv 1 \pmod 5.$$

More generally, an induction argument shows that

$$81^n \equiv 1 \pmod 5 \tag{$\star$}$$

for every positive integer exponent n. Recall that we chose 81 because $81 = 3^4$. Writing $81^n = \left(3^4\right)^n = 3^{4n}$, we can rewrite the congruence $(\star)$ above as

$$3^{4n} \equiv 1 \pmod 5.$$

This congruence holds for every positive integer n.

In order to find the least nonnegative residue of 3^{99} modulo 5, we can rewrite 3^{99} as $3^{96} \times 3^3 = 3^{4 \cdot 24} \times 3^3$ and apply Proposition 4.3 one more time. Using

$$3^{96} = 3^{4 \cdot 24} \equiv 1 \pmod 5$$

and

$$3^3 \equiv 2 \pmod 5,$$

we obtain

$$3^{99} \equiv 2 \pmod 5.$$

Thus 2 is the least nonnegative residue of 3^{99} modulo 5.

As another example, suppose we want to calculate the least nonnegative residue of 12^{85} modulo 7. We can use Proposition 4.3 to replace 12 by 5. Since $12 \equiv 5 \pmod{7}$, Proposition 4.3 implies that $12^2 \equiv 5^2 \pmod{7}$, that $12^3 \equiv 5^3 \pmod{7}$, and ultimately, by induction, that $12^{85} \equiv 5^{85} \pmod{7}$. Thus it suffices to calculate the least nonnegative residue of 5^{85} modulo 7. Let us determine the least nonnegative residue of each of the first few powers of 5 modulo 7. Once again, we seek a power of 5 that is congruent to 1 modulo 7. We can check that $5^2 \equiv 4 \pmod{7}$ and $5^3 \equiv 6 \pmod{7}$. Since $5^4 = 5^3 \times 5$, we see that

$$5^4 = \left(5^3 \times 5\right) \equiv (6 \times 5) \equiv 30 \equiv 2 \pmod{7}.$$

Continuing in this way, we find that $5^5 \equiv 3 \pmod{7}$ and $5^6 \equiv 1 \pmod{7}$. Therefore, using Proposition 4.3 yet again, we obtain the congruence $5^{6n} \equiv 1 \pmod{7}$ for every positive integer n. In particular, $5^{84} = 5^{6 \cdot 14} \equiv 1 \pmod{7}$, and so $5^{85} \equiv 5 \pmod{7}$. The least nonnegative residue of 5^{85} modulo 7 is therefore 5.

Exercise 4.4. In calculating the least nonnegative residues of powers of integers we made implicit use of the following corollary to Proposition 4.3: If $a \equiv b \pmod{m}$, then $a^n \equiv b^n \pmod{m}$ for every positive integer n. Prove this corollary using mathematical induction.

Exercise 4.5. Using Proposition 4.3, find the least nonnegative residues below. In addition to the answer, show the calculations you made to get it:

1. $2^{82} \pmod{5}$;
2. $3^{1502} \pmod{13}$;
3. $5^{1004} \pmod{7}$;
4. $6^{13\,334\,451} \pmod{7}$.

Propositions 4.2 and 4.3 illustrate ways in which congruences behave like equalities. Here is another way.

Proposition 4.4 *Fix an integer $m > 1$. Suppose a and b are integers satisfying*

$$a \equiv b \pmod{m}.$$

Then for every integer r, the congruence

$$ra \equiv rb \pmod{m}$$

holds.

Thus just as with an equality, you can multiply both sides of a congruence by an integer and preserve the congruence.

Proof. The proof is simple. The congruence $a \equiv b \pmod{m}$ is equivalent to the statement that m divides $a - b$. But then m divides $r(a - b) = ra - rb$, and this statement is equivalent to the desired congruence.

For integers a, b, and r, with $r \neq 0$, another familiar fact about equalities is that if $ra = rb$, then $a = b$. We often summarize this statement by saying that we can "cancel" r. Alas, the analogous statement about congruences need not hold. If $ra \equiv rb \pmod{m}$, we cannot conclude that $a \equiv b \pmod{m}$. For example, $10 \equiv 12 \pmod{2}$, or equivalently,

$$2 \times 5 \equiv 2 \times 6 \pmod{2}.$$

However, canceling the factor 2 from both sides yields

$$5 \equiv 6 \pmod{2},$$

which is false.

However, all is not lost, and we shall see that there are important cases in which cancellation in a congruence is possible. We note first that there is one case in which cancellation is not permitted in an equality: You cannot divide both sides of an equality by zero. The reason that cancellation failed in the congruence above has to do with the fact that 2 is congruent to 0 modulo 2, and in some sense we were trying to cancel out a "zero" (modulo 2). We will now show that cancellation in congruences is problematic when the integer we wish to cancel has a common factor (greater than 1) with the modulus.

With this in mind, let us fix a modulus $m > 1$ and suppose that r is an integer such that $(r, m) > 1$. We assert that in this case r cannot in general be canceled from congruences modulo m. To see why, let $d = (r, m)$. Then there are integers r' and m' with $r = dr'$ and $m = dm'$. Notice that $rm' = dr'm' = r'dm' = r'm$, so that m divides rm'. This yields the congruence $rm' \equiv 0 \pmod{m}$. If we try to cancel r from this congruence, we obtain $m' \equiv 0 \pmod{m}$, which means that m divides m'. But $0 < m' < m$, so m cannot divide m'. Thus r cannot be canceled from the congruence $rm' \equiv 0 \pmod{m}$.

That is the bad news. However, the good news is that if $(r, m) = 1$, then r can be canceled from congruences modulo m:

Theorem 4.5. *Let r and m be relatively prime integers, with $m > 1$. If a and b are integers for which*

$$ra \equiv rb \pmod{m},$$

then

$$a \equiv b \pmod{m}.$$

Exercise 4.6. Prove Theorem 4.5. (Hint: Observe that what you must prove is that if m divides $ra - rb$, then m divides $a - b$. Review Theorem 3.4 and observe that you have essentially proved this already.)

4.2 Solving Congruences

In algebra one studies equations involving an unknown, such as

$$3x + 4 = 5.$$

Similarly, one can study congruences involving an unknown, such as the congruence

$$3x + 4 \equiv 5 \ (\text{mod } m)$$

for some modulus m. To solve $3x + 4 = 5$ is to find a value of x, or perhaps all values of x, for which the equation is true. To *solve* $3x + 4 \equiv 5 \ (\text{mod } m)$ is to find the integers x for which the congruence is true. If a solution exists, the congruence is *solvable*; otherwise, the congruence is not solvable.

Suppose, for example, that $m = 13$, so that the congruence we want to solve is

$$3x + 4 \equiv 5 \ (\text{mod } 13).$$

Proposition 4.3 ensures that subtracting 4 from both sides of the congruence yields the new, equivalent, congruence

$$3x \equiv 1 \ (\text{mod } 13).$$

By testing integers, we find that $x = 9$ is a solution. If we keep going, we will find other solutions, such as $x = 22$ and $x = 35$. In fact, for every integer u, you can check that the integer $x = 9 + 13u$ is a solution of the congruence.

Consider the following question: For what positive integers a and m with $m > 1$ is the congruence

$$ax \equiv 1 \ (\text{mod } m)$$

solvable? Such congruences are certainly not solvable for all values of a and m. For instance, consider the congruence

$$2x \equiv 1 \ (\text{mod } 2).$$

Solving this means finding an integer x with the property that 2 divides $2x - 1$. But for every integer x, the integer $2x - 1$ odd, so 2 cannot divide $2x - 1$. In other words, the congruence is not solvable.

More generally, consider the congruence

$$2x \equiv 1 \ (\text{mod } m).$$

If m is even, we run into the same difficulty. For every integer x, the integer $2x - 1$ is odd, so the even integer m cannot divide $2x - 1$. Thus, for every even m, the congruence $2x \equiv 1 \ (\text{mod } m)$ is not solvable. However, if m is odd, you should be able to see that $2x \equiv 1 \ (\text{mod } m)$ is solvable (every odd integer m can be written in the form $m = 2r - 1$ for some integer r).

Let us try another example. Consider the congruence

$$3x \equiv 1 \ (\text{mod } 3).$$

A solution is an integer x with the property that 3 divides $3x - 1$. Can there be a solution? No. After all, 3 divides $3x$ for every integer value of x, so 3 cannot divide $3x - 1$. Therefore, the congruence is not solvable. More generally, if m is a positive integer divisible by 3, the congruence

$$3x \equiv 1 \ (\text{mod} \ m)$$

is not solvable, for m cannot divide $3x - 1$. On the other hand, if m is not divisible by 3, then

$$3x \equiv 1 \ (\text{mod} \ m)$$

is solvable. This is not so obvious, but it will follow as a special case of a more general theorem below. Let us first work out some more examples.

Exercise 4.7. In this problem, we will consider congruences in the special form $ax \equiv 1 \ (\text{mod} \ m)$. The problems below can be solved simply by trying all possibilities modulo m.

1. For each integer a from 1 to 4, find a solution to the congruence $ax \equiv 1 \ (\text{mod} \ 5)$.
2. For each integer a from 1 to 5, either find a solution to the congruence $ax \equiv 1 \ (\text{mod} \ 6)$, or show that there is no solution.
3. For each integer a from 1 to 6, find a solution to the congruence $ax \equiv 1 \ (\text{mod} \ 7)$.

The following theorem tells us when the congruence $ax \equiv 1$ can be solved modulo m.

Theorem 4.6. *Let a and m be positive integers with $m > 1$. The congruence*

$$ax \equiv 1 \ (\text{mod} \ m)$$

is solvable if and only if $(a, m) = 1$.

Exercise 4.8. Prove Theorem 4.6. Since it is an "if and only if" statement, it consists of two separate statements, so two proofs are needed.

1. Begin by proving that if the congruence

$$ax \equiv 1 \ (\text{mod} \ m)$$

is solvable, then $(a, m) = 1$. To do so, use the assumption that there is a solution to show that every positive integer d dividing both a and m must divide 1.
2. Then prove that if $(a, m) = 1$, the congruence

$$ax \equiv 1 \ (\text{mod} \ m)$$

is solvable. To do so, you can use Bézout's theorem.

Theorem 4.6 is a special case of the following more general theorem, one of the most important we will discuss.

Theorem 4.7. *Let a, e, and m be positive integers, with m > 1. The congruence*

$$ax \equiv e \pmod{m}$$

is solvable if and only if (a, m) divides e.

Notice that Theorem 4.7 really is a generalization of Theorem 4.6: By setting e equal to 1 in Theorem 4.7, we recover Theorem 4.6 as a special case. Since Theorem 4.7 is an "if and only if" statement, it too is two statements in one. The two statements can be proved by following the pattern used in the preceding exercise: One statement follows from elementary divisibility considerations, the other from Bézout's theorem. If you were to carry out the proof, you would find that the arguments are familiar, and for good reason. We have proved Theorem 4.7 already. It is Theorem 3.11 in disguise. Recall what Theorem 3.11 said: For integers a, b, and e, the equation $ax + by = e$ has an integer solution if and only if (a, b) divides e.

Exercise 4.9. Prove Theorem 4.7 by showing that it is a consequence of the already proved Theorem 3.11.

Let us try solving some congruences. Suppose we wish to solve

$$30x \equiv 16 \pmod{84}.$$

According to Theorem 4.7, this is solvable if and only if the greatest common divisor of 30 and 84, which is 6, divides 16. Since 6 does not divide 16, there is no solution. Suppose instead that we wish to solve

$$30x \equiv 18 \pmod{84}.$$

Since 6 does divide 18, there is a solution. A solution is an integer x such that 84 divides $30x - 18$. For 84 to divide $30x - 18$, there must be an integer y such that $30x - 18 = 84y$, and we can rewrite this equation as $30x - 84y = 18$. Thus solving the congruence $30x \equiv 18 \pmod{84}$ amounts to solving the Diophantine equation $30x - 84y = 18$.

We learned how to solve $30x - 84y = 18$ in Section 3.5, using the Euclidean algorithm. The algorithm, with input 30 and 84, produces the equalities

$$84 = (30 \times 2) + 24,$$
$$30 = (24 \times 1) + 6,$$
$$24 = (6 \times 4) + 0.$$

From this we obtain the equalities

$$6 = 30 - (24 \times 1),$$
$$24 = 84 - (30 \times 2),$$

and from these we find that

$$6 = 30 - (24 \times 1) = 30 - (84 - (30 \times 2)) = (30 \times 3) - 84.$$

Therefore,

$$18 = 6 \times 3 = (30 \times 9) - (84 \times 3).$$

This shows that $x = 9$ and $y = 3$ is a solution to the equation $30x - 84y = 18$. It follows that $x = 9$ is a solution to the congruence $30x \equiv 18 \pmod{84}$.

Exercise 4.10. For each of the congruences below, decide whether there is a solution. If there is one, find a solution using the Euclidean algorithm:

1. $8x \equiv 1 \pmod 6$;
2. $8x \equiv 4 \pmod 6$;
3. $15x \equiv 1 \pmod{21}$;
4. $15x \equiv 6 \pmod{21}$.

4.3 Congruence Classes and McNuggets

In Chapter 1 we introduced the problem of determining what quantities of Chicken McNuggets can be bought if the nuggets are sold in boxes of particular sizes. For example, if nuggets are sold in boxes of sizes 3 and 7, we found that we can buy every possible amount of nuggets larger than 11. To prove this, we partitioned the positive integers into three classes according to the remainder upon dividing by 3:

Class 0: 3, 6, 9, 12, 15, ...;
Class 1: 1, 4, 7, 10, 13, ...;
Class 2: 2, 5, 8, 11, 14, ...

Let us call these classes $\mathcal{C}(0)^+$, $\mathcal{C}(1)^+$, and $\mathcal{C}(2)^+$. Thus $\mathcal{C}(0)^+$ is the collection of positive integers whose remainder upon division by 3 is 0, $\mathcal{C}(1)^+$ is the collection of positive integers whose remainder upon division by 3 is 1, and $\mathcal{C}(2)^+$ is the collection of positive integers whose remainder upon division by 3 is 2.

We observed that if a number n of nuggets can be bought by buying boxes of sizes 3 and 7, then we can also buy $n + 3$ nuggets, or $n + 6$ nuggets, or any number of nuggets after n in the class $\mathcal{C}(i)^+$ that contains n. Since we can buy 3 nuggets, we can buy any number in the class $\mathcal{C}(0)^+$; since we can buy 7 nuggets, we can buy any number after 7 in $\mathcal{C}(1)^+$; since we can buy 14 nuggets, we can buy any number after 14 in $\mathcal{C}(2)^+$. The only quantities we cannot buy are 1, 2, 4, 5, 8, and 11.

In the real world, nuggets come in boxes of sizes 6, 9, and 20. Our first step in handling this situation was to divide all the positive integers into six classes, depending on their remainder upon division by 6. The class of remainder 1 is

$$1, 7, 13, 19, 25, 31, 37, 43, 49, \ldots.$$

We now recognize the numbers in this class as containing all the positive integers that are congruent to 1 modulo 6.

Let us introduce some terminology for collections of integers such as these, with negative integers included as well. Once we have the terminology and accompanying notation that we need, we can analyze the general two-box Chicken McNugget problem.

For an integer $m > 1$, the collection of all integers congruent, modulo m, to a given integer a is called a *congruence class modulo m*, or simply a *congruence class* if m is understood. For instance, if $m = 2$, the even integers are a congruence class: the set of integers congruent to 0 modulo 2. Similarly, the odd integers are a congruence class: the set of integers congruent to 1 modulo 2. If $m = 3$, there are three congruence classes: the class of integers congruent to 0 modulo 3, the class of integers congruent to 1 modulo 3, and the class of integers congruent to 2 modulo 3. The collections $C(0)^+$, $C(1)^+$, and $C(2)^+$ above contain the positive integers in these congruence classes.

Fix an integer $m > 1$ and write $C(i)$ for the congruence class that contains the integer i. For example, $C(0)$ is the class consisting of the integers

$$\ldots, \ -2m, \ -m, \ 0, \ m, \ 2m, \ 3m, \ 4m, \ \ldots,$$

and $C(1)$ is the class consisting of

$$\ldots, \ 1 - 2m, \ 1 - m, \ 1, \ 1 + m, \ 1 + 2m, \ 1 + 3m, \ 1 + 4m, \ldots.$$

The congruence class $C(i)$ is the collection of all integers that are congruent to i modulo m. Explicitly, $C(i)$ consists of all the integers in the list

$$\ldots, \ i - 2m, \ i - m, \ i, \ i + m, \ i + 2m, \ i + 3m, \ i + 4m, \ldots.$$

The statement in Theorem 4.1 that every integer is congruent modulo m to exactly one of the integers $0, 1, \ldots, m - 1$ can be rephrased to say that every integer is in exactly one of the congruence classes

$$C(0), \ C(1), \ \ldots, \ C(m - 1)$$

modulo m.

For a concrete example, suppose $m = 5$. Every integer is in exactly one of the five congruence classes modulo 5,

$$C(0), \ C(1), \ C(2), \ C(3), \ C(4).$$

The class $C(2)$ is the collection of integers congruent to 2 modulo 5:

$$\ldots, \ -13, \ -8, \ -3, \ 2, \ 7, \ 12, \ 17, \ 22, \ 27, \ 32, \ \ldots.$$

Each congruence class has many names, infinitely many in fact. This is not unusual; frequently, objects can be described by a variety of names. For example, the United States of America, U.S.A., U.S., Les Etats-Unis,

Die Vereinigten Staaten, and los Estados Unidos de América are names of the same country. Or to take an example from literature, the nineteenth-century Russian novel is known for its proliferation of names, and in Tolstoy's *Anna Karenina* the youngest daughter of Prince Shcherbatsky is referred to as "Princess Shcherbatskaya," "Kitty," "Kitty Shcherbatskaya," "Katerina Alexandrovna," and "Katya." Likewise, in the example above, the congruence class $\mathcal{C}(2)$ modulo 5 has additional names, including $\mathcal{C}(-8)$, $\mathcal{C}(12)$, $\mathcal{C}(17)$, and $\mathcal{C}(34\,489\,372)$, all, of course, modulo 5. After all, the integers congruent to 2 modulo 5 are the same as the integers congruent to -8 modulo 5, the integers congruent to 12 modulo 5, and the integers congruent to $34\,489\,372$ modulo 5, and if you think of $\mathcal{C}(r)$ as "the congruence class containing r," then it is clear that any of $\mathcal{C}(2)$, $\mathcal{C}(-8)$, $\mathcal{C}(12)$, $\mathcal{C}(17)$, and $\mathcal{C}(34\,489\,372)$ serves perfectly well as the name for this congruence class.

In general, if $i \equiv j \pmod{m}$, then every integer congruent to i modulo m is congruent to j modulo m, and vice versa. This means that the integers in the congruence class $\mathcal{C}(i)$ coincide with the integers in the congruence class $\mathcal{C}(j)$. In other words, if $i \equiv j \pmod{m}$, then $\mathcal{C}(i) = \mathcal{C}(j)$. Both $\mathcal{C}(i)$ and $\mathcal{C}(j)$ are names for the same congruence class.

Take m to be 5 again. We know that there are five congruence classes modulo 5, and that they are

$$\mathcal{C}(0), \ \ \mathcal{C}(1), \ \ \mathcal{C}(2), \ \ \mathcal{C}(3), \ \ \mathcal{C}(4).$$

Alternatively, we could describe these five congruence classes as

$$\mathcal{C}(15), \ \ \mathcal{C}(31), \ \ \mathcal{C}(47), \ \ \mathcal{C}(98), \ \ \mathcal{C}(24).$$

These are the same congruence classes, listed in the same order. Or we could describe them as

$$\mathcal{C}(18), \ \ \mathcal{C}(34), \ \ \mathcal{C}(82), \ \ \mathcal{C}(6), \ \ \mathcal{C}(120).$$

This time the classes are written in a different order, but all five classes are listed. The first one is the same as $\mathcal{C}(3)$, the second is the same as $\mathcal{C}(4)$, and so on.

Any one of the integers in a given congruence class modulo m can serve to "represent" that class, just as members of the U.S. House of Representatives represent the citizens of their congressional districts. For this reason, the members of a congruence class are called *representatives* of that congruence class. Each representative gives us enough information to determine all the other integers in the congruence class.

A list of m integers, one from each of the m congruence classes, serves as a complete list of names for all m congruence classes. For example, the list 18, 34, 82, 6, 120 given above constitutes a complete set of names of the five congruence classes modulo 5. We call such a set a *complete set of distinct congruence class representatives modulo m*. The set of integers from 0 to $m-1$

is always a complete set of distinct congruence class representatives modulo
m. However, we can choose other complete sets of distinct congruence class
representatives as well, and sometimes it is convenient to do so.

Let us consider the case of $m = 5$ and use the integers

$$0, \ 1, \ 2, \ 3, \ 4$$

as a complete set of distinct congruence class representatives. Let us multiply
every integer in the set by 3 to get the set

$$0, \ 3, \ 6, \ 9, \ 12.$$

This is also a complete set of distinct congruence class representatives modulo 5. If instead we take the initial set and multiply each element by 4, we
get

$$0, \ 4, \ 8, \ 12, \ 16.$$

This too is a complete set of congruence class representatives modulo 5.

Does this procedure always work? In other words, for an integer $m > 1$
and an integer a, do the multiples

$$0, \ a, \ 2a, \ 3a, \ \ldots, \ (m-1)a$$

of a form a complete set of congruence class representatives modulo m?

Another example shows that they need not. Suppose $m = 6$. Then

$$0, \ 1, \ 2, \ 3, \ 4, \ 5$$

is a complete set of congruence class representatives modulo 6. When we
multiply this set by 4, we get

$$0, \ 4, \ 8, \ 12, \ 16, \ 20.$$

These six integers do not lie in distinct congruence classes: 12 and 0 are
congruent modulo 12, as are 4 and 16, and 8 and 20. Only three congruence
classes are represented. If instead we multiply the initial set by 5 rather than
4, we get

$$0, \ 5, \ 10, \ 15, \ 20, \ 25,$$

and you can check that this is a complete set of congruence class representatives modulo 6.

Notice that with respect to 6, one difference between 4 and 5 is that 4
and 6 are not relatively prime, while 5 and 6 are relatively prime. Perhaps 5
worked and 4 did not for this reason. Indeed, this is the case:

Theorem 4.8. *Let $m > 1$ be an integer. If a is an integer relatively prime to
m, then the integers*

$$0, \ a, \ 2a, \ 3a, \ \ldots, \ (m-1)a$$

form a complete set of congruence class representatives modulo m.

We shall prove Theorem 4.8 using Theorem 4.5. Let us repeat the statement of Theorem 4.5 for easy reference, but with the notation changed slightly.

Theorem 4.9. *Let a and m be relatively prime integers, with $m > 1$. If r and s are integers for which*

$$ar \equiv as \pmod{m},$$

then

$$r \equiv s \pmod{m}.$$

Exercise 4.11. Let m be an integer greater than 1.

1. Using Theorem 4.9, observe that if r and s are integers such that $r \not\equiv s \pmod{m}$ and a is relatively prime to m, then $ar \not\equiv as \pmod{m}$.
2. Deduce that no two integers in the set

$$0, \quad a, \quad 2a, \quad 3a, \quad \ldots, \quad (m-1)a$$

   are congruent to each other modulo m.
3. Argue that therefore no two integers in the set lie in the same congruence class.
4. Prove Theorem 4.8.

We are ready to analyze the Chicken McNuggets problem. Suppose a and m are positive integers with $a < m$ and McDonald's sells Chicken McNuggets in boxes of sizes a and m. Then we can buy $ra + sm$ nuggets for every choice of nonnegative integers r and s. In Chapter 1 we called positive integers of this form (a, m)-*accessible* and positive integers that cannot be expressed in this form (a, m)-*inaccessible*. We would like to determine the (a, m)-accessible positive integers.

If a and m have a common divisor d, then for any integers r and s, we know that $ra + sm$ is divisible by d. In particular, if $d > 1$, then every (a, m)-accessible positive integer is divisible by d, and every positive integer that is not divisible by d is (a, m)-inaccessible. We obtain in this way infinitely many (a, m)-inaccessible positive integers.

Suppose instead that a and m are relatively prime. We studied several examples of this situation in Chapter 1, finding each time that there exists a positive integer N such that every integer n satisfying $n > N$ is (a, m)-accessible. In other words, we found in these examples that we can buy n nuggets for every value of n larger than N. We now have enough tools to prove the following general result.

Theorem 4.10. *Let a and m be relatively prime integers greater than 1, and let*

$$N = am - a - m.$$

Then N is (a, m)-inaccessible, but every integer n satisfying $n > N$ is (a, m)-accessible. Thus, one cannot buy N nuggets in box sizes a and m, but one can buy n nuggets for every integer $n > N$.

Exercise 4.12. Prove Theorem 4.10. You can proceed as follows:

1. Observe that ra is (a, m)-accessible for every positive integer r. In particular, the integers

$$a, \quad 2a, \quad 3a, \quad \ldots, \quad (m-1)a$$

   are all (a, m)-accessible.
2. Deduce from this that $ra + sm$ is also (a, m)-accessible for every positive integer s.
3. For an integer r between 0 and $m - 1$, the congruence class $C(ra)$ consists of all the integers congruent to ra modulo m. Observe that these are the integers of the form

$$\ldots, \quad ra - 2m, \quad ra - m, \quad ra, \quad ra + m, \quad ra + 2m, \quad ra + 3m, \quad \ldots.$$

   Deduce that all the integers in this congruence class that are greater than or equal to ra are (a, m)-accessible.
4. Deduce that in the congruence class $C(ra)$, the largest integer that might fail to be (a, m)-accessible, assuming that it is positive, is $ra - m$.
5. Conclude that the largest integer that may fail to be (a, m)-accessible is $(m-1)a - m$, and that this is just $am - a - m$, or N. Conclude in particular that every integer $n > N$ is (a, m)-accessible.
6. Finally, show that N is not (a, m)-accessible. In fact, show that for each r between 1 and $m - 1$ the integer ra is the smallest positive integer in its congruence class $C(ra)$ that is (a, m)-accessible. This is an even stronger statement than that of Theorem 4.10, since it tells us precisely which positive integers in each congruence class $C(ra)$ are (a, m)-inaccessible, namely, all the positive integers in $C(ra)$ that are strictly smaller than ra. Thus the (a, m)-inaccessible positive integers are all the integers of the form $ra - km$, where $1 \leq r \leq m - 1$ and $0 < k < ra/m$.

The idea underlying the solution to the two-box Chicken McNugget problem is the fact that for relatively prime integers a and m, multiplying the integers from 0 to $m - 1$ by a produces a new set of integers representing all m congruence classes modulo m. Multiplication by a "shuffles" the congruence classes. This is a special case of a result that we will study in Section 7.1.

5

Prime Numbers

5.1 Prime Numbers and Generalized Induction

One approach to the calculation of greatest common divisors is factoring, but we found that this approach takes too long. The Euclidean algorithm gave us a more efficient way to proceed, as well as a means of solving certain Diophantine equations. We would like, however, to return to the calculation of greatest common divisors by factoring, for this method still has theoretical importance. To discuss factoring, we must first discuss prime numbers.

Every positive integer n can be factored as $1 \times n$ and $n \times 1$. These factorizations are not particularly interesting; they are called the *trivial* factorizations of n. In contrast, a factorization $n = rs$ with $1 < r, s < n$ is called a *nontrivial* factorization of n. An integer $n > 1$ whose only factorizations as a product of two positive integers are the two trivial factorizations is called a *prime number*. Equivalently, an integer $n > 1$ is prime if and only if the only positive integers dividing n are 1 and n. An integer $n > 1$ that is not prime is called *composite* (since it is *composed* of at least two nontrivial factors).

The smallest prime number is 2. Its only possible factorizations as a product of two positive integers are 1×2 and 2×1. Similarly, 3 is prime, but 4 is not, since 4 has the nontrivial factorization 2×2. And 5 is prime, while 6 is not, since $6 = 2 \times 3$.

By this time in your mathematical career you are probably familiar with the notion of prime number, and you are probably also familiar with the fact that every integer greater than 1 factors as a product of prime numbers. As familiar as this fact is, it requires proof. We cannot prove it by verifying its truth directly for each and every positive integer, since there are too many, indeed infinitely many, of them. Instead, an induction argument is required. The form of induction we will use is called *generalized induction*.

Suppose we have a sequence of statements that we wish to prove, indexed by integers, the first statement having the integer a as its index. Typically, a is 1, or perhaps 0, but it may be larger, or it could even be negative. The principle of induction says that all of the statements can be proved by our performimg two acts. First, we must prove Statement a. Second, we need to show that for every integer k with $k \geq a$, if Statement k is true, then Statement $k+1$ is true. The principle of generalized induction is a little different. It says that we will succeed in proving the sequence of statements by our performing two acts. the first act is the same as in "regular" induction, and the second is new. First, we must prove Statement a, as before. Second, we need to show that for every integer k with $k \geq a$, if Statements $a, a+1, \ldots, k-1, k$ are all true, then Statement $k+1$ is true. If we do this, we will have proved all the statements.

Let us see right away an example of generalized induction at work.

Theorem 5.1. *Every integer $n \geq 2$ is divisible by a prime number.*

Proof. We begin with $n = 2$. Since 2 is prime, and 2 divides itself, 2 is divisible by a prime number. We have thus easily proved Statement 2. To apply generalized induction, we must prove the theorem for an integer $k+1$ on the assumption that the theorem is true for all integers between 2 and k, that is, for all integers in the range $2, \ldots, k$. If $k+1$ is prime, then the prime number $k+1$ itself divides $k+1$, and we will have thus shown that $k+1$ is divisible by a prime number. If $k+1$ is not prime, then $k+1 = rs$ for some positive integers r and s, neither of which is 1 or $k+1$. In particular, r is an integer between 2 and k. By our assumption, namely, that *every* integer between 2 and k is divisible by a prime number, we must have that r is divisible by some prime number; call it p. Since p divides r, and r divides $k+1$, it follows that p divides $k+1$. Thus in this case as well $k+1$ is divisible by a prime number. Thus we have shown that if each of the integers $2, \ldots, k$ is divisible by a prime number, then so is $k+1$, and therefore, by the principle of generalized induction, the theorem has been proved.

Induction and generalized induction are logically equivalent to each other, in the sense that if one of them is a valid principle of logic, so is the other. We will add generalized induction to our collection of tools.

Exercise 5.1. Use generalized induction to prove Theorem 5.2 below.

Theorem 5.2. *Every integer $n \geq 2$ is either a prime number or a product of a finite number of prime numbers.*

Mathematicians have put a great deal of ingenuity into finding efficient ways to determine whether a positive integer n is prime. In principle, of course, one can do so simply by dividing n by all the integers from 2 to $n-1$ to see whether any of them divides n. For instance, to decide whether 44 497 is a prime number, we can try dividing 44 497 by each of the integers from 2 to 44 496. But before we undertake so daunting a task, let us use some of our own ingenuity to simplify the problem:

Theorem 5.3. *Let n be an integer greater than 1. If n is not prime, then n has a prime divisor p such that $p \leq \sqrt{n}$.*

Exercise 5.2. Prove Theorem 5.3. (Hint: Suppose n factors as rs and $r > \sqrt{n}$. What can you say about the size of s?)

Using Theorem 5.3, we can test the primality of n by dividing n by the positive integers from 2 to $\sqrt{n}$. To decide whether 44 497 is a prime number, we need only try dividing it by all the integers from 2 to 210. This is much, much better than trying to divide 44 497 by all the integers up to 44 496. Still, it is a considerable amount of work to perform by hand or even with a pocket calculator. (Go ahead, try it.)

Even with the use of high-speed computers, testing the primality of integers with many digits by the method of trial division can take a prohibitively long time. However, mathematicians have developed tests for primality that use congruences and avoid the necessity of numerous trial divisions. This fact has practical importance in cryptography. The basic problem of cryptography is to encode a message that your intended recipient can decode easily but that someone who intercepts it is very unlikely to be able to decode. The great difficulty of factoring integers with many digits in comparison to the relative ease with which like-sized integers can be tested for primality is the basis for modern cryptography. The best-known of these encryption techniques is a system called RSA encryption, named for its inventors, Ronald Rivest, Adi Shamir, and Leonard Adleman. We shall return to this fascinating topic in Section 7.5.

For now, we return to our investigation of the prime numbers with a fundamental question: How many prime numbers are there? While there are infinitely many positive integers, it is conceivable that a finite set of prime numbers could produce the infinite set of positive integers. However, a famous theorem of Euclid tells us that there are, in fact, infinitely many primes.

Theorem 5.4 (Euclid). *There are infinitely many prime numbers.*

Exercise 5.3. Euclid's theorem can be rephrased in the following form, which though a bit awkward is admirably suited to proof by induction:

For each positive integer n, there are at least n distinct prime numbers.

Prove this by induction. You can follow the outline below.

1. First prove a little lemma that you will need later: If a positive integer m divides integers a and $a + 1$, then m equals 1.
2. Show that there is at least one prime number. (Statement 1)
3. Assume that there are at least k distinct prime numbers, say $p_1, \ldots, p_k$ (Statement k). Now prove Statement $k + 1$, that there are at least $k + 1$ primes, as follows: Consider the number $(p_1 p_2 \cdots p_k) + 1$, which we will call ℓ. Use the lemma above to show that none of $p_1, \ldots, p_k$ can be a divisor of ℓ and deduce that there is a prime number p_{k+1} distinct from $p_1, \ldots, p_k$.

4. That completes the proof by induction, and you may conclude that for every natural number n there exist at least n distinct prime numbers. In other words, there is no bound on the number of primes. Therefore, there must be infinitely many of them.

It is nice to know that there are infinitely many prime numbers, but that fact should whet our appetites for more precise knowledge about these special numbers. Is there some way that we can measure the proportion of positive integers that the prime numbers make up? To start with an easier example, consider the even positive integers. There are infinitely many of them, yet they are distributed regularly throughout the totality of positive integers: Every second positive integer is even. This suggests that in some sense, roughly one-half of the set of positive integers consists of even positive integers. We can make this more precise by saying that the number of even integers less than a positive integer n is approximately $n/2$.

The question of the distribution of the prime numbers among the integers is much more difficult. They do not appear regularly, like the even integers; rather, they are scattered among the integers seemingly at random. However, a famous theorem allows one to estimate how many positive integers less than a given positive integer n are prime. Following mathematical tradtion, we write $\pi(n)$ for the number of positive integers less than n that are prime. (This π has nothing whatsoever to do with the number π that appears throughout mathematics and has the value $\pi \approx 3.14159265$; the π in $\pi(n)$ is simply a Greek "p" for "prime.") For instance, the primes less than 10 are 2, 3, 5, and 7, so $\pi(10) = 4$. There are 25 prime numbers less than 100, so $\pi(100) = 25$.

If the only way to determine $\pi(n)$ is to write down all the numbers from 1 to n and then count how many are prime, we are not going to be able to compute $\pi(n)$ for large n. One of the most famous theorems in the history of mathematics, the prime number theorem, allows one to estimate $\pi(n)$ without knowing the prime numbers from 1 to n explicitly. The prime number theorem states that $\pi(n)$ has roughly the same size as $n/\ln(n)$, where ln denotes the natural logarithm function. More precisely, it says that as n gets larger, the ratio of $\pi(n)$ and $n/\ln(n)$ gets closer and closer to 1.

For small enough n we can directly calculate $\pi(n)$ and $n/\ln(n)$, to see what this ratio is. For instance, $\pi(100) = 25$ and $100/\ln(100) = 21.71$. The ratio of the two is 1.151. Or, $\pi(10\,000) = 1229$ and $10\,000/\ln(10\,000) = 1085.74$; their ratio is 1.132. For n equal to one billion, the ratio of

$$\pi(1\,000\,000\,000)$$

and $1\,000\,000\,000/\ln(1\,000\,000\,000)$ is 1.054. As these calculations suggest, the ratios get steadily closer to 1 as n increases. It is in this sense that $\pi(n)$ is approximated by $n/\ln(n)$.

The French mathematician Legendre guessed in 1780 that a result along the lines of the prime number theorem was true. Gauss made an improved conjecture in 1792, and mathematicians tried to prove the theorem throughout the nineteenth century. Finally, in 1896, proofs were found independently by Jacques Hadamard and C. J. de la Vallée Poussin. Their work represents one of the great moments in the history of mathematics, a moment whose hundredth anniversary was celebrated recently. Hadamard and de la Vallée Poussin both lived into the 1960s, dying in their nineties. Their long lives led to speculation, prior to their deaths, that their achievement had brought them immortality. Alas, it did not.

5.2 Uniqueness of Prime Factorizations

Suppose two students, Ilya and Anya, are given the same integer $a > 1$ and asked to factor it as a product of prime numbers. Assuming that they manage not to make any mistakes, will their answers be the same? There is a trivial reason why their answers may differ: Ilya and Anya may write their answers in different orders. For instance, if a is 42, Anya might factor 42 as $3 \times 2 \times 7$, while Ilya might come up with $42 = 2 \times 7 \times 3$. We are not going to care much about this difference, since these factorizations are essentially the same, differing only in regard to the order of the factors. We can eliminate this potential ambiguity by agreeing to write the factors in order of size, starting with the smallest primes and continuing in ascending order. Under this convention, the lone factorization of 42 is $42 = 2 \times 3 \times 7$. We hereby adopt this convention. Then a prime factorization of a will have the form $a = p_1 p_2 \cdots p_m$, where the p_i's are prime numbers that satisfy $p_1 \leq p_2 \leq \cdots \leq p_m$. Of course, the prime numbers in such a factorization may occur more than once. This happens, for instance, if $a = 45$, since $45 = 3 \times 3 \times 5$.

We can now address the fundamental issue. Suppose Anya factors a as $a = p_1 p_2 \cdots p_m$ with prime numbers $p_1 \leq p_2 \leq \cdots \leq p_m$, while Ilya factors a as $a = q_1 q_2 \cdots q_n$ with prime numbers $q_1 \leq \cdots \leq q_n$. Will these factorizations necessarily be the same? In other words, must it be true that $m = n$ and that $p_i = q_i$ for each index i?

Your experience with integers and your intuition may suggest that the answer is yes, but experience is not a sufficient guarantee. How much experience do you *really* have? Can you rely one hundred percent on your intuition? Can you be absolutely sure that there is not some really large N lurking among the integers that can be factored in two essentially different ways? Perhaps $N = p_1 p_2$ for two large prime numbers p_1 and p_2, while at the same time this N has the factorization $N = q_1 q_2 q_3$ for three other large prime numbers q_1, q_2, and q_3. These numbers might all be larger than any you have encountered in your mathematical journeys, so your experience would not be broad enough to rule out this possibility. If we are to show that *every* integer $a > 1$ has a unique factorization as a product of prime numbers, then we must come up with a *proof*. Let us state formally exactly what we want to prove:

Theorem 5.5. *Let a be an integer greater than 1. Suppose that $p_1 p_2 \cdots p_m$ and $q_1 q_2 \cdots q_n$ are two prime factorizations of a, arranged so that $p_1 \leq p_2 \leq$*

$\cdots \leq p_m$ and $q_1 \leq q_2 \leq \cdots \leq q_n$. Then $m = n$, and $p_i = q_i$ for each $i = 1, \ldots, m$.

Theorem 5.5 is called the fundamental theorem of arithmetic. In order to prove the fundamental theorem of arithmetic, we need the following important result about prime numbers:

Theorem 5.6. *Suppose a prime number p divides the product bc of integers b and c. Then p divides b, or p divides c.*

The word *or* is used in the statement of Theorem 5.6 in a special way. In logic and mathematics, if $\mathcal{A}$ and $\mathcal{B}$ are mathematical statements or logical propositions, then the composite statement "$\mathcal{A}$ or $\mathcal{B}$" asserts that at least one of $\mathcal{A}$ and $\mathcal{B}$ is true. The possibility that both statements are true is not excluded. For example, "3 divides 6 or 3 divides 14" is true, since 3 divides 6. But "3 divides 6 or 3 divides 15 is also true, since 3 divides both 6 and 15. The fact that 3 divides both is fine; the more the merrier. The statement of Theorem 5.6 will be satisfied if p divides one or both of b and c.

In proving a statement of the form "$\mathcal{A}$ or $\mathcal{B}$" it is often convenient to adopt the following logical point of view: We wish to prove that at least one of $\mathcal{A}$ and $\mathcal{B}$ is true. If $\mathcal{A}$ is true, then we are done, for then certainly "$\mathcal{A}$ or $\mathcal{B}$" is true. If $\mathcal{A}$ is not true, then the only way for "$\mathcal{A}$ or $\mathcal{B}$" to be true is for $\mathcal{B}$ to be true. Thus proving the statement "$\mathcal{A}$ or $\mathcal{B}$" is equivalent to proving the statement "if $\mathcal{A}$ is false, then $\mathcal{B}$ is true." To put it another way, the statements "$\mathcal{A}$ or $\mathcal{B}$" and "If not $\mathcal{A}$, then $\mathcal{B}$" are logically equivalent.

Exercise 5.4. Prove Theorem 5.6. (As just observed, you can restate the theorem as follows: Suppose that a prime number p divides the product bc of integers b and c. Then if p does not divide b, it divides c. This is essentially a special case of Theorem 3.4.)

Exercise 5.5. Use induction to prove the following generalization of Theorem 5.6:

Theorem 5.7. *Suppose a prime number p divides the product*

$$a_1 a_2 \cdots a_n$$

of integers $a_1, \ldots, a_n$. Then p divides at least one of the factors a_i.

We can now use Theorem 5.7 to prove the fundamental theorem of arithmetic, and we do so in the following two exercises.

Exercise 5.6. Suppose

$$a = p_1 p_2 \cdots p_m$$

and

$$a = q_1 q_2 \cdots q_n$$

are two factorizations of an integer $a > 1$ as a product of prime numbers. Use Theorem 5.7 to prove that there is an index j for which $p_m = q_j$. Deduce that

$$p_1 p_2 p_3 \cdots p_{m-1} = q_1 q_2 \cdots q_{j-1} q_{j+1} \cdots q_n.$$

Next obtain a slight rephrasing of this result. Suppose

$$p_1 p_2 \cdots p_m$$

and

$$q_1 q_2 \cdots q_n$$

are two factorizations of an integer $a > 1$ as a product of prime numbers, but this time assume that $p_1 \leq p_2 \leq \cdots \leq p_m$ and $q_1 \leq q_2 \leq \cdots \leq q_n$. Prove that $p_m = q_n$. (Hint: From the first version, you know that $p_m = q_j$ for some index j. Observe that a similar argument yields $q_n = p_i$ for some index i. Now use the inequalities to deduce that $p_m \leq q_n$ and $q_n \leq p_m$ and deduce from this that $p_m = q_n$.) Conclude also that $p_1 p_2 p_3 \cdots p_{m-1} = q_1 q_2 \cdots q_{n-1}$.

Exercise 5.7. Prove the fundamental theorem of arithmetic. Recall that you are given two prime factorizations $p_1 p_2 \cdots p_m$ and $q_1 q_2 \cdots q_n$ of an integer $a > 1$ arranged so that $p_1 \leq p_2 \leq \cdots \leq p_m$ and $q_1 \leq q_2 \leq \cdots \leq q_n$. You may as well assume that $m \leq n$. You wish to show that in fact $m = n$ and that $p_i = q_i$ for each i between 1 and m. Proceed by induction on m, the length of the possibly shorter of the two factorizations.

1. Deal with the case $m = 1$. In this case, $a = p_1$, so that a is a prime number. You must prove that the second factorization $q_1 \leq q_2 \leq \cdots \leq q_n$ of a consists of a single prime number and that this prime number is p_1. This will follow from the definition of primality.
2. Perform the inductive step. On the assumption of Statement m, which states that factorization is unique for every positive integer that has a prime factorization of length m, prove statement $m + 1$, which is the uniqueness result for an integer a whose possibly shorter prime factorization has length $m + 1$. You can use the last part of the preceding exercise for this.

5.3 Greatest Common Divisors Revisited

When greatest common divisors were introduced in Section 3.1, you computed several examples, such as the greatest common divisor of 6 and 9 and the greatest common divisor of 25 and 40. You may have obtained the answer $(6, 9) = 3$ by factoring 6 and 9 as 2×3 and 3×3, from which you read off that 3 is the greatest common divisor; and then you proceeded similarly to see that $(25, 40) = 5$, by factoring 25 and 40 as 5×5 and $5 \times 2 \times 2 \times 2$, leading to the answer 5. We also computed the greatest common divisor of 30 and 84. The same procedure would work:

$$30 = 2 \times 3 \times 5$$

and

$$84 = 2 \times 2 \times 3 \times 7.$$

From this we see that 30 and 84 have a factor of 2 in common and a factor of 3 in common, and so $(30, 84) = 2 \times 3 = 6$.

Ultimately, we rejected the prime factorization approach to computing greatest common divisors because factoring integers as products of prime numbers can take too long. For instance, for the pair 227 761 and 661 643, the prime factorizations are

$$227\,761 = 421 \times 541, \quad 661\,643 = 541 \times 1223,$$

and therefore 541 is the greatest common divisor. But it takes a long time to find these prime factorizations. In contrast, when you apply the Euclidean algorithm to find the greatest common divisor of 227 761 and 661 643, you get the same answer in just a couple of minutes. Thus prime number factorization is not the best approach for calculating greatest common divisors. Nevertheless, using prime factorizations to describe greatest common divisors is important theoretically, and for this reason we shall discuss the approach further.

Let us make another calculation of a greatest common divisor using the prime factorization approach, as a guide to formulating precisely what the approach is. We will compute $(8316, 19\,800)$. The prime factorizations are

$$8316 = 2^2 \times 3^3 \times 7 \times 11$$

and

$$19\,800 = 2^3 \times 3^2 \times 5^2 \times 11.$$

Let us rewrite these factorizations as

$$8316 = 2^2 \times 3^3 \times 5^0 \times 7^1 \times 11^1$$

and

$$19\,800 = 2^3 \times 3^2 \times 5^2 \times 7^0 \times 11^1.$$

Doing so allows us to line up the prime numbers occurring in these two factorizations, so that we can compare exponents at a glance. Now we can see that 8316 and 19 800 have in common the following factors: two 2's, two 3's, no 5's, no 7's, and one 11. From this we conclude that $(8316, 19\,800) = 2^2 \times 3^2 \times 11 = 396$.

In general, prime factorizations can be used to compute greatest common divisors just as in the last example. We want to formulate a precise statement of this procedure and then prove that the statement is true. For this purpose we need to develop some notation.

A typical way to write a prime factorization of an integer $a > 1$ is by writing the prime factors with their exponents, as we did in the example above. For instance, we can write

$$a = p_1^{e_1} p_2^{e_2} \cdots p_r^{e_r},$$

where the integers $p_1, \ldots, p_r$ are distinct prime numbers and the exponents $e_1, \ldots, e_r$ are positive integers. If we wish to compare a to another integer b, we might allow 0 as an exponent, as we did in the example above. For instance, if $p_1, \ldots, p_r$ is a complete list of the distinct prime numbers dividing one or both of a or b, then we can write

$$a = p_1^{e_1} p_2^{e_2} \cdots p_r^{e_r}$$

and

$$b = p_1^{f_1} p_2^{f_2} \cdots p_r^{f_r},$$

where the exponents e_i and f_i are nonnegative integers. Let us adopt this notation.

The following divisibility criterion is one whose truth seems obvious. Nonetheless, it requires proof. Notice the role of the fundamental theorem of arithmetic in its proof.

Proposition 5.8 *Suppose a and b are integers greater than 1 with prime factorizations*

$$a = p_1^{e_1} p_2^{e_2} \cdots p_r^{e_r}$$

and

$$b = p_1^{f_1} p_2^{f_2} \cdots p_r^{f_r},$$

where the exponents are nonnegative integers and $p_1, \ldots, p_r$ are distinct prime numbers. Then a divides b if and only if for each index i the inequality $e_i \leq f_i$ holds.

Proof. Suppose a divides b. Then there is a positive integer k such that $b = ka$. Since k divides b, every prime divisor of k is also a prime divisor of b. Therefore, the prime factorization of k has the form

$$p_1^{d_1} p_2^{d_2} \cdots p_r^{d_r}$$

for some nonnegative integer exponents d_i, $i = 1, \ldots, r$. Substituting this expression for k in the equality $b = ka$, we obtain

$$b = p_1^{d_1 + e_1} p_2^{d_2 + e_2} \cdots p_r^{d_r + e_r}.$$

The fundamental theorem of arithmetic yields the equality $d_i + e_i = f_i$ for each index i. Since d_i is nonnegative,

$$e_i = f_i - d_i \leq f_i.$$

This proves one half of the proposition.

For the converse, assume that $e_i \leq f_i$ for each i. Then there are nonnegative integers d_i, $i = 1, \ldots, r$, such that $f_i = d_i + e_i$. Hence,

$$b = p_1^{f_1} p_2^{f_2} \cdots p_r^{f_r} = p_1^{d_1 + e_1} p_2^{d_2 + e_2} \cdots p_r^{d_r + e_r} = p_1^{d_1} p_2^{d_2} \cdots p_r^{d_r} a,$$

and b is divisible by a.

Using Proposition 5.8 and the fundamental theorem of arithmetic, we can state and prove the result we want on the calculation of greatest common divisors using prime factorizations.

Theorem 5.9. *Suppose a and b are integers greater than 1 with prime factorizations*

$$a = p_1^{e_1} p_2^{e_2} \cdots p_r^{e_r}$$

and

$$b = p_1^{f_1} p_2^{f_2} \cdots p_r^{f_r},$$

where the exponents are nonnegative integers and $p_1, \ldots, p_r$ are distinct prime numbers. For each index i between 1 and r, let g_i equal the smaller of the two exponents e_i and f_i. Then the greatest common divisor of a and b is given by the formula

$$(a, b) = p_1^{g_1} p_2^{g_2} \cdots p_r^{g_r}.$$

Exercise 5.8. Prove Theorem 5.9.

A *common multiple* of integers a and b is an integer c that is divisible by both a and b. For example, ab is a common multiple of a and b, but there may be smaller common multiples. The *least common multiple* of a and b is the smallest positive integer that is a common multiple of a and b. It is written $[a, b]$. For instance, the least common multiple of 3 and 7 is 21, and the least common multiple of 4 and 6 is 12.

In principle, we can always compute the least common multiple of two positive integers a and b. For instance, if $a < b$, one way to proceed would be to search through the list $b, 2b, 3b, \ldots$ of positive multiples of b until we find the first one that is divisible by a. This must be their least common multiple. Another way to proceed is given by the following theorem, a companion to Theorem 5.9.

Theorem 5.10. *Suppose a and b are integers greater than 1 with prime factorizations*

$$a = p_1^{e_1} p_2^{e_2} \cdots p_r^{e_r}$$

and

$$b = p_1^{f_1} p_2^{f_2} \cdots p_r^{f_r},$$

where the exponents are nonnegative integers and $p_1, \ldots, p_r$ are distinct prime numbers. For each i between 1 and r, let h_i equal the larger of the two exponents e_i and f_i. Then the least common multiple of a and b is given by the formula

$$[a, b] = p_1^{h_1} p_2^{h_2} \cdots p_r^{h_r}.$$

Exercise 5.9. Prove Theorem 5.10.

Exercise 5.10. Let a and b be positive integers.

1. Using the expressions for (a, b) and $[a, b]$ in terms of prime factorizations of a and b, prove that

$$ab = (a, b) \times [a, b].$$

Conclude that

$$[a, b] = \frac{ab}{(a, b)}.$$

2. Explain how this equality can be combined with the Euclidean algorithm to provide an efficient means of computing the least common multiple of a and b without using prime factorizations.

6

Rings

6.1 Numbers

In this chapter we begin the study of some new number systems. But before we consider these new systems, we would do well to look at the number systems that we already know. Our story begins with the numbers that we use to count: $1, 2, 3, \ldots$. For millennia these satisfied all of humankind's mathematical requirements, serving to keep track of whatever objects needed to be counted. In the third century B.C.E. the Babylonians invented a symbol for the number zero, and zero was reinvented by the Mayans in the fourth century C.E. and again in India in the fifth century. By now, sixteen hundred years later, zero seems a natural enough number, and so we adjoin it to our original set of counting numbers and call this new set the *natural numbers*. We shall denote the set of natural numbers by the symbol $\mathbb{N}$:

$$\mathbb{N} = \{ 0, 1, 2, 3, 4, 5, \ldots \}.$$

The natural numbers allow us to count, and they also allow us to add and multiply numbers: The sum of two natural numbers is again a natural number, and likewise, the product of two natural numbers is a natural number. We say that the natural numbers are *closed* under the operations of addition and multiplication.

However, the natural numbers are incomplete in the sense that they are *not* closed under subtraction: If you try to subtract a natural number from another natural number, the result is not necessarily another natural number. When we were children, we simply declared that one *cannot* subtract a larger (natural) number from a smaller number, and indeed, the human race faced the same difficulty in its childhood. But the desire to express the result of such subtractions demanded that this gap in the natural numbers be filled, and so negative numbers were invented.

In order to be able to subtract *any* natural number from *any other* natural number we are going to have to augment our set of numbers. To do so, we simply invent the numbers that we need by introducing a new symbol, the negative sign ("−"). When the negative sign is placed in front of a natural number n it gives us another natural number, called "negative n" and written $-n$, which we understand to mean something along the lines of "the quantity n, but in a negative sense." For example, we can now express the concept of "net worth" applied to individuals whose liabilities exceed their assets: "If I have 12 000 dollars in assets and 15 000 dollars in liabilities, then my net worth is −3000 dollars." The *amount* is 3000, but alas, it is in the negative sense.

We are now ready to create a set of numbers that is closed under the operation of subtraction. For each natural number $n \geq 1$ we adjoin the number $-n$ to the set of natural numbers. This gives us the set $\{\ldots, -4, -3, -2, -1, 0, 1, 2, 3, 4, \ldots\}$. We call this new collection of numbers the *integers*. The word "integer" comes from the Latin adjective *integer*, which means "whole" or "complete," and indeed, the integers are a *complete* collection of *whole* numbers. The set of integers is denoted by $\mathbb{Z}$:

$$\mathbb{Z} = \{\ldots, -4, -3, -2, -1, 0, 1, 2, 3, 4, \ldots\}.$$

The reason for the letter "Z" is that for historical reasons the Germans obtained the naming rights for the integers, and the German word for "number" is *Zahl* (pronounced "tsahl").

To go along with our increasing mathematical sophistication, let us introduce some fancy terminology to describe what we have done in creating the negative integers. Zero has the property that when it is added to any other natural number n, the result is n again. Because of this property, 0 is called an *additive identity*: When 0 is added to n, the "identity" of n is preserved.

The numbers $-n$ that we created also have an important property: When an integer n is added to $-n$, the result is 0, the additive identity. For this reason, $-n$ is called the *additive inverse* of n. For instance, −3 is the additive inverse of 3, and 18 is the additive inverse of −18. In this language, the additive inverses of positive numbers in $\mathbb{N}$ are not in $\mathbb{N}$; the set $\mathbb{N}$ is not closed under additive inverses. We can see the creation of the set $\mathbb{Z}$ of integers as a successful attempt to fix this defect in $\mathbb{N}$. The integers have the pleasant property that the additive inverse of every integer is an integer. Thus the integers are closed under addition and additive inverses. This allows us to do addition and subtraction in the integers.

Before computer typesetting there was a limited collection of typographical symbols for printing mathematical texts, and perhaps it is for that reason that we tend to use the same symbol for a number of different functions. For example, the negative sign "−" that we use to indicate an additive inverse is nothing other than the familiar minus sign that we use to indicate subtraction. Now that we have additive inverses, we could abandon the idea of subtraction altogether and use addition of additive inverses instead. For example, instead of writing $7 - 3$ we could write $7 + (-3)$, and in fact, we use both notations in our daily mathematical lives. In some of the new number systems that we are about to create, we will define addition, but not subtraction, but then we will be a bit sloppy and write $a - b$ when what we really mean is $a + (-b)$, namely, the sum of a and the additive inverse of b.

The attentive reader may have noticed a subtle shift in the meaning of the negative sign. We introduced this symbol to mean "in the negative sense," but now we are using it to mean "additive inverse." This latter sense is much more flexible, because we can place "−" in front of any integer to indicate its additive inverse. Thus all of the following uses of "−" make sense:

1. -11, which represents the additive inverse of 11, which we call "negative eleven" for short;
2. $-(-12)$, which represents the additive inverse of the additive inverse of 12, which is simply 12;
3. In general, for an integer n, its additive inverse can be written $-n$, regardless of whether n is positive, negative, or zero.

By the way, the attentive reader may also have wondered why in creating the negative integers we defined $-n$ only for $n \geq 1$. We did *not* define -0. In the language of additive inverses the reason is clear: Zero is its own additive inverse, $0 + 0 = 0$, and therefore we did not need to define -0, since $-0 = 0$.

Now that we have the integers, we could congratulate ourselves on a job well done and not worry about anything further. However, mathematicians are worriers by nature, and what we are going to worry about now is the question of *multiplicative* inverses. But first let us discuss the notion of *multiplicative identity*. The number 1 has the same property with respect to multiplication that 0 does with respect to addition: 1 times any number n is n, and so in multiplication, 1 preserves a number's identity. For this reason we call 1 a *multiplicative identity* for $\mathbb{Z}$.

A *multiplicative inverse* of a number n is a number m with the property that $n \times m = 1$. We note that the number 0 has no multiplicative inverse, and we are not going to attempt to invent one. There are, however, some integers with multiplicative inverses: The multiplicative inverse of 1 is 1, and the multiplicative inverse of -1 is -1. However, no other integer has another integer as its multiplicative inverse, and if we are to have a number system that is closed under multiplicative inverses, which amounts to having a number system in which we can do division, we are going to have to invent them. Therefore, for each nonzero integer n we *define* its multiplicative inverse to be $\frac{1}{n}$. That is, $\frac{1}{n}$ has, by definition, the property that $n \times \frac{1}{n} = 1$. Thus, for example, the multiplicative inverse of 2 is $\frac{1}{2}$, and the multiplicative inverse of 3 is $\frac{1}{3}$.

If we adjoin these new numbers to the set of integers, we get the set $\{\ldots, -3, -\frac{1}{3}, -2, -\frac{1}{2}, -1, 0, 1, 2, \frac{1}{2}, 3, \frac{1}{3}, \ldots\}$. Observe that we have created a new number for each integer except -1, 0, and 1. Now, except for 0, which will remain a special case, this augmented set is closed under both additive and multiplicative inverses. However, we definitely have something to worry about:

This augmented set is no longer closed under addition and multiplication. Therefore, we have to make the set bigger.

We start out by enlarging the set of integers and their multiplicative inverses by adjoining all products of integers and multiplicative inverses of integers, and we write such a product $m \times \frac{1}{n}$ as $\frac{m}{n}$. We will need such numbers if we want closure under multiplication. (We also declare that $-\frac{m}{n} = \frac{-m}{n} = \frac{m}{-n}$.) Since another way of thinking about such fractions is as *ratios* of integers, we call our new set the *rational numbers*. To put it formally, the *rational numbers* comprise all numbers of the form $\frac{m}{n}$, with m and n integers and n nonzero. Traditionally, this collection of numbers is denoted by $\mathbb{Q}$. The letter "Q" is used because it is the first letter of "quotient."

Let us examine the set of rational numbers to see whether it satisfies, or fails to satisfy, closure under addition, multiplication, and additive inverses. Every rational number has a rational number as its additive inverse, and every rational number besides 0 has a rational number as its multiplicative inverse, and furthermore, the set of rational numbers is closed under addition and multiplication. This makes $\mathbb{Q}$ a good setting for carrying out the arithmetic operations of addition, subtraction, multiplication, and division, and indeed, $\mathbb{Q}$ is the setting for most of our daily contact with numbers.

However, the set $\mathbb{Q}$ of rational numbers is still not large enough. The realization of the inadequacy of the rational numbers for geometry was a shock to the ancient Greeks known as the Pythagoreans. In the sixth century B.C.E. they discovered that not all line segments that can be constructed with a straightedge and compass have lengths that are rational numbers. For example, consider the diagonal of a 1×1 square. Let us call its length a (see Figure 6.1). If we recall what the Pythagorean theorem says about the sides

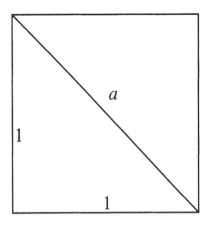

Figure 6.1. A diagonal of a 1×1 square.

of a right triangle, we see that an $a \times a$ square has area equal to the sum of the areas of two 1×1 squares. That is, a is a number such that $a \times a = 1^2 + 1^2 = 2$. What the Pythagoreans discovered is that a is not a rational number; that is, there is no way to write a as a quotient of two integers. Therefore, to do the most elementary geometry we will have to enlarge the set $\mathbb{Q}$ of rational numbers to include numbers like a. First we need to invent a new symbol for such numbers, and so we do, writing $a = \sqrt{2}$. We read this as "the square root of 2," to indicate that a is a number that when multiplied by itself gives the result 2.

We will prove that $\sqrt{2}$ is not a rational number using a method that is a powerful logical weapon in the mathematician's armamentarium. The idea is to reason from certain premises that we believe to be false to derive an obviously false conclusion, thereby establishing that the premises were indeed false. This method of proof is called "proof by contradiction" or "reductio ad absurdum," since the goal is to obtain a logical contradiction or establish an absurd conclusion. In the next exercise we will show that the premise that $\sqrt{2}$ is rational is false by deducing, based on that premise, that two relatively prime integers have a factor greater than 1, an absurd conclusion and a contradiction to the definition of relative primality.

Exercise 6.1. Prove that $\sqrt{2}$ is not a rational number. Proceed as follows, assuming that what we are trying to prove is false (that is, assuming that $\sqrt{2}$ is rational) and reaching a contradiction:

1. Suppose that $\sqrt{2}$ is rational, so that it equals $\frac{m}{n}$ for some integers m and n with $n \neq 0$. Explain why it can be assumed that m and n are relatively prime; then make this assumption.
2. Square both sides of the equation $\sqrt{2} = \frac{m}{n}$, clear denominators, and use Theorem 5.6 to deduce that m is divisible by 2.
3. Use Theorem 5.6 again to deduce that n is divisible by 2.
4. Observe that this contradicts the assumption that m and n are relatively prime.
5. Conclude that $\sqrt{2}$ is not rational.

Prove also, using a similar argument, that $\sqrt{3}$ is not a rational number.

Since there are not enough rational numbers to measure length, we enlarge our system of numbers to include all the numbers that lie on the number line. The positive real numbers are the numbers we need for measuring lengths, and we include the negative real numbers too so that we can be sure to have additive inverses. We call this set the *real numbers*, and we denote this set by the symbol $\mathbb{R}$.

We have not given a precise mathematical definition of the real numbers because to do so would take us too far afield. However, we can say something about our new and improved set $\mathbb{R}$. We included "negative lengths" in $\mathbb{R}$ so that we would be guaranteed additive inverses. But what about multiplicative inverses? It is reassuring to know that every nonzero real number r has another real number s as its multiplicative inverse, and we can see this by a nice geometric argument. Figure 6.2 is the result of the following construction: We are given a positive real number r. To obtain its multiplicative inverse, we construct the right triangle ABC with base r and altitude 1. We then extend the line segment AC to the right and construct a perpendicular to segment AB at B, calling the point where these two lines intersect D. If we denote the length of the segment CD by s, then $r \times s = 1$, and therefore, s is a multiplicative inverse of r.

Exercise 6.2. Prove that s in Figure 6.2 is a multiplicative inverse of r.

We denote the multiplicative inverse of a nonzero real number r by $\frac{1}{r}$ or $1/r$. We have actually shown only that *positive* real numbers have multiplicative inverses. What about multiplicative inverses for negative real numbers?

Exercise 6.3. Prove that every negative real number has a multiplicative inverse.

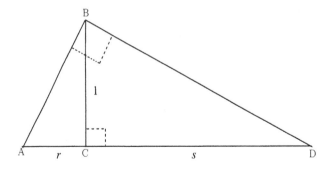

Figure 6.2. Geometric demonstration that every positive real number r has a multiplicative inverse s.

The laying of a proper foundation for the real numbers is an interesting and important story. The starting point is the idea that real numbers are numbers that can be approximated arbitrarily closely by rational numbers. This is what a decimal expansion of a real number r is; the expansion records a list of rational numbers that are closer and closer to r. To make this precise requires a well-developed theory of *convergence* for sequences of numbers. The underlying ideas also serve as the foundation of calculus. This is neither the time nor place to develop these ideas, and we shall not say much more than that the set of real numbers is an interesting set indeed. It contains a vast menagerie of different types of numbers.

An enormous benefit produced by the introduction of the real numbers is that we have obtained enough numbers to solve many equations that are unsolvable using only integers or rational numbers. We can regard the passage from $\mathbb{N}$ to $\mathbb{Z}$ as a means of providing the numbers needed to solve equations of the form $x + m = 0$ when m is a positive integer. The passage from $\mathbb{Z}$ to $\mathbb{Q}$ enables us to find solutions to equations of the form $nx - m = 0$ when m and n are arbitrary integers, with $n \neq 0$. What about the equation $x^2 - 2 = 0$? It has no integer or rational solution, but it has the real solutions $x = \pm\sqrt{2}$. Introducing real numbers allows us to solve this equation and many others as well. In fact, for every *positive* real number r, we can solve the equation $x^2 - r = 0$.

Even the system of real numbers is not large enough to solve all algebraic equations, and we quickly run into trouble. For example, the equation $x^2 + 1 = 0$ has no real-number solution, since the square of every real number is 0 or positive. This motivates one more expansion of our system of numbers.

We will do as we have done before when our set of numbers needed to be enlarged: We simply invent a new symbol. We do this now by introducing a special number i with the property that $i^2 = -1$. In other words, i is a square root of -1, a solution to the equation $x^2 = -1$. The letter "i" is used because it is the first letter of the word "imaginary." After all, if i is not a *real* number, it must be *imaginary*. We are not going to worry about what i is just yet. We just declare that it is a new number with the property that its square is -1. Indeed, we are never going to worry about what i actually is, because it is nothing more or less than our invention, a number that we have made up that has the property that $i \times i = -1$.

Once we adjoin i to the set of real numbers, we are going to have to adjoin many more numbers to ensure that our new set of numbers is closed under addition, multiplication, additive inverses, and (except for zero) multiplicative inverses. Therefore, we include the additive inverse of i, denoted by $-i$. Notice that $-i$ is also a square root of -1, for $(-i)(-i) = (-1)(-1)(i)(i) = i^2 = -1$. For closure under addition we will also need such numbers as $1 + i$, $3i$, and $42i - 17$. In fact, we require the entire collection of all numbers of the form $r + si$, with r and s real. This collection is called the set of *complex numbers*, and it is denoted generally by $\mathbb{C}$.

If you are not already familiar with complex numbers, try the following calculations:

1. $(2 + 3i) + (5 - i) =$
2. $(2 + 3i)(5 - i) =$
3. More generally, suppose that a, b, c, and d are real numbers.
   (a) Calculate $(a + bi) + (c + di)$.
   (b) Verify that the additive inverse of $a + bi$ is $-a - bi$.
   (c) Calculate $(a + bi) \times (c + di)$.
   (d) Conclude that the sum and product of every two complex numbers is another complex number.

The property of the complex numbers that you checked in the last item was a surprise bonus. We enlarged the set of real numbers by a single imaginary number i; then we threw in everything that we needed for closure under addition. And then it turned out that we got closure under multiplication at no extra charge.

What about closure under multiplicative inverses? To hope for closure under multiplicative inverses "for free" is madness, is it not? Surely, we are going to have to enlarge our set yet again. Or are we? Here is an even bigger surprise: The set $\mathbb{C}$ of complex numbers is closed (except, as always, for zero) under multiplicative inverses. Getting started in proving that this is so is easy: We see at once that the multiplicative inverse of i is $-i$, since $(-i) \times i = -(i \times i) = -(-1) = 1$. Similarly, for every nonzero real number r, the multiplicative inverse of ri is $-\frac{1}{r}i$. But now the going gets a bit rough. If $a + bi$ is a nonzero complex number (that is, a and b are real, not both zero), is there another complex number $r + si$ that is its multiplicative inverse?

If $a + bi$ is to have a multiplicative inverse $r + si$, then we are going to have to obtain the product $(a + bi)(r + si) = 1$. It is not clear in advance that *in general* we can multiply two complex numbers together and end up with a *real number*, let alone end up with the real number 1.

Exercise 6.4. Find a complex number $r + si$ such that $(3 + 5i) \times (r + si)$ is a real number.

After a bit of experimentation you may have discovered that $(3 + 5i) \times (3 - 5i) = 34$, and 34 is a real number. This trick works in general, since it is based on the familiar identity $x^2 - y^2 = (x + y)(x - y)$, and therefore, we now introduce the notion of *complex conjugation*. For a complex number $a + bi$, the complex number $a - bi$ is called its *complex conjugate*, or simply its *conjugate*. You should write down some complex numbers and multiply them by their conjugates. For instance, calculate $(1 + i)(1 - i)$ and $(2 + 3i)(2 - 3i)$.

Exercise 6.5. Suppose a and b are real numbers.

1. Calculate the product $(a + bi)(a - bi)$.
2. Show that for $a + bi$ not equal to 0 this product is a positive real number.
3. Using the fact that the product is nonzero and real, find a multiplicative inverse $r + si$ to $a + bi$. In other words, find explicit real numbers r and s, expressed in terms of a and b, so that

$$(a + bi)(r + si) = 1.$$

(You should start with $a - bi$ and recognize that although it is not a multiplicative inverse, it almost is. Make an adjustment to it to obtain an inverse.)

4. Conclude that every nonzero complex number has a multiplicative inverse.

We have gone from the natural numbers to the integers to the rational numbers to the real numbers to the complex numbers. Each time we enlarged our system of numbers, it became possible to solve additional algebraic equations. How much further must we go? A theorem called the *fundamental theorem of algebra* states that for the purposes of solving algebraic equations, we need go no further. More precisely, every polynomial equation with complex-number coefficients has a complete set of solutions in the complex numbers. This famous theorem was first proved by Gauss over two hundred years ago. We will return to the fundamental theorem in Section 10.7.

6.2 Number Rings

We have considered four number systems that are closed under addition and multiplication: the integers $\mathbb{Z}$, the rational numbers $\mathbb{Q}$, the real numbers $\mathbb{R}$, and the complex numbers $\mathbb{C}$. The traditional word used in algebra for a system of numbers that is closed under addition and multiplication is *ring*. A precise definition will be given later, but for now think of a ring as a collection of numbers you can add and multiply, with the further properties that there is an additive identity 0, there is a multiplicative identity 1, and every number in the ring has an additive inverse. Thus, $\mathbb{Z}$, $\mathbb{Q}$, $\mathbb{R}$, and $\mathbb{C}$ are rings, but $\mathbb{N}$ is not. The rings $\mathbb{Q}$, $\mathbb{R}$, and $\mathbb{C}$ have the additional property that every nonzero number in the ring has a multiplicative inverse in the ring. The integers do not have this property. A ring in which every nonzero number has a multiplicative inverse is called a *field*. Thus, $\mathbb{Q}$, $\mathbb{R}$, and $\mathbb{C}$ are fields, but $\mathbb{Z}$ is not.

A number in a ring is called a *unit* in that ring if it has a multiplicative inverse in the ring.[1] This is a convenient terminology, allowing us to abbreviate the statement "r is a number with a multiplicative inverse in the ring" by the shorter statement "r is a unit." The only units in $\mathbb{Z}$ are 1 and -1. In contrast, since $\mathbb{Q}$, $\mathbb{R}$, and $\mathbb{C}$ are fields, every nonzero element in them is a unit. If a number u is a unit, we will sometimes write its multiplicative inverse $1/u$ as u^{-1}.

We would like now to study two new rings that like $\mathbb{Z}$ have only a few units. One way to produce such rings is to adjoin numbers to the integers, but not too many, making sure that what we get is closed under addition and multiplication. For instance, inside $\mathbb{R}$, consider the collection of real numbers of the form $a + b\sqrt{2}$, where a and b are integers. We will give this set the name $\mathbb{Z}\left[\sqrt{2}\right]$, since we have adjoined $\sqrt{2}$ to $\mathbb{Z}$. Some elements of $\mathbb{Z}\left[\sqrt{2}\right]$ are $11 + 32\sqrt{2}$ and $-5 + 21\sqrt{2}$.

Practice arithmetic in $\mathbb{Z}\left[\sqrt{2}\right]$ by doing the following sample calculations, and more if you wish:

1. $\left(3 - 4\sqrt{2}\right) + \left(5 + 2\sqrt{2}\right) =$

---

[1] Why are these elements called "units"? In $\mathbb{Z}$, the ring of integers, our archetypal model of a ring, the units are 1 and -1. The Latin word for the number 1 is *unum*, and in a sense, the units of a ring play the role played by 1 and -1 in the ring of integers.

2. $\left(3 - 4\sqrt{2}\right)\left(5 + 2\sqrt{2}\right) =$
3. $\left(3 - 4\sqrt{2}\right)\left(3 + 4\sqrt{2}\right) =$
4. $\left(3 + \sqrt{2}\right)\left(3 - \sqrt{2}\right) =$

Exercise 6.6. Let us do some more calculations in $\mathbb{Z}\left[\sqrt{2}\right]$.

1. Let $a + b\sqrt{2}$ represent a generic number in $\mathbb{Z}\left[\sqrt{2}\right]$. Show that

$$\left(a + b\sqrt{2}\right)\left(a - b\sqrt{2}\right) = a^2 - 2b^2.$$

   In particular, notice that the product $(a + b\sqrt{2})(a - b\sqrt{2})$ is an integer.
2. Show that $\mathbb{Z}\left[\sqrt{2}\right]$ is closed under addition and multiplication; that is, the sum and product of every pair of numbers $a + b\sqrt{2}$ and $c + d\sqrt{2}$ with a, b, c, and d integers is another number $e + f\sqrt{2}$ of the same form.

Exercise 6.7. Now let us look for units in $\mathbb{Z}\left[\sqrt{2}\right]$. Certainly, every nonzero number $a + b\sqrt{2}$ in $\mathbb{Z}\left[\sqrt{2}\right]$ has some real number as its multiplicative inverse, but this does not mean that $a + b\sqrt{2}$ is a unit in $\mathbb{Z}\left[\sqrt{2}\right]$. All it tells us is that $a + b\sqrt{2}$ is a unit in the ring of real numbers. The question is, does the real number that is the multiplicative inverse of $a + b\sqrt{2}$ lie in $\mathbb{Z}\left[\sqrt{2}\right]$?

1. Suppose that $a + b\sqrt{2}$ is a unit in $\mathbb{Z}\left[\sqrt{2}\right]$ and that its multiplicative inverse is $c + d\sqrt{2}$. Show that $ac + 2bd = 1$ and $ad + bc = 0$.
2. Using these equations, deduce that $a - b\sqrt{2}$ is also a unit in $\mathbb{Z}\left[\sqrt{2}\right]$, and that its inverse is $c - d\sqrt{2}$.
3. Continuing with these numbers, show that the product

$$\left(a + b\sqrt{2}\right)\left(a - b\sqrt{2}\right)\left(c + d\sqrt{2}\right)\left(c - d\sqrt{2}\right)$$

   equals 1 and deduce that $a^2 - 2b^2$ must equal 1 or -1.
4. You have proved that if $a + b\sqrt{2}$ is a unit in $\mathbb{Z}\left[\sqrt{2}\right]$, then $a^2 - 2b^2 = \pm 1$.
5. Conversely, suppose a and b are integers satisfying either $a^2 - 2b^2 = 1$ or $a^2 - 2b^2 = -1$. Prove that $a + b\sqrt{2}$ is a unit in $\mathbb{Z}\left[\sqrt{2}\right]$. What is its multiplicative inverse?
6. Conclude that the units in $\mathbb{Z}\left[\sqrt{2}\right]$ correspond to solutions to the Diophantine equations

$$x^2 - 2y^2 = 1; \quad x^2 - 2y^2 = -1.$$

7. Observe that $(1, 1)$ is a solution to one of these equations, as is $(3, 2)$. Deduce that $\sqrt{2} + 1$ is a unit, with inverse $\sqrt{2} - 1$, and $3 + 2\sqrt{2}$ is a unit with inverse $3 - 2\sqrt{2}$.
8. Observe that $3 + 2\sqrt{2}$ is just $\left(\sqrt{2} + 1\right)^2$, and the inverse of $3 + 2\sqrt{2}$ is $\left(\sqrt{2} - 1\right)^2$. More generally, show that $\left(\sqrt{2} + 1\right)^n$ is a unit for every positive integer n by describing its inverse. Observe that we get in this way infinitely many units in $\mathbb{Z}\left[\sqrt{2}\right]$.

9. Prove that if a is an even integer and b is an integer, the number $a + b\sqrt{2}$ cannot be a unit in $\mathbb{Z}\left[\sqrt{2}\right]$. (Is it possible for such a pair (a, b) to satisfy the equation $x^2 - 2y^2 = \pm 1$?) Conclude that there are infinitely many numbers in $\mathbb{Z}\left[\sqrt{2}\right]$ that are not units.

Exercise 6.8.

In fact, we can show that in some sense the units are few and far between in the ring $\mathbb{Z}\left[\sqrt{2}\right]$.

1. For N a positive integer, how many numbers $a + b\sqrt{2}$ are there in $\mathbb{Z}\left[\sqrt{2}\right]$ with $-N \leq a, b \leq N$?
2. Observing that for a given a there is at most one integer b that satisfies the Diophantine equation $x^2 - 2y^2 = 1$ and at most one integer b that satisfies the Diophantine equation $x^2 - 2y^2 = -1$, deduce an upper bound on the number of units $a + b\sqrt{2}$ of $\mathbb{Z}\left[\sqrt{2}\right]$ with $-N \leq a, b \leq N$.
3. Argue that the proportion of units among the numbers in $\mathbb{Z}\left[\sqrt{2}\right]$ with $-N \leq a, b \leq N$ is at most $1/N$ for large N.

The proportion of units among the numbers in $\mathbb{Z}\left[\sqrt{2}\right]$ is actually much less than the bound given in Exercise 6.8. You have shown that the powers of $\sqrt{2} + 1$ and $\sqrt{2} - 1$ are units in $\mathbb{Z}\left[\sqrt{2}\right]$. One can prove, though it is difficult, that these are the *only* units in $\mathbb{Z}\left[\sqrt{2}\right]$. Since the units correspond to the solutions to the Diophantine equations $x^2 - 2y^2 = 1$ and $x^2 - 2y^2 = -1$, this gives a recipe for writing down a complete list of the solutions to these equations. We need only compute all the powers of $\sqrt{2} \pm 1$ and read off the integer coefficients.

More generally, for a positive integer D that is not divisible by the square of an integer, one can introduce the ring $\mathbb{Z}\left[\sqrt{D}\right]$ of real numbers of the form $a + b\sqrt{D}$, where a and b are integers. Then one can ask what the units of $\mathbb{Z}\left[\sqrt{D}\right]$ are. Once again, it turns out that there is a special unit such that every other unit is a power of this special unit or its inverse. This fact allows one to write down all solutions to the Diophantine equations

$$x^2 - Dy^2 = 1$$

and

$$x^2 - Dy^2 = -1.$$

These are known as *Pell's equations*,[2] and they have been studied in some form going back to the ancient Greeks. The first general method of solution was

---

[2] John Pell (1611–1685) apparently did little to merit having these equations named after him. Many were more deserving, such as the Indian mathematicians Brahmagupta (598–670) and Bhaskara (1114–1185), who studied these equations, and Fermat, who in 1657 challenged the European community of mathematicians to come up with a method of solving these Diophantine equations (from which we may infer that Fermat had solved the problem himself). Fermat's challenge was successfully met by Bernard Frénicle de Bessy, William Brouncker, and John Wal-

found in 1657. By introducing the rings $\mathbb{Z}[\sqrt{D}]$, we are able to translate the number-theoretic problem of solving the Diophantine equation $x^2 - Dy^2 = 1$ into an algebraic problem about units in $\mathbb{Z}[\sqrt{D}]$.

Let us look at another ring of a similar type, the one obtained by replacing the 2 in $\mathbb{Z}[\sqrt{2}]$ by -1. Choosing a negative integer forces us into the domain of complex numbers. This complicates matters, but the analysis of units becomes easier because the corresponding Pell's equation is trivial to solve, as we shall see.

Consider the set of all complex numbers of the form $a + b\sqrt{-1}$, where a and b are integers. We can write this set as $\mathbb{Z}[\sqrt{-1}]$, in analogy with our notation $\mathbb{Z}[\sqrt{2}]$ and $\mathbb{Z}[\sqrt{D}]$. However, since we have already introduced the abbreviation i for $\sqrt{-1}$, we may as well keep using this. Thus, we are considering the complex numbers of the form $a + bi$ with a and b integers, and we will write $\mathbb{Z}[i]$ for the collection of these numbers. They are called the *Gaussian integers*.

Practice arithmetic in $\mathbb{Z}[i]$ by calculating:

1. $(3 - 4i) + (5 + 2i)$.
2. $(3 - 4i)(5 + 2i)$.
3. $(3 - 4i)(3 + 4i)$.

Now do the following exercises.

Exercise 6.9. Show that $\mathbb{Z}[i]$ is closed under addition and multiplication; that is, the sum and product of every pair of numbers $a + bi$ and $c + di$ in $\mathbb{Z}[i]$ is another number of this form.

Exercise 6.10. Find the units in $\mathbb{Z}[i]$.

1. Suppose that $a + bi$ is a unit in $\mathbb{Z}[i]$ and that its multiplicative inverse is $c + di$. Show that $ac - bd = 1$ and $ad + bc = 0$. Using these equations, deduce that $a - bi$ is also a unit in $\mathbb{Z}[i]$, and that its inverse is $c - di$.
2. Continuing with these numbers, show that the product

$$(a + bi)(a - bi)(c + di)(c - di)$$

equals 1 and deduce that $a^2 + b^2 = 1$. Show that this means that the only possibilities for units in $\mathbb{Z}[i]$ are 1, -1, i, and $-i$, and then check that these four numbers are indeed units in $\mathbb{Z}[i]$.
3. Conclude that ± 1 and $\pm i$ are the *only* units in $\mathbb{Z}[i]$, since $(\pm 1, 0)$ and $(0, \pm 1)$ are the only integer solutions to $x^2 + y^2 = 1$.

lis. Perhaps the equations should have been named after one of these gentlemen. But apparently, Leonhard Euler (pronounced "oiler"), the greatest mathematician of the eighteenth century, thought that Pell had done major work on the problem and dubbed them "Pell's equations." The name stuck.

We have found that the units in $\mathbb{Z}[i]$ are ± 1 and $\pm i$. In contrast, we found infinitely many units in $\mathbb{Z}\left[\sqrt{2}\right]$. The reason for this difference in the number of units is that $x^2 + y^2 = 1$ has only finitely many integer solutions, while $x^2 - 2y^2 = 1$ has infinitely many.

For every integer n, not just the integer $n = 1$, the problem of finding integer solutions to the equation $x^2 + y^2 = n$ is related to algebraic questions in $\mathbb{Z}[i]$. We will discuss these connections soon, but first let us study the equation $x^2 + y^2 = n$.

Exercise 6.11. Let n be a positive integer. Consider the equation $x^2 + y^2 = n$.

1. We have already dealt with the case $n = 1$. Now suppose $n = 2$. Find all integer solutions to the equation $x^2 + y^2 = 2$.
2. Next suppose that n is a positive integer congruent to 3 modulo 4. Prove that the equation $x^2 + y^2 = n$ has no such integer solutions. You can proceed as follows.
   (a) Show that if r is an even integer, then r^2 is divisible by 4, or equivalently, r^2 is congruent to 0 modulo 4.
   (b) Show that if r is an odd integer, then r^2 is congruent to 1 modulo 4.
   (c) Deduce that for every pair of integers a and b, $a^2 + b^2$ is congruent to 0, 1, or 2 modulo 4 but never congruent to 3 modulo 4.
   (d) Conclude as desired that $x^2 + y^2 = n$, for $n \equiv 3 \pmod 4$, has no integer solutions.
   (e) Deduce as an example that the equation

$$x^2 + y^2 = 334\,257\,891\,443\,112\,355$$

has no integer solutions.

Notice how the simple argument you have made just above allows you to conclude effortlessly that

$$334\,257\,891\,443\,112\,355$$

is not the sum of two integer squares. It is not necessary to proceed by trial and error, adding together all the squares that are smaller.

Exercise 6.12. Let p be each of the prime numbers 5, 13, 17, 29, 37, 41, 53, 61, 73, and 89 in turn. These are the first few odd prime numbers that are congruent to 1 modulo 4. For each such p find an integer solution to the equation $x^2 + y^2 = p$.

It is possible to show that the problem of solving the equation $x^2 + y^2 = n$ for an integer n can be reduced to the problem of solving $x^2 + y^2 = p$ for the prime divisors p of n. This focuses attention on the equations $x^2 + y^2 = p$ for prime numbers p. We have found that $x^2 + y^2 = p$ has a solution if $p = 2$, that there is no solution for the prime numbers p congruent to 3 modulo 4, and that there are solutions for the first few prime numbers p congruent to 1 modulo 4. Every odd prime number is congruent to 1 or 3 modulo 4. (Why is this true?) Therefore, in analyzing the equation $x^2 + y^2 = p$ for all prime numbers p, all that remains to be considered are the larger prime numbers p congruent to 1 modulo 4. In the 1600s, Fermat settled the general question of the solvability of the equation $x^2 + y^2 = p$ by proving the following result. We have already proved the second part. We shall prove the first part in Section 16.3.

Theorem 6.1 (Fermat). *Suppose p is an odd prime number.*

1. *If $p \equiv 1 \pmod 4$, then the equation $x^2 + y^2 = p$ has an integer solution.*
2. *If $p \equiv 3 \pmod 4$, then the equation $x^2 + y^2 = p$ has no solution in integers.*

Suppose we are working in some ring, say $\mathbb{Z}$ or $\mathbb{Z}[i]$ or $\mathbb{Q}$, and we are interested in how numbers factor. If u is a unit, then an element n of the ring factors as $(u)(u^{-1}n)$. For instance, in $\mathbb{Z}$, the number 5 factors as 1×5 and as $(-1) \times (-5)$. In $\mathbb{Q}$, the number 2 factors, for example, as $7 \times (\frac{1}{7} \cdot 2)$, that is, as $7 \times \frac{2}{7}$. We regard all such factorizations as uninteresting and call them *trivial* factorizations. A factorization not of this type is called a *nontrivial* factorization; that is, a factorization of n as ab in a ring is nontrivial if neither a nor b is a unit in the ring and trivial if one of a or b is a unit. In $\mathbb{Z}$, these are exactly the factorizations we called nontrivial and trivial earlier. In $\mathbb{Z}[i]$, the units are ± 1 and $\pm i$. Therefore, in addition to the trivial factorizations of 5 as 1×5 and $(-1) \times (-5)$, there are the trivial factorizations $i \times (-5i)$ and $(-i) \times 5i$. For the rings $\mathbb{Q}$, $\mathbb{R}$, and $\mathbb{C}$, since every nonzero number is a unit, all factorizations are trivial, and factorization questions are not interesting. That is why we do not study arithmetic questions in $\mathbb{Q}$ or $\mathbb{R}$.

A positive integer is prime in $\mathbb{Z}$ by definition if the only possible factorizations of it in $\mathbb{Z}$ are trivial ones. In general, we use the term *irreducible* for this: A nonzero number in a ring is *irreducible* if it is not a unit and its only factorizations in the ring are the trivial ones. We can study factorization questions in rings such as $\mathbb{Z}\left[\sqrt{2}\right]$ and $\mathbb{Z}[i]$, asking what the irreducible numbers are and how numbers factor as products of irreducible numbers. Consider, for example, the number 5 in $\mathbb{Z}[i]$. Check that

$$5 = (2 + i)(2 - i).$$

This factorization is not trivial, since $2 + i$ and $2 - i$ are not units in $\mathbb{Z}[i]$. Thus, even though 5 is irreducible in $\mathbb{Z}$, it is *not* irreducible in $\mathbb{Z}[i]$.

Exercise 6.13. Let us look further at factorization in the ring $\mathbb{Z}[i]$ of Gaussian integers.

1. Prove that if a pair of integers a and b is a solution to the equation $x^2 + y^2 = n$, then n factors in $\mathbb{Z}[i]$ as the product $(a + bi)(a - bi)$.
2. Describe a nontrivial factorization of 2 in $\mathbb{Z}[i]$.
3. Let p be each of the prime numbers 5, 13, 17, 29, 37, 41, 53, 61, 73, and 89 in turn. Describe a nontrivial factorization of p in $\mathbb{Z}[i]$.

4. Using Fermat's theorem, show that every prime number p congruent to 1 modulo 4 has a nontrivial factorization in $\mathbb{Z}[i]$.

Exercise 6.13 shows that the prime number 2 and the odd prime numbers p congruent to 1 modulo 4 are no longer irreducible in $\mathbb{Z}[i]$. By allowing the use of i, we obtain nontrivial factorizations of these numbers in $\mathbb{Z}[i]$.

What about the prime numbers of $\mathbb{Z}$ congruent to 3 modulo 4? Are these irreducible in $\mathbb{Z}[i]$ too, or do they have nontrivial factorizations in $\mathbb{Z}[i]$? We shall prove in Chapter 16 that these numbers are irreducible in $\mathbb{Z}[i]$ as well as in $\mathbb{Z}$. In doing so, we shall show that the number-theoretic problem of finding integer solutions to the equation

$$x^2 + y^2 = p$$

is equivalent to the algebraic problem of factoring p in the ring $\mathbb{Z}[i]$. This is just one example of the close relation between problems in number theory and algebraic problems for rings such as $\mathbb{Z}[i]$ and $\mathbb{Z}\left[\sqrt{2}\right]$.

6.3 Fruit Rings

So far, a ring has been a collection of numbers that one can add and multiply without leaving the collection (closure), with the further properties that there is an additive identity 0, there is a multiplicative identity 1, and every number in the ring has an additive inverse. We have looked at the rings of integers, rational numbers, real numbers, complex numbers, numbers of the form $a + b\sqrt{2}$, and the Gaussian integers. We also will be interested in rings that consist of objects other than integers, rational numbers, real numbers, or complex numbers.

What sorts of objects does it make sense to add and multiply? Fruit? No, of course not. Furniture? No again. Only numbers, then?

You already know one example of objects besides numbers that can be added and multiplied: polynomials. For instance, you can add or multiply together

$$3x^2 - 4x + 2$$

and

$$x^7 - 14400x^5 + 17x.$$

In fact, the set of all polynomials behaves just like a ring of numbers. It is closed under addition and multiplication, the constant polynomial 0 is an additive identity, the constant polynomial 1 is a multiplicative identity, and every polynomial has an additive inverse, obtained by changing the signs of all the coefficients. The set of polynomials is a ring. It is not a ring of numbers, but it is a ring nonetheless.

We will study rings of polynomials in Chapter 12. Seeing them as an example here helps to make the point that we can add and multiply objects other than numbers, so that it makes perfect sense to consider rings whose objects are not numbers. We shall see later that the study of polynomials as a ring provides a framework in which to investigate questions about factorization of polynomials and roots of polynomials.

We can add and multiply numbers. We can add and multiply polynomials. What else? We agreed a moment ago that fruit will not do. Why not?

The obvious difficulty with adding and multiplying fruit is that we have no idea what addition or multiplication could possibly mean. How do we multiply an orange and a banana? It appears to make no sense. But what if we decide to invent addition and multiplication rules for fruit, just for fun? Is there a reason that we cannot do so? Who is to stop us? Let us try.

Suppose we have an orange, a banana, and a pear, and we want to invent a rule that tells us the sum of each pair of these. This would produce answers to questions such as the following:

$$\text{orange} + \text{banana} = ?$$
$$\text{banana} + \text{pear} = ?$$
$$\text{pear} + \text{pear} = ?$$

We also want to invent product rules that would let us answer questions such as

$$\text{pear} \times \text{orange} = ?$$

To save space, and to make the formulas we want to write look a little less strange, let us introduce symbols for the three fruits. We shall write *0* for the orange, since it is shaped a bit like a zero, *1* for the banana, since it is shaped a bit like a one, and *2* for the pear, since pear sounds like pair. Keep in mind throughout this discussion that *0*, *1*, and *2* are being used as symbols for the orange, the banana, and the pear, not as symbols for the numbers zero, one, and two.

Let us make addition and multiplication tables for the orange, banana, and pear, or for their symbols *0*, *1*, and *2*. We shall follow the same format traditionally used in elementary school in writing the addition and multiplication tables for the numbers from 1 to 10. Each table will list the three kinds of fruit along the top and along the side, and will have nine entries summarizing the results of adding or multiplying a given pair of fruits. There is no obvious way to construct natural addition and multiplication tables. For the sake of experiment, let us try the ones below.

+	0	1	2
0	0	1	2
1	1	2	0
2	2	0	1

×	0	1	2
0	0	0	0
1	0	1	2
2	0	2	1

To illustrate how the tables are used, notice that they tell us that

$$1 + 1 = 2$$

and
$$2 \times 2 = 1.$$

In words,
$$\text{banana} + \text{banana} = \text{pear}$$

and
$$\text{pear} \times \text{pear} = \text{banana}.$$

Remember, we are not trying to make physical sense of the tables or the equations that they encode. Think of the tables for now as the data for a weird game.

Exercise 6.14. Verify the following:

1. The orange, 0, is an additive identity.
2. The banana, 1, is a multiplicative identity.
3. Each fruit has an additive inverse. Specify what the additive inverse of each fruit is.

What game are we playing? Let us call it the *ring game*. At the beginning of the game we are given a collection of objects, such as the orange, banana, and pear. Our first step is to create two tables, an addition table and a multiplication table. Each table is indexed by the objects. If there are infinitely many objects, the tables will be hard to picture, so let us restrict ourselves for now to games with only finitely many objects, n, say. We then have n^2 entries in each table. The objects may be fruit, furniture, numbers, leaves, or whatever you wish. The entry in the addition table that is in the row marked with an object x and the column marked with an object y is the sum $x + y$. Similarly, the entry in this position in the multiplication table is the product $x \times y$.

Without further rules, the ring game is too easy to play. We are free to insert any objects we wish in any positions of the table. The game becomes more difficult if additional rules are imposed. Let us impose the rules below:

1. The addition and multiplication tables must be filled out in such a way that the operations of addition and multiplication are *commutative*. This means that the order in which we add or multiply two objects does not matter. For objects x and y, we are required to have $x + y = y + x$ and $x \times y = y \times x$.
2. The tables must be filled out in such a way that the operations are associative. This means that for every three objects x, y, and z, we have

$$x + (y + z) = (x + y) + z,$$

   and similarly for multiplication.
3. The operations must also satisfy the distributive law. This means that for three objects x, y, and z, we have $x \times (y + z) = x \times y + x \times z$.

4. There must be an additive identity, an object $\mathcal{A}$ such that for every object x, we have $\mathcal{A} + x = x$. Usually, when we play the game we shall arrange to label the additive identity 0.

5. There is a multiplicative identity, an object $\mathcal{M}$ such that for every other object x, we have $\mathcal{M} \times x = x$. Usually, when we play the game we shall arrange to label the multiplicative identity 1.

6. Every object x has an additive inverse, an associated object y such that $x + y = 0$.

A *solution* to the ring game for a collection of objects now consists of an addition table and a multiplication table satisfying all these rules. If we succeed in finding a solution to the game for a given collection of objects, we shall say that we have made the collection into a *ring*.

Some objects in a ring may have multiplicative inverses and some may not. We have not imposed the requirement that they all do. An object that has a multiplicative inverse will be called a *unit* of the ring. (Notice that by our rules, every ring will have at least one unit, namely, the multiplicative identity.) *Unit* is the same name we used for numbers that have multiplicative inverses in our earlier rings. Let us continue the previous exercise.

Exercise 6.15. Verify that the addition and multiplication tables for the orange, banana, and pear are solutions to the ring game. You have already verified that there are additive and multiplicative identities, and that every fruit has an additive inverse. Here is what remains:

1. Verify the associative law. This can be tedious. Do not check every case of this, but do check a few.
2. Verify the commutative laws for addition and multiplication.
3. Verify a few cases of the distributive law.
4. Verify also that each nonzero fruit is a unit. Specify what the multiplicative inverse of each nonzero fruit is. (The fact that each nonzero fruit is a unit is an "extra." It is not required of a solution to the ring game.)

Let us play the ring game with a new collection of objects. We shall keep the orange, banana, and pear, but add a strawberry. Since strawberry has three syllables, we shall use the number *3* as a symbol for it. So we write *0*, *1*, *2*, and *3* for our four pieces of fruit. Forget the ring game we just played, with the addition and multiplication tables above. We are starting over again. Keep in mind that the symbols *0*, *1*, *2*, and *3* are not numbers. They are symbols for the four fruits, and we are free to make up any addition and multiplication tables we wish for them, provided that the rules of the ring game are satisfied. Let us use the following tables for these four pieces of fruit:

+	0	1	2	3
0	0	1	2	3
1	1	2	3	0
2	2	3	0	1
3	3	0	1	2

×	0	1	2	3
0	0	0	0	0
1	0	1	2	3
2	0	2	0	2
3	0	3	2	1

Exercise 6.16. Verify that the addition and multiplication tables above are solutions to the ring game. Specifically, answer the following questions.

1. What is the additive identity?
2. What is the multiplicative identity?
3. Does every fruit have an additive inverse?

Also, decide which nonzero fruits have multiplicative inverses, and for each such fruit, specify what its inverse is.

Let us play one more version of the game. For our starting collection we shall keep the orange and banana of the last game, but we shall replace the pear and strawberry with an apple and some other kind of berry. The berry is an unfamiliar one, so we will not be more specific about what type of berry it is. Continue to write 0 for the orange and 1 for the banana. In the absence of any additional clever ways to assign number symbols to fruits, let us just write a for the apple and b for the berry. Our four objects, then, are 0, 1, a, and b. Here are proposed addition and multiplication tables for this collection of fruit:

+	0	1	a	b
0	0	1	a	b
1	1	0	b	a
a	a	b	0	1
b	b	a	1	0

×	0	1	a	b
0	0	0	0	0
1	0	1	a	b
a	0	a	b	1
b	0	b	1	a

Verify that 0 is an additive identity, that 1 is a multiplicative identity, and that the commutative laws are satisfied.

A variant of the ring game is the field game. A solution to the field game for a given collection of objects is an addition table and a multiplication table that satisfy all the rules of the ring game plus the following rule:

Every object x other than the additive identity 0 has a multiplicative inverse, that is, an object y with $x \times y = 1$. In other words, every object in the ring other than 0 is a unit.

A solution to the field game is called a *field*. Thus, a field is a ring with the additional property that every object except for the additive identity is a unit. Let us consider whether the three solutions to the ring game that we have just examined are fields.

Exercise 6.17. Decide which of the three fruit rings constructed above are fields.

1. For the first two, we have already determined which nonzero fruits are units. State whether or not the rings are fields and explain why or why not.
2. For the third fruit ring, answer the following questions:
   (a) What are a^2 and b^2?
   (b) What are a^3 and b^3?
   (c) Is a a unit? If so, what is its multiplicative inverse?
   (d) Is b a unit? If so, what is its multiplicative inverse?
   (e) Is the ring a field? Why or why not?

6.4 Modular Arithmetic Rings

Rings of fruit have been introduced in order to illustrate the idea that one can add and multiply objects, symbols, or numbers in ways that are not the familiar ones. Next we will study rings that behave like the rings of fruit.

Let us start with some calendar arithmetic. If today is Friday and your friend Alice tells you that she is coming to your house in two weeks and three days, you know after a moment's thought that she will be there on a Monday. You might calculate this by adding 3 days to Friday, to get Monday, and then ignoring the 2 weeks, since they do not change the day. You add 17 days to Friday, but do not count 14 of those 17 days, so you reduce the problem of adding 17 days to that of adding 3 days. In fact, from this point of view, adding 3 days, 10 days, 17 days, 24 days, 31 days, or, more generally, $3 + 7n$ days to Friday does not change the result. It certainly changes the actual day on which Alice shows up, but not the name of the day. We are all familiar with this process. It leads to a new ring, the *days-of-the-week ring*.

The objects in the days-of-the-week ring will be the symbols 0, 1, 2, 3, 4, 5, and 6. These are not the usual integers zero, one, two, three, four, five, and six. Rather, they are symbols, pieces in a game, just as *0, 1, 2* were symbols in the fruit ring games. Since they are not the usual numbers, we are free to make them behave in ways other than the usual way. We can introduce addition and multiplication rules as we please, just as we did for fruit. Do not let the symbols fool you. These are new objects, which will not be added or multiplied according to the old rules.

We shall add two objects of the days-of-the-week ring as follows: If the sum of two such objects as ordinary integers is less than 7, then that is their sum in our new ring. If their sum as ordinary integers is 7 or more, subtract 7 from the usual sum to get the sum in the days-of-the-week ring. For example,

$$2 + 3 = 5,$$
$$3 + 5 = 1,$$
$$1 + 6 = 0.$$

More generally, we shall add a string of objects in the days-of-the-week ring by adding in the usual way, and then subtracting as many 7's as necessary to get down to a number between 0 and 6. For example,

$$3 + 6 + 4 + 2 + 2 + 1 + 6 + 5 = 1.$$

In other words, add all the symbols in the usual way, and then find the remainder when you divide the result by 7. This is the least nonnegative residue of the ordinary sum modulo 7.

We take the same approach in defining multiplication in the days-of-the-week ring: To multiply two objects together, first multiply them in the usual way, and then subtract as many 7's as necessary to get down to a number between 0 and 6, the least nonnegative residue modulo 7 of the usual product. For example,

$$4 \times 6 = 3,$$
$$2 \times 3 = 6,$$
$$5 \times 5 = 4.$$

Let us write $\mathbb{Z}_7$ for the days-of-the-week ring. Another way to think about the rules for addition and multiplication in the days-of-the-week ring is to regard them as *rewrite rules*. We perform addition and multiplication as usual, and then we apply the rewrite rule that has us replace the provisional sum or product with its least nonnegative residue modulo 7.

Exercise 6.18. Do the following:

1. Calculate the following sums and products in $\mathbb{Z}_7$.
   (a) $2 + 5$,
   (b) $3 + 1 + 4 + 2$,
   (c) 4×3,
   (d) 2×2,
   (e) $3 \times 4 \times 6$.
2. Write a multiplication table for $\mathbb{Z}_7$. In other words, make a 7×7 table with the rows and columns labeled by the numbers 0 to 6. In the square of the grid whose row is marked by r and whose column is marked by s, enter the product $r \times s$, as calculated in $\mathbb{Z}_7$. For instance, since $4 \times 6 = 3$, we would enter the symbol 3 in the square in row 4 and column 6. (You can simplify the task by remembering that multiplication is commutative in $\mathbb{Z}_7$.)
3. Using your multiplication table, check that 1 is a multiplicative identity for the ring $\mathbb{Z}_7$.
4. Using your multiplication table, decide which numbers in $\mathbb{Z}_7$ are units; that is, which numbers have multiplicative inverses. Specify what the multiplicative inverse of each unit is.

For each integer $m > 1$, there is a ring $\mathbb{Z}_m$ analogous to the ring $\mathbb{Z}_7$. It consists of the objects $0, 1, 2, \ldots, m - 2, m - 1$. These are not the usual integers 1 through $m - 1$. They are new objects, just as in the days-of-the-week ring the symbols $0, 1, 2, 3, 4, 5, 6$ are new objects and in the rings of fruit, 0, 1, 2, 3 are new objects. We will play the ring game with the objects $0, 1, 2, \ldots, m - 2, m - 1$; that is, we will establish addition and multiplication formulas for these objects that satisfy the rules of the ring game.

The sum and product of two objects a and b in $\mathbb{Z}_m$ are defined according to the rules of *modular arithmetic*. We proceed in two steps: First, view a and b as if they were ordinary integers and compute their sum $a + b$ or product $a \times b$ as usual. Suppose n is the result. Then subtract enough copies of m from n to obtain a number in the range 0 to $m - 1$; in other words, find the least nonnegative residue of n modulo m. The resulting number, viewed as one of our new objects, is the sum or product. For instance, for $m = 4$ the objects in $\mathbb{Z}_4$ are $0, 1, 2, 3$. The addition and multiplication rules yield $2 + 3 = 1$ and $2 \times 3 = 2$. In case $m = 12$, the addition in $\mathbb{Z}_m$ is just the familiar addition we use in telling time, the addition of *clock arithmetic*.

Exercise 6.19. Let us take a look at the rings $\mathbb{Z}_m$ for small m.

1. Suppose $m = 2$.
   (a) Write the multiplication table for $\mathbb{Z}_2$.
   (b) Using the multiplication table, check that 1 is a multiplicative identity for $\mathbb{Z}_2$.
   (c) Which numbers in $\mathbb{Z}_2$ are units? What are their multiplicative inverses? Is $\mathbb{Z}_2$ a field?
2. Suppose $m = 3$.
   (a) Write the multiplication table for $\mathbb{Z}_3$. Have you seen this table before?
   (b) Using the multiplication table, check that 1 is a multiplicative identity for $\mathbb{Z}_3$.
   (c) Which numbers in $\mathbb{Z}_3$ are units? What are their multiplicative inverses? Is $\mathbb{Z}_3$ a field?
3. Suppose $m = 4$.
   (a) Write the multiplication table for $\mathbb{Z}_4$. Have you seen this table before?
   (b) Using the multiplication table, check that 1 is a multiplicative identity for $\mathbb{Z}_4$.
   (c) Which numbers in $\mathbb{Z}_4$ are units? What are their multiplicative inverses? Is $\mathbb{Z}_4$ a field?

Exercise 6.20. Repeat the steps of Exercise 6.19 yet again, for each of the rings $\mathbb{Z}_m$ as m runs from 5 to 12.

The object 3 in the ring $\mathbb{Z}_4$ is not the number 3 in the ring $\mathbb{Z}$ of ordinary integers. Sometimes it is convenient to use a notation that makes this clear. For this purpose, we can write [3] for the object we call 3 in $\mathbb{Z}_4$. Using this notation for all the objects in $\mathbb{Z}_4$, we can write statements of the following sort:

$$[3] \times [2] = [2]$$

and

$$[3] \times [3] = [1].$$

In general, when we wish to avoid possible ambiguity, we will write $[a]$ for the object in $\mathbb{Z}_m$ that so far we have written simply as a.

We can take this notational convention one step further if we need to make clear which ring $\mathbb{Z}_m$ we are studying. After all, the object 3 in $\mathbb{Z}_4$ is different from the object 3 in $\mathbb{Z}_5$ or the object 3 in every other $\mathbb{Z}_m$. If we need a notation that allows us to keep track of which 3 we are talking about, we will write $[3]_4$ for the object we call 3 in $\mathbb{Z}_4$, we will write $[3]_5$ for the object we call 3 in $\mathbb{Z}_5$, and so on. The general convention will be to write $[a]_m$ for the object in $\mathbb{Z}_m$ that so far we have written as a. This notation is useful when we are working with several rings at once. Notice, as examples of this notation, that we have the following equalities:

$$[3]_4 \times [2]_4 = [2]_4$$

and

$$[3]_5 \times [2]_5 = [1]_5.$$

In Exercises 6.19 and 6.20 we determined all the units in the rings $\mathbb{Z}_m$ for $2 \leq m \leq 12$. These examples may suggest the following general result.

Theorem 6.2. *Let a and m be positive integers with $a < m$. The element $[a]_m$ is a unit in $\mathbb{Z}_m$ if and only if $(a, m) = 1$.*

Exercise 6.21. Prove Theorem 6.2 by translating the statement that $[a]_m$ is a unit in $\mathbb{Z}_m$ into a statement about the solvability of certain congruences modulo m and quoting a result from Section 4.2. Then describe which rings $\mathbb{Z}_m$ are fields. In other words, for which integers $m > 1$ is every nonzero object in $\mathbb{Z}_m$ a unit?

6.5 Congruence Rings

We can gain further insight into the modular arithmetic rings $\mathbb{Z}_m$ by relating the objects in them to congruence classes. Let us begin with the special case $m = 5$. Choose an integer a from $\mathcal{C}(2)$, the congruence class of 2, and an integer b from $\mathcal{C}(4)$. Then $a \equiv 2 \pmod 5$ and $b \equiv 4 \pmod 5$. Recall Proposition 4.3: Suppose a, b, e, and f are integers satisfying $a \equiv e \pmod 5$ and $b \equiv f \pmod 5$. Then $a + b \equiv e + f \pmod 5$, and $ab \equiv ef \pmod 5$. Applying Proposition 4.3 to our situation, we find that $a + b \equiv 6 \equiv 1 \pmod 5$ and $ab \equiv 8 \equiv 3 \pmod 5$. Thus if we add a number a from $\mathcal{C}(2)$ to a number b from $\mathcal{C}(4)$, then no matter what choice of numbers a and b we make from $\mathcal{C}(2)$ and $\mathcal{C}(4)$, the sum $a + b$ is in the congruence class $\mathcal{C}(1)$. Similarly, in multiplying a number a from $\mathcal{C}(2)$ by a number b from $\mathcal{C}(4)$ we obtain a product ab in the congruence class $\mathcal{C}(3)$. We thus have a reasonable way of talking about addition and multiplication of congruence classes, and we may summarize the statements above in the equations

$$\mathcal{C}(2) + \mathcal{C}(4) = \mathcal{C}(1)$$

and

$$\mathcal{C}(2) \times \mathcal{C}(4) = \mathcal{C}(3).$$

Pursuing this idea further, we might try to define addition and multiplication on the five congruence classes $\mathcal{C}(0)$, $\mathcal{C}(1)$, $\mathcal{C}(2)$, $\mathcal{C}(3)$, and $\mathcal{C}(4)$ and make a new ring. In fact, we can do this. More generally, for an integer $m > 1$, Proposition 4.3 allows us to define addition and multiplication on congruence classes modulo m, and we thereby obtain a ring.

Exercise 6.22. Let us treat the easiest case first. Let $m = 2$. The two congruence classes modulo 2 are $\mathcal{C}(0)$ and $\mathcal{C}(1)$.

1. Observe that the sum of an even integer and an odd integer is odd, the sum of an even integer and an even integer is even, and the sum of an odd integer and an odd integer is even.
2. Rephrase this to say that the sum of an element from $\mathcal{C}(0)$ and an element from $\mathcal{C}(1)$ is in $\mathcal{C}(1)$, the sum of an element from $\mathcal{C}(0)$ and an element from $\mathcal{C}(0)$ is in $\mathcal{C}(0)$, and the sum of an element from $\mathcal{C}(1)$ and an element from $\mathcal{C}(1)$ is in $\mathcal{C}(0)$.
3. These three statements can be summarized in equations. For example, the first statement can be written as

$$\mathcal{C}(0) + \mathcal{C}(1) = \mathcal{C}(1).$$

Write the two other statements as equations of the same type.
4. Perform analogous steps for multiplication: What is the product of an even integer and an odd integer? An even integer and an even integer? An odd integer and an odd integer? Summarize your answers in three equations of the form
$$\mathcal{C}(i) \times \mathcal{C}(j) = \mathcal{C}(k).$$
5. Observe that your sum and product rules turn the collection

$$\{\, \mathcal{C}(0),\ \mathcal{C}(1) \,\}$$

into a ring.
6. Explain how this ring can be considered as equivalent to the ring $\mathbb{Z}_2$ that we studied earlier, with the only difference being that different symbols are used for the objects.

Exercise 6.23. Let $m > 1$ be an integer.

1. Suppose $C(i)$ and $C(j)$ are two congruence classes modulo m. Choose an integer a from $C(i)$ and an integer b from $C(j)$. Show that $a + b$ is in the congruence class $C(i + j)$ and that ab is in the congruence class $C(ij)$.

2. Deduce that this allows you to define addition and multiplication rules for the set of m different congruence classes

$$\{ \ C(0), \ C(1), \ C(2), \ \ldots, \ C(m - 2), \ C(m - 1) \}$$

modulo m. In other words, you can form sums $C(i) + C(j)$ and products $C(i) \times C(j)$.

3. Show that with respect to these rules of addition and multiplication, $C(0)$ is an additive identity and $C(1)$ is a multiplicative identity. Check that the commutative laws hold for addition and multiplication. The other rules of the ring game hold as well, as you can check. Thus the collection of congruence classes modulo m forms a ring.

4. Explain how the ring of congruence classes modulo m can be regarded as equivalent to the ring $\mathbb{Z}_m$, the only difference being that different symbols are used for the objects. Do so by identifying each object in $\mathbb{Z}_m$ with a congruence class and showing that under this identification, the addition and multiplication rules correspond.

We have found that for an integer $m > 1$ the collection of congruence classes modulo m forms a ring and that this ring is the ring $\mathbb{Z}_m$ with different names given to the objects. Therefore, we can regard the objects of $\mathbb{Z}_m$, if we wish, not as the symbols $0, 1, \ldots, m-1$ with funny addition and multiplication rules but as the congruence classes $C(0)$, $C(1)$, $\ldots$, $C(m - 1)$ with natural addition and multiplication rules.

The identification of the objects of $\mathbb{Z}_m$ with congruence classes of integers modulo m also clarifies the meaning of the bracket notation $[a]$ for objects in $\mathbb{Z}_m$. We can think of $[a]$ as a symbol for the congruence class $C(a)$ that contains a. Once we do this, we realize that we need not restrict a to the integers $0, 1, \ldots, m - 1$; rather, the symbol a can now represent any integer. For example, in $\mathbb{Z}_5$, the object $[2]$ is really the congruence class $C(2)$ of integers congruent to 2 modulo 5. We could just as well write this as $[7]$ or $[82]$. The fact that $[3] \times [4] = [2]$ in $\mathbb{Z}_5$ could just as well be written as the rule $[3] \times [4] = [12]$, since $[2]$ and $[12]$ represent the same object of $\mathbb{Z}_5$.

7

Euler's Theorem

7.1 Units

One reason to introduce the ring $\mathbb{Z}_m$ is that congruence statements about integers modulo m turn into equalities in $\mathbb{Z}_m$. In particular, Theorem 4.9 on cancellation of factors in congruences turns into a statement about cancellation of factors in equations in $\mathbb{Z}_m$. This is a special case of a still more general result about cancellation of units in rings, as we will soon see.

Suppose we wish to solve the equation

$$3x = 6.$$

We all know how to proceed. We divide both sides by 3 to get $x = 2$. What about

$$3x = 2?$$

Again we know what to do. We divide both sides by 3 again to get the solution $x = \frac{2}{3}$. But wait! Are solutions in rational numbers allowed? In posing the question we need to specify whether we are looking only for integer solutions, or allowing rational or real or even complex solutions. If we want only integer solutions, dividing by 3 is not a good idea, but if we want solutions in $\mathbb{Q}$ or in $\mathbb{R}$, then dividing by 3 is fine.

More generally, suppose we allow arbitrary numbers on the right-hand side of the equation, so that the equation we now want to solve is

$$3x = m$$

for different choices of m. If we are working in $\mathbb{Q}$, the general solution will be $x = m/3$. If we are working in $\mathbb{Z}$, the equation is solvable if m is divisible by 3, but it is not solvable otherwise.

The reason we can solve $3x = m$ in $\mathbb{Q}$ no matter what value of m is chosen is that we can always divide by 3. It is helpful to rephrase what we are doing by saying not that we are dividing by 3 but that we are multiplying by the multiplicative inverse of 3, that is, by $\frac{1}{3}$. We can divide any number in $\mathbb{Q}$ by 3, or multiply any number in $\mathbb{Q}$ by $\frac{1}{3}$, because 3 is a unit in $\mathbb{Q}$. In contrast, 3 is not a unit in $\mathbb{Z}$, so that $\frac{1}{3}$ is not in $\mathbb{Z}$, and we do not always remain in $\mathbb{Z}$ when we multiply a number in $\mathbb{Z}$ by $\frac{1}{3}$.

An earlier result on congruences can be reinterpreted in a similar way. Suppose we wish to solve the congruence

$$3x \equiv 2 \ (\text{mod } 11).$$

We have proved a theorem that tells us that we can solve this congruence if and only if the greatest common divisor $(3, 11)$ of 3 and 11 divides 2. Since $(3, 11) = 1$, and 1 divides 2, there is a solution. We can use the Euclidean algorithm to find a solution, as we have seen, or we can just experiment until we find a solution, such as $x = 8$.

The idea of solving the congruence $3x \equiv 2 \ (\text{mod } 11)$ by dividing by 3, or multiplying by $\frac{1}{3}$, does not appear to make sense. It would make sense if in contrast we were to replace 2 by 24. Since $24 \equiv 2 \ (\text{mod } 11)$, solutions to the congruence $3x \equiv 2 \ (\text{mod } 11)$ are the same as solutions to the congruence

$$3x \equiv 24 \ (\text{mod } 11).$$

However, we can solve the new congruence by dividing by 3, obtaining $x = 8$.

To understand better what is happening, we can pass from *congruences* in $\mathbb{Z}$ to *equations* in $\mathbb{Z}_{11}$. Let us use the notation $[a]$ in representing elements of $\mathbb{Z}_{11}$, in order to avoid confusion with ordinary integers. With this notation, the elements of $\mathbb{Z}_{11}$ are

$$[0], \ [1], \ [2], \ \ldots, \ [10].$$

Suppose we wish to solve the equation

$$[3]\, x = [2]$$

in $\mathbb{Z}_{11}$. To solve this equation we would like to divide by $[3]$. Since we have not defined what it means to divide congruence classes, a much better way to say this is that we want to multiply both sides of the equation by the multiplicative inverse of $[3]$, if, in fact, $[3]$ has a multiplicative inverse in $\mathbb{Z}_{11}$. Some experimentation shows that it does, and that the inverse is $[4]$:

$$[4] \times [3] = [1].$$

We can solve the equation

$$[3]\, x = [2]$$

by multiplying both sides by $[4]$. Doing so yields

$$[4] \times [3]\, x = [4] \times [2],$$

or
$$[1]\, x = [8].$$

Since [1] is the multiplicative identity in $\mathbb{Z}_{11}$, we can write this as $x = [8]$.

Let us repeat this without the bracket notation, keeping in mind at all times that our numbers represent not integers but elements of $\mathbb{Z}_{11}$. We wish to solve $3x = 2$, where 3 and 2 are not integers but elements of $\mathbb{Z}_{11}$. We notice that 3 has 4 as an inverse: $4 \times 3 = 1$. We then multiply both sides of the equation $3x = 2$ by 4 to obtain $x = 8$. We are done.

This is analogous to what we did earlier in solving $3x = 2$ in $\mathbb{Q}$. The multiplicative inverse of the integer 3 in $\mathbb{Q}$ is $\frac{1}{3}$, and we multiplied both sides of the equation $3x = 2$ by $\frac{1}{3}$ to get $x = \frac{2}{3}$. In $\mathbb{Z}_{11}$ the multiplicative inverse of 3 is 4, and we multiply both sides of $3x = 2$ by 4 to obtain the solution $x = 8$.

Exercise 7.1. For each of the following equations, find a solution or explain why there is no solution:

1. $3x = 7$ in $\mathbb{Z}_{11}$.
2. $5x = 4$ in $\mathbb{Z}_7$.
3. $6x = 4$ in $\mathbb{Z}_{12}$.
4. $5x = 4$ in $\mathbb{Z}_{12}$.

Suppose a and e are elements of an arbitrary ring R and we wish to solve the equation $ax = e$ in R; that is, we want to find a value of x in R that solves the equation. The above examples suggest a way to solve the equation if a is a unit in R. We find its multiplicative inverse a^{-1}, multiply both sides of the equation by a^{-1}, and obtain $x = a^{-1}e$. If a is not a unit, the equation may or may not be solvable; in this case solvability depends on the specific choice of a and e. But if a is a unit, we see that the equation is solvable and that we have a procedure for finding a solution. This is summarized in the following proposition.

Proposition 7.1 *Suppose a is a unit in a ring R, and e is an arbitrary element of R. Then the equation*

$$ax = e$$

is solvable in R, with the solution $x = a^{-1}e$.

The ability to multiply equations by the multiplicative inverse of a unit also yields the following result.

Proposition 7.2 *Suppose a is a unit in a ring R, and suppose that b and c are also in R. If $ab = ac$, then $b = c$.*

Exercise 7.2. Prove Proposition 7.2. Then show that the conclusion of Proposition 7.2 can fail if a is not a unit. In other words, it is possible for a ring R to contain elements a, b, and c satisfying $ab = ac$ but not $b = c$. To show this, describe a specific ring R and specific elements a, b, and c of this type. (The case $a = 0$ is trivial. Find an example with $a \neq 0$.)

The conclusion of Proposition 7.2 can be summarized by saying that the unit a can be *canceled* in equations, or that *cancellation* holds for units. In contrast, cancellation need not hold for nonunits. Combining Proposition 7.1 and Theorem 6.2, we conclude that for an integer a relatively prime to an integer $m > 1$, the equation

$$[a] \, x = [e]$$

can always be solved in $\mathbb{Z}_m$. To do so, we need only multiply both sides of the equation by the multiplicative inverse of $[a]$ in $\mathbb{Z}_m$. It follows also, for such an a, that we can always solve the congruence

$$ax \equiv e \pmod{m},$$

as we already know.

Let us discuss an important property of units. This property was used in the congruence setting to solve the Chicken McNugget problem. What we saw is that for relatively prime positive integers a and m with $a < m$ and, of course, $m > 1$, multiplying a complete set of congruence class representatives modulo m by a yields another complete set of congruence class representatives modulo m, possibly in a new, "shuffled," order. This can be restated in $\mathbb{Z}_m$ as the fact that multiplying by a unit shuffles the elements.

The word "shuffle" reminds us of playing cards. When we perform a shuffle on a standard deck of 52 playing cards, what we do is to keep the set of 52 playing cards intact while changing their order. Multiplication by a unit performs a similar "shuffle" on the deck of elements of a ring. This shuffle idea makes sense for rings of arbitrary size, but in order to avoid discussing shuffles of infinite sets, let us restrict ourselves to situations involving a shuffle of a finite set.

Exercise 7.3. Suppose u is a unit in a ring R.

1. Recall that units can be canceled: If r and s are elements of R satisfying $ur = us$, then $r = s$. Observe that the logically equivalent contrapositive takes the following form: If r and s are elements of R satisfying $r \neq s$, then $ur \neq us$.
2. Suppose that R has only finitely many elements, and that

$$r_1, \ldots, r_t$$

is a complete list of them, with no repetitions. Deduce that

$$ur_1, \ldots, ur_t$$

is also a complete list of the elements of R, with no repetitions. We might describe this result by saying that multiplication by u *shuffles* R.
3. Give an example to show that in contrast, if a is an arbitrary nonzero element of R, the set

$$ar_1, \ldots, ar_t$$

may not be a complete list of elements of R. (Again note that $a = 0$ is a trivial example.) Instead, some elements may be repeated in this list, and some may be omitted. You need only produce a single ring R and a single nonzero element a of R in order to demonstrate this. Thus, multiplication by an arbitrary element may not shuffle R.

Let us connect shuffles of the ring $\mathbb{Z}_m$ with the congruence class shuffle that we used in Section 4.3.

Exercise 7.4. Suppose $m > 1$ is an integer and a is an integer relatively prime to m. Recall the discussion of bracket notation at the end of Section 6.5.

1. Recall that since $(a, m) = 1$, it follows that $[a]$ is a unit in $\mathbb{Z}_m$. Deduce that multiplication of all the elements of $\mathbb{Z}_m$ by $[a]$ produces a shuffle of $\mathbb{Z}_m$.
2. Conclude that the set

$$[0], \ [a], \ [2a], \ [3a], \ldots, \ [(m-1)a]$$

   is a complete list of the m elements of $\mathbb{Z}_m$.
3. Reinterpret this, in terms of congruences, to say that every integer is congruent, modulo m, to exactly one of the integers

$$0, \ a, \ 2a, \ 3a, \ \ldots, \ (m-1)a.$$

4. Conclude that we have shown again that

$$0, \ a, \ 2a, \ 3a, \ \ldots, \ (m-1)a$$

   is a complete set of distinct congruence class representatives modulo m. Thus, if $(a, m) = 1$, then multiplication by $[a]$ shuffles congruence classes modulo m.

7.2 Roots of Unity

A special kind of unit is a root of unity. An element a of a ring R is called an *nth root of unity* if $a^n = 1$, and a *root of unity* if it is an nth root of unity for some positive integer n. Here the word "unity" refers to the identity element 1. Notice that every root of unity is a unit in the ring, since if $a^n = 1$, then $a \cdot a^{n-1} = 1$, and therefore a^{n-1} is a multiplicative inverse of a. For many rings, the roots of unity are just a small minority of all the units in the ring. Two extremes are $\mathbb{Z}$ and $\mathbb{R}$. In $\mathbb{Z}$, the roots of unity are 1 and -1, and these are the only units of $\mathbb{Z}$. By contrast, in $\mathbb{R}$, the roots of unity are still only 1 and -1, as you will prove in the next exercise, but every nonzero number is a unit.

Exercise 7.5. Suppose x is a real number.

1. Assume first that $x > 1$. Recall that this implies that $x^2 > x$. Use this to prove that there is no positive integer n for which $x^n = 1$.
2. Assume that $0 < x < 1$. Recall that this implies that $x^2 < x$. Use this to prove that there is no positive integer n for which $x^n = 1$.
3. Deduce that the only positive real number x satisfying $x^n = 1$ is $x = 1$.
4. Suppose $x < 0$. Show that if $x^n = 1$, then n must be even. Deduce from this and the previous part that $x^2 = 1$, and then deduce that $x = -1$.
5. Conclude that 1 and -1 are the only roots of unity in $\mathbb{R}$. We see that even though $\mathbb{R}$ has infinitely many units, only two of them are roots of unity.

Next let us consider roots of unity in the ring $\mathbb{C}$ of complex numbers. It is clear that $\mathbb{C}$ has more roots of unity than $\mathbb{R}$ does, since i and $-i$ are roots of unity: Their squares are -1, so they satisfy $i^4 = (-i)^4 = 1$. Thus there are four fourth roots of unity in $\mathbb{C}$, namely, 1, i, -1, and $-i$. In fact, for each positive integer n, there are n different nth roots of unity in $\mathbb{C}$, namely, the n complex numbers

$$\cos \frac{2\pi k}{n} + i \sin \frac{2\pi k}{n}$$

obtained by letting k take on the n integer values from 0 to $n - 1$. We will not prove this in general, but we will check its truth in a few special cases. For a general proof, one can use trigonometric identities or the exponential function.

Exercise 7.6. Verify the following statements:

1. The two numbers 1 and -1 are second roots of unity. They can be written as $\cos 0 + i \sin 0$ and $\cos \pi + i \sin \pi$.
2. The two complex numbers

$$-\frac{1}{2} \pm \frac{\sqrt{3}}{2} i$$

   are third roots of unity, as is 1. These three numbers can be written as

$$\cos \frac{2\pi k}{3} + i \sin \frac{2\pi k}{3}$$

   for k equal to 0, 1, and 2.
3. The four complex numbers ± 1 and $\pm i$ are fourth roots of unity. These four numbers can be written as

$$\cos \frac{2\pi k}{4} + i \sin \frac{2\pi k}{4}$$

   for k equal to 0, 1, 2, and 3.

4. The two complex numbers

$$\frac{1}{2} \pm \frac{\sqrt{3}}{2} i$$

are sixth roots of unity, as are ± 1 and $-\frac{1}{2} \pm \frac{\sqrt{3}}{2}i$. These six numbers can be written as

$$\cos\frac{2\pi k}{6} + i\sin\frac{2\pi k}{6}$$

for k equal to $0, 1, 2, 3, 4, 5$.

If r is a root of unity in a ring R, the smallest positive integer n for which $r^n = 1$ is called the *order* of r. For example, in $\mathbb{C}$, the order of 1 is 1, the order of -1 is 2, and the order of each of i and $-i$ is 4. The roots of unity in $\mathbb{C}$ have many important applications.

Our larger concern is not roots of unity in $\mathbb{R}$ or $\mathbb{C}$, but roots of unity in the rings $\mathbb{Z}_m$. Typically, the roots of unity of a ring R form only a small subset of the entire set of units of R. However, if R has only a finite number of elements, as is the case for the rings $\mathbb{Z}_m$, then all the units in R are roots of unity. More generally, if a ring R has only a finite number of units, then all the units of R are roots of unity. This statement is the last part of Proposition 7.3 below.

Proposition 7.3 *Let R be a ring with multiplicative identity 1.*

1. *Suppose u and v are units in R. Then uv is a unit too.*
2. *More generally, if $u_1, \ldots, u_t$ are units in R, then the product $u_1 u_2 \cdots u_t$ is a unit.*
3. *In particular, if u is a unit and t is a positive integer, then u^t is a unit.*
4. *Suppose R has only finitely many units; say, R has t units. Then every unit u in R satisfies $u^n = 1$ for some integer n less than or equal to t.*

Exercise 7.7. Prove Proposition 7.3. For the first part, you can write down the inverse explicitly, in terms of the inverses of u and v. A similar approach handles the next two parts. For the last part, can the units $1, u, u^2, u^3, \ldots, u^t$ all be distinct in R? If not, two of them must be equal. Write down an equation displaying the equality of two of these powers of u and deduce that $u^n = 1$ for some $n \leq t$.

7.3 The Theorems of Fermat and Euler

We saw in Section 4.1 that for an integer $m > 1$ and a positive integer a, we can determine the least nonnegative residue of a power a^t modulo m with relative ease if we are able to find an exponent e with the property that

$$a^e \equiv 1 \pmod{m}.$$

Must such an exponent e exist? Let us prove that such an exponent can exist only if $(a, m) = 1$.

Suppose that such an e exists. Factoring a^e as $a \times a^{e-1}$, we find that

$$a \times a^{e-1} \equiv 1 \pmod{m}.$$

Thus a^{e-1} is a solution to the congruence

$$ax \equiv 1 \pmod{m}.$$

We know from Theorem 4.6 that this congruence is solvable if and only if $(a, m) = 1$. Thus, if there is to be an exponent e satisfying $a^e \equiv 1 \pmod{m}$, then we must have $(a, m) = 1$. Or to state the contrapositive, if $(a, m) > 1$, then there can be no such exponent e. Therefore, we are led to ask the following question:

Question 7.4 *If a and $m > 1$ are integers with $(a, m) = 1$, is there an exponent e such that $a^e \equiv 1 \pmod{m}$?*

Let us translate this question into a multiplication question in $\mathbb{Z}_m$. We will use the bracket notation for elements of $\mathbb{Z}_m$. To find an e such that $a^e \equiv 1 \pmod{m}$ is the same as to find an e such that

$$[a]^e = [1]$$

in $\mathbb{Z}_m$. The condition that $(a, m) = 1$ is the same as the condition that $[a]$ is a unit in $\mathbb{Z}_m$. Therefore, we are asking the following question:

Question 7.5 *If $m > 1$ is an integer and u is a unit in $\mathbb{Z}_m$, is there an exponent e such that $u^e = 1$ in $\mathbb{Z}_m$?*

We have answered this question already, and more. We proved in Proposition 7.3 that every unit in a ring with only finitely many units is a root of unity. Since $\mathbb{Z}_m$ has only finitely elements altogether, it certainly has only finitely many units. Therefore, every unit u in $\mathbb{Z}_m$ is a root of unity, and this is all that the question is asking.

We can even say a bit more. Suppose u and v are units in $\mathbb{Z}_m$, with $u^i = 1$ and $v^j = 1$ for some positive integers i and j. Then $u^{ij} = (u^i)^j = 1$ and $v^{ij} = v^{ji} = (v^j)^i = 1$. Similarly, if $u_1, \ldots, u_t$ is a complete list of the units of $\mathbb{Z}_m$ and $i_1, \ldots, i_t$ are positive integers such that $u_r^{i_r} = 1$ for each index r, then there is a positive integer e such that every unit u_r satisfies $(u_r)^e = 1$. For instance, e can be the product of the i_r's, or their least common multiple. We have proved the following proposition:

Proposition 7.6 *Suppose R is a ring with only finitely many units. Then there is a positive integer e such that every unit u in R satisfies $u^e = 1$. In particular, for an integer $m > 1$, there is a positive integer e such that every unit u in $\mathbb{Z}_m$ satisfies $u^e = 1$. Equivalently, there is a positive integer e such that every integer a relatively prime to m satisfies the congruence*

$$a^e \equiv 1 \pmod{m}.$$

We would like to refine Proposition 7.6 for the ring $\mathbb{Z}_m$ in order to obtain an exponent e satisfying two special properties: e is not large (compared to m) and e is explicitly computable (in terms of m). Let us look back over our examples to see what choice of e can be made for specific values of m.

Exercise 7.8. For each of the values of m from 2 to 13 answer the following questions:

1. What are the units in $\mathbb{Z}_m$?
2. What is the order of each unit u in $\mathbb{Z}_m$? In other words, for each unit u, what is the smallest positive integer f such that $u^f = 1$?
3. What is the smallest integer e such that the every unit u of $\mathbb{Z}_m$ satisfies $u^e = 1$?
4. Let ϕ_m stand for the number of units in $\mathbb{Z}_m$. Does every unit u in $\mathbb{Z}_m$ satisfy $u^{\phi_m} = 1$? In particular, for prime values of m in the range from 2 to 13, does every unit u in $\mathbb{Z}_m$ satisfy $u^{m-1} = 1$?

Presumably, you have found that the answer to the last question in Exercise 7.8 is yes. Thus, for the prime numbers p that you have checked, those between 2 and 13, every unit u in $\mathbb{Z}_p$ satisfies

$$u^{p-1} = 1.$$

Equivalently, for every integer a not divisible by the prime p,

$$a^{p-1} \equiv 1 \pmod{p}.$$

This is true in general, and this fact was proved by Fermat:

Theorem 7.7 (Fermat). *Let p be a prime number and let a be a positive integer not divisible by p. Then*

$$a^{p-1} \equiv 1 \pmod{p}.$$

Equivalently, in the ring $\mathbb{Z}_p$, every nonzero element u satisfies

$$u^{p-1} = 1.$$

The argument we will use to prove Fermat's theorem actually works in much greater generality. Thus, we may as well state and prove the more general result.

Theorem 7.8. *Suppose R is a ring with only finitely many units, t, say. Then every unit u in R satisfies*

$$u^t = 1.$$

Exercise 7.9. Explain why Fermat's theorem follows as a special case of Theorem 7.8. Then prove Theorem 7.8 following the outline below.

1. Suppose that the ring R has precisely t units, $u_1, \ldots, u_t$. Suppose also that u a unit of R. Thus, u is one of the u_i's, but we do not care which one. Show that $uu_1, uu_2, \ldots, uu_t$ is also a complete list of the t units of R, possibly listed in a different order from the original list. (Thus, multiplication by u *shuffles* the set of units.)
2. Using the unit shuffle, prove that

$$u_1 u_2 \cdots u_t = (uu_1)(uu_2) \cdots (uu_t).$$

3. Rearrange the factors in this equality and use cancellation to deduce that

$$u^t = 1.$$

We can apply Theorem 7.8 to $\mathbb{Z}_m$ for every integer $m > 1$, not just for m prime. The result we obtain by doing so is known as Euler's theorem. It is usually stated using the *Euler ϕ-function*, which we shall now define.

If m is a positive integer, set $\phi(m)$ equal to the number of integers in the range from 1 to m that are relatively prime to m. Thus $\phi(1) = 1$, while for $m > 1$ we have $1 \leq \phi(m) < m$. (The reason that $\phi(m)$ is strictly less than m for $m > 1$ is that for $m > 1$, the integer m is never relatively prime to itself, and so there are at most $m - 1$ integers in the range from 1 to m that are relatively prime to m.) The value of $\phi(m)$ for $m > 1$ can also be described as the number of units in $\mathbb{Z}_m$. You should compute $\phi(m)$ for all integers $m \leq 13$, in order to check your understanding of its definition. You already have all the data you need.

Using the notation of the Euler ϕ-function, we can state Theorem 7.8 in the case of the ring $\mathbb{Z}_m$.

Theorem 7.9 (Euler's Theorem). *Let m be an integer greater than 1. In the ring $\mathbb{Z}_m$, every unit u satisfies*

$$u^{\phi(m)} = 1.$$

Equivalently, every integer a relatively prime to m satisfies

$$a^{\phi(m)} \equiv 1 \;(\text{mod } m).$$

The proof of Euler's theorem is simply a matter of quoting Theorem 7.8 and observing that $\phi(m)$ is the number of units in $\mathbb{Z}_m$. We shall discuss how to calculate $\phi(m)$ effectively in Section 7.4.

Let us discuss a surprising and important application of Fermat's theorem, a test for primality.

Proposition 7.10 *Let m be an integer greater than 1 and let a be a positive integer relatively prime to m. If the congruence*

$$a^{m-1} \equiv 1 \;(\text{mod } m)$$

does not hold, then m is not prime.

Exercise 7.10. Prove Proposition 7.10. Then illustrate its use by calculating the least nonnegative residue of 2^{510} modulo 511 and concluding that 511 is not prime.

This idea for testing whether a particular integer is prime is the beginning of a large and important subject. One way to test the primality of a positive integer m is trial factorization, in which you experiment by dividing m by all the integers up to $\sqrt{m}$. As we have discussed, this can take a long time. Alternatively, we can use Fermat's theorem to test the primality of m as follows: We can choose different integers a relatively prime to m and determine the least nonnegative residue of a^{m-1} modulo m. We have seen that these tests do not take a long time to perform. If any one of these residues is not 1, then m is not prime. This approach has the benefit that it may produce the nonprimality of m in a short time. It has the drawback, though, that when it tells us that a given integer m is not prime, it gives us no idea of the nontrivial factors of m.

There is another pitfall to the method of primality testing provided by Fermat's theorem: It is not guaranteed to give us conclusive information. Fermat's theorem allows us to conclude that an integer is not prime if it fails to pass a certain test, but it does *not* tell us anything about the primality of the integer if it in fact passes the test for all appropriate values of a. That is, Fermat's theorem does not preclude the possibility that there might exist some positive integer m that is not prime but that nonetheless "passes the Fermat test" for each a relatively prime to m. There may exist composite integers m that satisfy the condition that for *every* a relatively prime to m, we have $a^{m-1} \equiv 1 \pmod{m}$. Thus, even though m is not prime, the Fermat test would fail to reveal this.

In fact, such numbers do exist. They are called *Carmichael numbers* after R. D. Carmichael, who first noted the existence of such numbers in 1910. The first three of them are 561, 1105, and 1729. The Carmichael numbers are relatively few and far between (there are only 105 Carmichael numbers less than ten million), so if an integer m passes the Fermat test for every a relatively prime to m, we may conclude that m is "probably" prime. Fortunately, the idea behind this test can be refined to yield a better primality test, leading to techniques for determining with near certainty in a practical amount of time whether a large integer is prime. There are also, of course, tests to determine with absolute certainty whether an integer is prime, but the price of certainty is a slower test.

7.4 The Euler ϕ-Function

For Euler's theorem to be used effectively in practical situations, we need to be able to compute $\phi(m)$. We have already done so in the case that m is prime, obtaining the value $\phi(m) = m - 1$. For arbitrary m, the value of $\phi(m)$

can be computed in principle by checking each positive integer less than m to decide whether it is relatively prime to m, but this can take a long time. What we would like is a general formula for $\phi(m)$, and since we know the value of the Euler ϕ-function for prime numbers, we might suspect that such a formula would express $\phi(m)$ in terms of the prime numbers that are factors of m. Indeed, that is exactly the type of formula that we are going to deduce. Let us start with the simplest case.

Exercise 7.11. For a prime number p, you already know that $\phi(p) = p - 1$. More generally, prove that for every positive integer e, the value of the Euler ϕ-function applied to p^e is given by $\phi(p^e) = p^e - p^{e-1}$. (Hint: If an integer is not relatively prime to p^e, what are the only possible common factors that it can have with p^e? Use this idea to specify a set of p^{e-1} integers between 1 and p^e that are not relatively prime to p^e. Argue that the remaining $p^e - p^{e-1}$ integers between 1 and p^e are relatively prime to p^e.)

One way to compute the Euler ϕ-function for arbitrary positive integers is to combine the calculation of ϕ for prime powers from Exercise 7.11 with the following theorem.

Theorem 7.11. *Let a and b be relatively prime positive integers. Then*

$$\phi(ab) = \phi(a)\phi(b).$$

More generally, let $a_1, \ldots, a_r$ be positive integers satisfying the condition that any two of them are relatively prime. Then

$$\phi(a_1 a_2 \cdots a_r) = \phi(a_1)\phi(a_2) \cdots \phi(a_r).$$

Exercise 7.12. Let n be a positive integer.

1. Suppose $n = p^e q^f$ for distinct prime numbers p and q. Use Theorem 7.11 to show that
$$\phi(N) = \left(p^e - p^{e-1}\right) \cdot \left(q^f - q^{f-1}\right).$$

2. Calculate $\phi(1728)$.
3. Suppose $n = p_1^{e_1} p_2^{e_2} \cdots p_r^{e_r}$ for distinct prime numbers $p_1, \ldots, p_r$. Use Theorem 7.11 to write a formula for $\phi(n)$ in terms of the primes p_i and the exponents e_i.
4. Using this formula, calculate $\phi(60)$, $\phi(900)$, and $\phi(7875)$.

Notice that the second part of Theorem 7.11 follows from the first by induction. We will not prove Theorem 7.11 directly. Instead, we will take a different approach to the calculation of $\phi(N)$, one that will lead to the same formula for $\phi(N)$ given in the exercise above, and from this Theorem 7.11 will follow. The alternative approach depends on a general counting principle known as the *inclusion–exclusion principle*. Let us discuss an example in which this principle arises.

Suppose you are given a bag with 100 marbles. You are told that 30 are yellow (chipped or unchipped), that 15 are chipped (yellow or not yellow), and that 8 are both yellow and chipped. How many marbles are not yellow and not chipped?

Drawing a picture helps to see what is happening (see Figure 7.1). We can

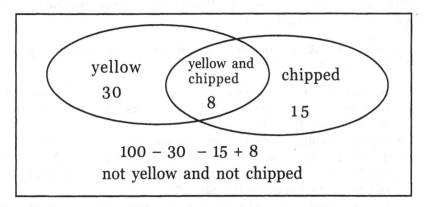

Figure 7.1. Counting marbles with the inclusion–exclusion principle.

draw a big rectangle representing the 100 marbles. Inside it we draw an ellipse representing the yellow marbles and another ellipse representing the chipped marbles. The two ellipses overlap, since some marbles are both chipped and yellow. The unchipped, nonyellow marbles are those lying outside the two ellipses. One way to count the number of unchipped, nonyellow marbles is to start with 100, subtract 30 to get rid of the yellow marbles, and subtract 15 to get rid of the chipped marbles. This gives us 55. But some marbles have been subtracted too often in this count: The chipped yellow ones have been subtracted twice. There are 8 of them, so we need to add 8 to 55, getting a count of 63 marbles that are neither yellow nor chipped. This is our answer. Let us generalize this idea and then apply it to the problem of calculating the Euler ϕ-function.

Suppose you have a set $\mathcal{X}$ containing N objects, for some positive integer N. The objects might be marbles, numbers, cars, or the integers from 1 to N. You are interested in two properties that the objects may satisfy. Let us call the properties P and Q. In the example above, property P might be the property of being yellow and Q the property of being chipped. An object in our collection may satisfy property P alone, property Q alone, both at once, or neither. Let us write N_P for the number of objects in the set $\mathcal{X}$ that satisfy property P. Any of these objects may satisfy Q or they may not. Similarly, write N_Q for the number of objects that satisfy property Q, and write N_{PQ} for the number of objects that satisfy both P and Q.

Exercise 7.13. Prove that the number of objects in $\mathcal{X}$ satisfying neither P nor Q is given by the formula

$$N - N_P - N_Q + N_{PQ}.$$

(Do not go into great detail. It would be sufficient to make a suitable diagram with various portions labeled and then to explain what the diagram means and how the formula follows.)

Exercise 7.14. We will apply this formula to the Euler ϕ-function for integers N of the form $p^e q^f$.

1. Suppose m and n are positive integers and m divides n, so that $n = bm$ for some positive integer b. List all the integers between 1 and n that are divisible by m. How many are there?
2. Suppose p and q are distinct prime numbers and N is a positive integer divisible by both p and q. How many integers between 1 and N are divisible by p? by q? by both p and q? (Hint: For the third part, if p and q divide an integer, does pq divide the integer?)
3. Suppose p and q are distinct prime numbers and N is an integer divisible by both p and q. How many integers between 1 and N are divisible by neither p nor q? In other words, how many integers between 1 and N are relatively prime to both p and q? (Hint: Use the result of the preceding problem.)
4. As a special case, suppose $N = p^e q^f$. Prove that

$$\phi(N) = N - \frac{N}{p} - \frac{N}{q} + \frac{N}{pq}.$$

5. Use this formula to deduce that

$$\phi(N) = \left(p^e - p^{e-1}\right)\left(q^f - q^{f-1}\right) = \phi\left(p^e\right)\phi\left(q^f\right).$$

Let us extend the idea above to integers divisible by three primes. First, begin with marbles again. Suppose you have a bag containing 100 marbles of two different sizes and you are given the following information about them:

(a) 22 marbles are green.
(b) 12 marbles are chipped.
(c) 14 marbles are large.
(d) 8 marbles are green and chipped.
(e) 3 marbles are green and large.
(f) 5 marbles are chipped and large.
(g) 2 marbles are green, chipped, and large.

How many marbles are not green, not chipped, and not large?

Begin by drawing a diagram, with a big rectangle for all the marbles and three small circles representing the green, the chipped, and the large marbles. The three circles need to be drawn so that they overlap in all possible ways. We want to count the number of marbles outside the three small circles. Our first estimate of this number is $100 - 22 - 12 - 14$, or 52. As in the earlier example, we have subtracted too much. Those that are green and chipped (but not large) have been subtracted twice. Similarly, we have subtracted the marbles that are green and large (but not chipped) twice and those that are chipped and large (but not green) also twice. So we add back $8 + 3 + 5$ to get a second, improved, estimate of 68. However, the marbles that are green, chipped, and large have been subtracted three times in the first estimate and added back three times in the second estimate, so the count of 68 includes them. And so we must finally subtract these 2 marbles from our count, getting 66. Our final calculation takes the form:

$$68 = 100 - (22 + 12 + 14) + (8 + 3 + 5) - 2.$$

What general principle covers this situation? Suppose you have a set $\mathcal{X}$ containing N objects, for some positive integer N, and now you are interested in three properties that the objects may satisfy. Let us call the properties P, Q, and R. Let us write N_P for the number of objects that satisfy property P, and then N_Q for the number satisfying Q, and N_R for the number satisfying R. Also write N_{PQ} for the number satisfying both P and Q, and write N_{PR} and N_{QR} for the analogous quantities. Finally, write N_{PQR} for the number of objects satisfying P, Q, and R simultaneously.

Exercise 7.15. Prove that the number of objects in $\mathcal{X}$ satisfying neither P nor Q nor R is given by the formula

$$N - N_P - N_Q - N_R + N_{PQ} + N_{PR} + N_{QR} - N_{PQR}.$$

Again, you do not need to go into great detail. Use a diagram.

Exercise 7.16. Apply this formula to the Euler ϕ-function.

1. Suppose p, q, and r are distinct prime numbers and N is a positive integer divisible by all three. How many integers between 1 and N are divisible simultaneously by p, by q, and by r? (Hint: If p, q, and r divide an integer, does pqr divide the integer?)
2. Suppose p, q, and r are distinct prime numbers and N is an integer divisible by all three. How many integers between 1 and N are divisible by neither p nor q nor r? In other words, how many integers between 1 and N are relatively prime to all three of p, q, and r?
3. As a special case, suppose that $N = p^e q^f r^g$. Prove that

$$\phi(N) = N - \frac{N}{p} - \frac{N}{q} - \frac{N}{r} + \frac{N}{pq} + \frac{N}{pr} + \frac{N}{qr} - \frac{N}{pqr}.$$

4. Prove for N as in the preceding part that

$$\phi(N) = \left(p^e - p^{e-1}\right)\left(q^f - q^{f-1}\right)\left(r^g - r^{g-1}\right) = \phi\left(p^e\right)\phi\left(q^f\right)\phi\left(r^g\right).$$

Let us consider the general situation. We are given N objects again, and this time we consider r properties, which we will call $p_1, p_2, \ldots, p_r$. We will again write N_{p_i} for the number of objects satisfying property p_i, and $N_{p_i p_j}$ for the number of objects satisfying p_i and p_j. We extend this notation in the obvious way. For instance, $N_{p_1 p_3 p_4 p_6 p_9}$ is the number of objects satisfying the five properties p_1, p_3, p_4, p_6, and p_9. We have proved in the special cases of $r = 2$ and $r = 3$ a result known in general as the *principle of inclusion and exclusion* or the *inclusion–exclusion principle*.

Theorem 7.12. *Let X be a set consisting of N objects. The number of objects in X that satisfy none of the properties $p_1, \ldots, p_r$ is given by the formula*

$$N - \sum_i N_{p_i} + \sum_{i \neq j} N_{p_i p_j} - \sum_{\substack{i,j,k \\ \text{distinct}}} N_{p_i p_j p_k} + \cdots + (-1)^r N_{p_1 p_2 \ldots p_r}.$$

The proof is not difficult, but it depends on some ideas that we have not yet discussed. These ideas and the proof are treated in Section 8.1. Theorem 7.12 is fundamental in many combinatorial and probabilistic situations.

Exercise 7.17. Use the inclusion–exclusion principle in the problems below.

1. Count how many integers between 1 and 63 000 are not divisible by 2, 3, 5, or 7. What is $\phi(63\,000)$?
2. Suppose N is a positive integer whose prime factorization is

$$p_1^{e_1} p_2^{e_2} \cdots p_r^{e_r},$$

where $p_1, \ldots, p_r$ are distinct prime numbers. Use the inclusion–exclusion principle to obtain a formula for $\phi(N)$.
3. Prove that this formula agrees with the product formula you obtained in Exercise 7.12 for $\phi(N)$.

7.5 RSA Encryption

Anya Krupskaya and Ilya Kuryakin are working as secret agents for a government whose name we need not reveal here. They are operating in different areas of one of the world's trouble spots, and they need to communicate in code. However, Mira the Malevolent, who works for another country's spy agency, is also operating in the area, and she is known to have devised a number of clever ways of intercepting communications between Ilya and Anya. Therefore, these two spies require a cryptographic system by which each of them can encode and send secret messages that the other can decode but that

Mira will be unable to decode should she chance to intercept a message. In fact, we would like to do something more: We want to be able to describe the encoding procedure publicly, so that anyone, even Anya's mother, or even Mira for that matter, can encode data in the same way and send it to Anya, but we want to ensure that anyone who might intercept the encoded data, even someone who knows the publicly available procedure for encoding data—even Ilya!—will nonetheless be unable to decode the message.

This may seem like rather a tall order, but in fact, it can be done. What makes it possible, according to one scheme known as RSA encryption, are three facts that we have already discussed:

1. It takes a *very* long time, even with the fastest computers available, to factor an integer that is the product of two unknown large prime numbers.
2. With the help of primality tests, along the lines of the one we introduced using Fermat's theorem, it is possible to *find* large prime numbers in a relatively short amount of time.
3. Euler's theorem: Every integer a relatively prime to an integer $m > 1$ satisfies $a^{\phi(m)} \equiv 1 \pmod{m}$.

RSA encryption depends on the asymmetry between the effort it takes to determine whether an integer is prime and the much greater effort it takes to find the prime factors of a number that is known to be composite. Therefore, RSA encryption will become obsolete if anyone ever finds a fast way to factor integers. Until such a method is found—and the consensus among mathematicians is that such a method will never be found, because factorization is an inherently labor-intensive enterprise—the RSA encryption scheme appears to be secure.

RSA encryption works as follows. First we find two distinct large prime numbers, p and q, and multiply them to obtain a number n. How large p and q need to be depends on the latest state of the art in factoring large integers. For medium security they might each have about 75 digits, for high security, about 150. One can deduce from the prime number theorem (see Chapter 5) that the odds of a random integer n being prime are roughly $1/\ln n$. Therefore, since $\ln 10^{75} \approx 173$, we can find our large primes in a reasonable amount of time by simply testing 75-digit integers chosen at random for primality until we find two of them.

The knowledge of p and q is what we must keep secret, but we do not mind if n is publicly known, because we anticipate that no one will be able to factor it to discover the secret p and q. Since we know p and q, we also know how to calculate $\phi(n)$:

$$\phi(n) = (p-1)(q-1) = pq - p - q + 1.$$

This number $\phi(n)$ is information that we also keep to ourselves and that someone without the knowledge of the factorization of n will be unable to compute in a reasonable amount of time.

We now choose an integer d relatively prime to $\phi(n)$. Since we know $\phi(n)$, such a d is easily found. Bézout's theorem ensures that integers e and f exist satisfying

$$de - \phi(n)f = 1.$$

The Euclidean algorithm allows us to find e and f quickly, and we now have integers d and e such that

$$de \equiv 1 \pmod{\phi(n)}.$$

We publicize the numbers e and n. We keep to ourselves the numbers d and $\phi(n)$. If one could factor n, then one would know $\phi(n)$, and one could use $\phi(n)$ and e to determine d, at least up to congruence modulo $\phi(n)$, which turns out to be good enough to break the code. Our system depends on the assumption that no one will succeed in factoring n.

How does one use e and n to encode data? First the data must be converted in some standardized way to a string of positive integers. For instance, we could assign to each of the 52 uppercase and lowercase letters an integer between 10 to 99, and assign additional such integers to the ten Arabic numerals, standard punctuation symbols, and maybe a few other symbols like a space symbol. It does not matter how we do this. By concatenating all these two-digit numbers, we obtain a huge string of digits, and that is what we want to encode and send.

Going back and forth between the string of two-digit integers and the original message is a routine matter, and the procedure is public. The problem is to send the string in a secure way. We will break the string into blocks of digits of some length k, each block giving us a k-digit number a that we wish to encode and send. We could take $k = 2$ and use our original two-digit numbers, but it might be more efficient to make k larger. In any case, it is important that k be shorter than the number of digits in the prime numbers p and q. This ensures that the k-digit numbers a that we encode and send are less than p and q. Therefore, a is relatively prime to both p and q, implying that a is relatively prime to n as well. We wish for the sender to use the public data of e and n to encode a into a new number b and send b. Then we, the receiver, knowing d and $\phi(n)$, must be able, on receiving b, to use d and $\phi(n)$ in some way to recover a.

Here is what we do. The sender raises a to the power e to get a new integer a^e and then finds the least nonnegative residue b of a^e modulo n. As we have learned, as large as a^e may be, the calculation of the residue b can be done quickly. The number a^e itself never needs to be found explicitly. This b is the number sent to the receiver. The receiver takes b, raises it to the power d, and finds the least nonnegative residue of b^d modulo n. Again, this can be done quickly, without calculating b^d explicitly. The reason this system works is that the least nonnegative residue of b^d modulo n turns out to be a, as we will check in a moment. The receiver has decoded the message, and has done so using d, which no one else knows. Anyone who intercepts the message, even

Mira the Malevolent, will be unable to decode it, for the interceptor would need to know d, which depends on knowing $\phi(n)$, which in turn depends on knowing the prime factorization of n.

Why is the least nonnegative residue of b^d modulo n equal to a? This is where Euler's theorem comes in. We have

$$b \equiv a^e \pmod{n}.$$

Therefore,

$$b^d \equiv (a^e)^d \pmod{n}.$$

But $(a^e)^d = a^{de}$, and de is congruent to 1 modulo $\phi(n)$, by the choice of d and e. This means, as we have seen, that there is an integer f such that $de - \phi(n)f = 1$. Let us rewrite this as $de = 1 + \phi(n)f$. Then we find that

$$(a^e)^d = a^{de} = a^{1+\phi(n)f} = a \cdot a^{\phi(n)f} = a \cdot \left(a^{\phi(n)}\right)^f.$$

Now we use Euler's theorem. Since a has fewer digits than both p and q, it is relatively prime to n. Therefore, Euler's theorem tells us that

$$a^{\phi(n)} \equiv 1 \pmod{n}.$$

This yields

$$\left(a^{\phi(n)}\right)^f \equiv 1 \pmod{n}$$

and

$$a \cdot \left(a^{\phi(n)}\right)^f \equiv a \pmod{n}.$$

Therefore,

$$(a^e)^d \equiv a \pmod{n}$$

and

$$b^d \equiv a \pmod{n}.$$

Raising a to the power e and reducing modulo n and then raising the result b to the power d and reducing modulo n has returned us to the original a. The receiver is able to recover a from b, but no one else can do so, because no one else knows what exponent d to use.

Notice the large number of results and ideas that we have treated in this book that have come into play in constructing RSA encryption schemes. These include the Euclidean algorithm, Bézout's theorem, congruences, and Euler's theorem. RSA encryption is one of the most important, and unexpected, applications of number theory and algebra to everyday life.

8

Binomial Coefficients

8.1 Pascal's Triangle

In Section 7.4 we used Theorem 7.12, the inclusion–exclusion principle, to prove that the Euler phi function satisfies $\phi(ab) = \phi(a)\phi(b)$ if a and b are relatively prime positive integers. Theorem 7.12 can be proved using the theory of *combinations*, which we introduce in this section. The theory is not itself a part of algebra, but it has important applications to algebra.

The theory of combinations begins with the classical question of how many different ways there are to choose r objects from a collection of n objects. Here r and n are positive integers with $r \leq n$. One can give an unending sequence of examples of situations in which such questions arise. Many such situations arise in games of chance, and indeed, the theory of combinations— or *combinatorics*, as the field of mathematics that studies the various ways in which objects can be counted is called—was developed in large measure to assist gamblers in figuring out the odds of the occurrence of a particular roll of the dice or a particular hand at cards.

The card game poker is a convenient source of examples. In the standard variant of this game several players are dealt five cards each, and they proceed to wager on the strength of their particular combinations of cards, which are ranked according to their rarity. The rarest poker hand of all is the "straight flush," in which all cards are of the same suit (clubs, diamonds, hearts, spades) and are adjacent in counting sequence. An example of a straight flush is the hand consisting of the eight, nine, ten, jack, queen of hearts. (The "face cards" are ranked above the two through ten, in the order jack, queen, king, and these, in turn, are followed by the ace.)

The *probability* of a player being dealt a particular poker hand, or a particular type of poker hand, is the number of ways of choosing that hand or type of hand divided by the number of all possible poker hands. How many different poker hands are there? Put differently, starting with a standard pack of 52 cards, how many *different* ways are there of choosing 5 cards from this set of 52? Each such choice is called a *combination*. We are going to have to be careful about just what we mean by *different*. If we are playing a hand of poker, we pick up our cards after they are dealt and are free to arrange them in any order: The *order* in which the cards are dealt is irrelevant. All we care about is what five cards we have. So when we ask how many different poker hands there are, we are asking how many different (unordered) subsets of five elements there are in a set of fifty-two elements.

In addition to the the question of just how rare a straight flush is (how rare? we will soon find out!), we can ask additional questions, such as how many of these combinations are full houses (three cards of one denomination and two of another, such as two kings and three fives), or flushes (all five cards of the same suit, but not a straight flush).

To proceed with the development of the elementary theory of combinations, we will introduce some notation. The standard mathematical notation for the number of combinations of r objects selected from a collection of n objects is $\binom{n}{r}$. For instance, if we are playing with a deck of n cards and deal everyone r cards, then $\binom{n}{r}$ is the number of possible hands you can be dealt. In particular, $\binom{52}{5}$ denotes the number of poker hands. The numbers $\binom{n}{r}$ are called *binomial coefficients* (because $\binom{n}{r}$ is the coefficient of x^r in the expansion of the binomial $(1+x)$ raised to the nth power; more on this later). When we read $\binom{n}{r}$ aloud, we say "n choose r." The number of different poker hands is "fifty-two choose five."

Notice that $\binom{n}{n}$ counts the number of ways of choosing n objects from n objects, which is 1 (since order is irrelevant!), while $\binom{n}{1}$ counts the number of ways to choose 1 object from n, which is n, since a collection of n objects has n one-element subsets. It is convenient to introduce the idea of choosing 0 objects. This has a certain sense, in that if we have before us, say, an ice cream cone, a chocolate bar, and a piece of chess pie, and we have taken a New Year's resolution not to eat any desserts, and we have decided to choose none of the items offered, there is only one way to do it: Just say no. And so $\binom{3}{0} = 1$. If we look at our interpretation of n choose r as the number of r-element subsets of an n-element set, then it is clear that $\binom{n}{0}$ should be set equal to 1, since every set has precisely one zero-element subset, namely, the set with no elements, which is called the *empty set* and denoted by $\varnothing$. But even if neither of these justifications appeals to you, it is convenient in formulas involving the binomial coefficients to have $\binom{n}{0}$ equal to 1, and so we may simply *define* $\binom{n}{0}$ to be 1. We make no exception for $n = 0$. We define $\binom{0}{0}$ to be 1, to be consistent (the empty set has a single zero-element subset, namely, itself, the empty set).

We define $\binom{n}{r}$ to be 0 if r is negative or if $r > n$, since there is no way in the physical world to choose a negative number of objects or to choose more objects than are available. Once again, the subset interpretation can help to lend support to these definitions: There are no r-element subsets of an n-element set if r is greater than n, and no set has subsets with a negative number of elements. Why we bother assigning numbers to these symbols will become clear as we develop mathematical expressions that relate different binomial coefficients.

Exercise 8.1. Determine the values of the following binomial coefficients:

1. $\binom{2}{1}$.

2. $\binom{3}{2}$.

3. $\binom{n}{n-1}$ (in terms of n).

Theorem 8.1. *Let n be a positive integer and suppose r is an integer with $0 \leq r \leq n$. Then*

$$\binom{n}{r} = \binom{n-1}{r-1} + \binom{n-1}{r}.$$

Before we proceed to the proof, which you will work out in Exercise 8.2, note that in the case $n = r$, the expression $\binom{n}{r}$ makes "real-world" sense, but that $\binom{n-1}{r}$ does not, and that if we hadn't defined $\binom{n-1}{n}$ to be zero, we would have had to include in our theorem the disclaimers that r must be at least 1 and no greater than $n - 1$, and that n must be at least 1. But with our extension of the definition of the binomial coefficients to embrace the cases of n equal to zero and of r negative or greater than n, Theorem 8.1 is true for *all* values of $n > 0$ and $0 \leq r \leq n$.

Exercise 8.2. Prove Theorem 8.1. You can proceed as follows:

1. First, verify the formula directly in the two extreme cases $r = 0$ and $r = n$. Assume in the remainder that $0 < r < n$.
2. Imagine that the n objects are lined up in a row with a group of $n - 1$ on the left, then a gap, then a single object on the right. Any choice of r objects from this lineup either includes the lone object on the right or does not include it. From this consideration, show that the desired equality holds.

Exercise 8.3. We have not defined the binomial coefficients for $n < 0$. We could proceed by intuition and say that $\binom{n}{r}$ is zero if n is negative. Show by citing a particular choice of n and r that with this proposed definition Theorem 8.1 is false.

Theorem 8.1 gives us a means of computing the binomial coefficients *recursively*; that is, we can compute binomial coefficients in terms of "previous" binomial coefficients. What is meant by previous is that the number in the upper position of the binomial coefficient is smaller. For instance, we know that $\binom{1}{0} = \binom{1}{1} = 1$. Using the theorem, we can then compute the binomial coefficients $\binom{2}{r}$, since, for example, $\binom{2}{1} = \binom{1}{0} + \binom{1}{1} = 1 + 1 = 2$ and $\binom{2}{0} = \binom{1}{-1} + \binom{1}{0} = 0 + 1 = 1$. We can then use the binomial coefficients for $n = 2$ to compute the binomial coefficients $\binom{3}{r}$, and so on as far as time and patience allow.

Exercise 8.4. Make a table displaying the values of the binomial coefficients $\binom{n}{r}$ for n running from 1 to 8 and r running from 0 to n. Your table should take the shape of a triangle, with the top row (which we will call row zero) containing $\binom{0}{0}$, the next (first) row containing $\binom{1}{0}$ and $\binom{1}{1}$, and so on. The entry $\binom{n}{r}$ should sit in the nth row with $\binom{n-1}{r-1}$ above it to the left and $\binom{n-1}{r}$ above it to the right.

The advantage of numbering the rows in the triangle of Exercise 8.4 from zero, rather than one, is that under this numbering scheme, the nth row contains the binomial coefficients $\binom{n}{r}$, for n fixed and r running from 0 to n. Thus there are $n+1$ binomial coefficients in row n. Notice that this triangle can be continued downward indefinitely. It is called *Pascal's triangle*, in honor of the philosopher and mathematician Blaise Pascal (1623–1662), who founded the modern science of combinatorics and probability in response to a gambler's question about the odds of rolling a double six at least once in twenty-four rolls of two dice. (He was betting even money. Should he have been?) The triangle has been discovered many times in the past two thousand years, by thinkers in many civilizations. It is a wondrous object.

In studying Pascal's triangle you may make the following empirical observations:

1. Each row is *palindromic*: it reads the same from left to right as from right to left.
2. The sum of the terms in row n is 2^n.
3. The alternating sum of the terms in any row is 0.

These empirical observations, made above only for n from 1 to 8, hold for every n:

Theorem 8.2. *Let n be a nonnegative integer.*

1. *For every integer r, we have*

$$\binom{n}{r} = \binom{n}{n-r}.$$

2. *The following equality holds:*

$$\binom{n}{0} + \binom{n}{1} + \binom{n}{2} + \cdots + \binom{n}{n-1} + \binom{n}{n} = 2^n.$$

3. *The following equality holds:*

$$\binom{n}{0} - \binom{n}{1} + \binom{n}{2} - \cdots + (-1)^{n-1}\binom{n}{n-1} + (-1)^n\binom{n}{n} = 0.$$

Exercise 8.5. Prove Theorem 8.2. Proceed as follows:

1. Interpret the first statement in terms of combinations and explain why it holds.

2. For the second statement, imagine that the n objects from which you are choosing are lined up in a row. Show how each possible choice of objects from these n can be identified with a sequence of 0's and 1's of length n, and that in turn, each such sequence specifies a particular combination. Then show by induction that there are 2^n sequences of 0's and 1's of length n.

3. For the third statement, use Theorem 8.1.

Surprisingly, we can prove the inclusion–exclusion principle, Theorem 7.12, using the result of Theorem 8.2 that the alternating sum of binomial coefficients in a row of Pascal's triangle is zero. We repeat the theorem here as Theorem 8.3. Recall the notation that, for example, $N_{p_i p_j}$ is the number of objects in a set of N objects that satisfy the properties p_i and p_j:

Theorem 8.3. *Let $\mathcal{X}$ be a set consisting of N objects. The number of objects in $\mathcal{X}$ that satisfy none of the properties $p_1, \ldots, p_r$ is given by the formula*

$$N - \sum_i N_{p_i} + \sum_{i \neq j} N_{p_i p_j} - \sum_{\substack{i,j,k \\ \text{distinct}}} N_{p_i p_j p_k} + \cdots + (-1)^r N_{p_1 p_2 \cdots p_r}.$$

Exercise 8.6. Prove Theorem 8.3 (7.12). Proceed as follows:

1. Focus your attention on a particular object in $\mathcal{X}$. Let us call it Grendel. Suppose that from among the r properties that Grendel could satisfy, he satisfies exactly k of them. Here k must be some integer between 0 and r. To see where the argument is heading, let us suppose that k is 2 and that the two properties Grendel satisfies are p_2 and p_5. Consider the alternating sum expression in the theorem. Call it $\mathcal{A}$. The N on the left represents a count of all the objects in $\mathcal{X}$, including Grendel, so Grendel contributes a 1 to the expression. We then pass to the term $-\sum_i N_{p_i}$. Since Grendel satisfies p_2 and p_5, he contributes a 1 to N_{p_2} and a 1 to N_{p_5}. But since Grendel satisfies no other property, he contributes 0 to the other N_{p_i}'s. Thus in the second term of $\mathcal{A}$, Grendel contributes -2. Now pass to the third term, $\sum_{i \neq j} N_{p_i p_j}$. Grendel contributes 1 to $N_{p_2 p_5}$ but 0 to every other summand $N_{p_i p_j}$, so he contributes 1 to the sum $\sum_{i \neq j} N_{p_i p_j}$. Grendel satisfies no set of more than 2 properties, so Grendel contributes 0 to every other summand. Taking inventory, we see that Grendel has contributed to $\mathcal{A}$ the following amount:

$$1 - 2 + 1,$$

which is equal to 0.

2. Repeat the same analysis, but now suppose that Grendel satisfies the properties p_4, p_6, and p_7. Observe that he gets counted in $\mathcal{A}$ when we look at the summand N; at the summands N_{p_4}, N_{p_6}, and N_{p_7}; at the summands $N_{p_4 p_6}$, $N_{p_4 p_7}$, and $N_{p_6 p_7}$; and at the summand $N_{p_4 p_6 p_7}$; but in

every other summand of $\mathcal{A}$ he contributes 0. His total contribution to $\mathcal{A}$, taking signs into account, is

$$1 - 3 + 3 - 1,$$

or 0. (Do these numbers look familiar?)

3. In general, if Grendel satisfies k properties and $k > 0$, show that Grendel appears in a given summand $N_{p_{i_1} p_{i_2} \cdots p_{i_j}}$ of $\mathcal{A}$ only when the set of properties $\{p_{i_1}, p_{i_2}, \ldots, p_{i_j}\}$ is a subset of the set of k properties that Grendel satisfies. Observe in this way that Grendel's contribution to the alternating sum $\mathcal{A}$ is

$$\binom{k}{0} - \binom{k}{1} + \binom{k}{2} - \cdots + (-1)^k \binom{k}{k},$$

which equals 0.

4. Suppose Grendel satisfies none of the properties $p_1, p_2, \ldots, p_r$. Observe that Grendel's contribution to $\mathcal{A}$ is 1.

5. Finally deduce that $\mathcal{A}$ counts precisely the number of objects in $\mathcal{X}$ that satisfy none of the properties $p_1, p_2, \ldots, p_r$, as desired.

8.2 The Binomial Theorem

We have used properties of binomial coefficients in order to prove the inclusion–exclusion principle. Next let us take a closer look at binomial coefficients, in order to obtain a formula for $\binom{n}{r}$. Suppose we want to compute $\binom{52}{5}$, the binomial coefficient that counts the number of different poker hands. So far, the only method we have is to calculate rows of Pascal's triangle recursively until we get to row 52. There must be a better way.

Recall that in the earlier discussion of poker hands, we observed that in asking how many ways there are to choose 5 cards from a deck of 52, we have no interest in the order in which the cards are dealt to us. But let us now suppose that we do care about the order in which the cards are dealt. In other words, suppose we distinguish as different two hands that consist of the same cards laid out in a different order. We will call each ordered set of five cards an *arrangement*. It is easy to count the number of arrangements. There are 52 different cards, so there are 52 different possibilities for the first card in a given arrangement. Once this card is chosen, there are 51 possibilities for the second card, then 50 possibilities for the third, 49 for the fourth, and 48 for the fifth. We see that the number of arrangements is

$$52 \times 51 \times 50 \times 49 \times 48.$$

Let us count the number of arrangements of 5 cards from a 52-card deck a different way. For each fixed choice or combination of 5 cards—what we have called a hand—in how many ways can this hand be arranged? In other words, for these this collection of five different cards, in how many ways can we order them to obtain different arrangements? We count these arrangements the same way we counted above, except that now we start with only five cards, instead of fifty-two. There are 5 choices for the first card in the arrangement, then 4 for the second, 3 for the third, 2 for the fourth, and 1 choice for the last, the last remaining card. Since there are $\binom{52}{5}$ different hands, and $5 \times 4 \times 3 \times 2 \times 1$ different arrangements of each hand, the total number of arrangements of 5 cards is

$$\binom{52}{5} \times 5 \times 4 \times 3 \times 2 \times 1.$$

We have counted the number of arrangements in two different ways. Equating the two answers, we obtain

$$\binom{52}{5} = \frac{52 \times 51 \times 50 \times 49 \times 48}{5 \times 4 \times 3 \times 2 \times 1} = \frac{311\,875\,200}{120} = 2\,598\,960.$$

Let us generalize this argument.

Exercise 8.7. Let n be a positive integer and suppose r is an integer between 0 and n.

1. Prove that the number of arrangements of r objects chosen from a collection of n objects is

$$n \times (n - 1) \times (n - 2) \times \cdots \times (n - r + 1).$$

2. Prove Theorem 8.4 below.

Theorem 8.4. *Let n be a positive integer and let r be an integer satisfying $0 \le r \le n$. Then*

$$\binom{n}{r} = \frac{n \times (n - 1) \times (n - 2) \times \cdots \times (n - r + 1)}{r \times (r - 1) \times (r - 2) \times \cdots \times 2 \times 1}.$$

We can rewrite these two results using factorial notation.

Exercise 8.8. Let n be a positive integer and let r be an integer satisfying $0 \le r \le n$.

1. Prove that the number of arrangements of r objects chosen from a collection of n objects is

$$\frac{n!}{(n - r)!}.$$

2. Prove that

$$\binom{n}{r} = \frac{n!}{r!\,(n - r)!}.$$

3. Prove the sum formula

$$\binom{n}{r} = \binom{n-1}{r-1} + \binom{n-1}{r}$$

of Theorem 8.1 by using the factorial formula for binomial coefficients rather than the combinatorial meaning of them.

Let us study another application of binomial coefficients, the application that gives them their name. Consider how you calculate $(x+y)^2$. You write it out as $(x+y)(x+y)$ and then add together the products of all possible choices of x or y from the first factor and x or y from the second factor. This produces $x^2 + xy + yx + y^2$, or $x^2 + 2xy + y^2$. Similarly, you can calculate $(x+y)^3$ by writing it as $(x+y)(x+y)(x+y)$ and adding together the products of all choices of x or y from the first, the second, and the third factors. Do this and you find that you get a single x^3, a single y^3 and 3 each of x^2y and xy^2. In other words, $(x+y)^3 = x^3 + 3x^2y + 3xy^2 + y^3$.

In general, for n a positive integer, we can calculate $(x+y)^n$ by writing a product of n copies of $(x+y)$ and then adding together the products of all choices of x or y from each factor. We can think of this process as follows. Line up the n copies of $x+y$ in a row. When we multiply them together, we obtain a term $x^{n-r}y^r$ each time we choose a y from r of the copies and an x from the other $n-r$ copies. There are $\binom{n}{r}$ ways to choose r copies of $x+y$, so the coefficient of $x^{n-r}y^r$ in the product must be $\binom{n}{r}$. This reasoning suggests the following result:

Theorem 8.5. *Let n be a positive integer. Then*

$$(x+y)^n = \sum_{r=0}^{n} \binom{n}{r} x^{n-r}y^r$$

$$= \binom{n}{0}x^n + \binom{n}{1}x^{n-1}y + \binom{n}{2}x^{n-2}y^2 + \cdots + \binom{n}{n-1}xy^{n-1} + \binom{n}{n}y^n$$

$$= x^n + \binom{n}{1}x^{n-1}y + \binom{n}{2}x^{n-2}y^2 + \cdots + \binom{n}{n-1}xy^{n-1} + y^n.$$

The argument that led us to write down this theorem does not amount to a proof. More needs to be said to make it precise:

Exercise 8.9. Prove the binomial theorem by the following argument:

1. First show that for each positive integer n, the expansion of $(x+y)^n$ has the form

$$\sum_{r=0}^{n} b(n,r)x^{n-r}y^r = x^n + b(n,1)x^{n-1}y + b(n,2)x^{n-2}y^2 + \cdots + y^n,$$

where $b(n,r)$ is an integer, with $b(n,0) = b(n,n) = 1$. Do this by induction on n. It is certainly true for $n = 1$. Show that it is true for $k+1$ on the assumption that it is true for a positive integer k.

2. Now show, by reexamining your induction argument with a little more care, that the integers $b(n, r)$ satisfy the formula

$$b(n, r) = b(n - 1, r - 1) + b(n - 1, r).$$

For this formula to make sense for every r between 0 and n, you may wish to define $b(n, -1) = b(n, n + 1) = 0$.

3. It now remains to prove that $b(n, r) = \binom{n}{r}$. Do so by induction on n. First verify the equality for $n = 1$. Then show that if it holds for $n = k$, it holds for $n = k + 1$.

Exercise 8.10. We can now prove some earlier results in a new way, as consequences of the binomial theorem:

1. Use the binomial theorem to show again that the sum of the binomial coefficients $\binom{n}{r}$, with n fixed and r running from 0 to n, is 2^n. (Hint: Substitute a specific choice of integers for x and y in the binomial theorem.)

2. Use the binomial theorem to show again that the alternating sum of the binomial coefficients $\binom{n}{r}$, with n fixed and r running from 0 to n, is 0. (Hint: Again, substitute a specific choice of integers for x and y in the binomial theorem.)

Exercise 8.11. The binomial theorem can be used to give another proof of Fermat's theorem, Theorem 7.7, which states that if p is a prime number and a a positive integer not divisible by p, then $a^{p-1} \equiv 1 \pmod{p}$. Alternatively, the theorem states that for every nonzero element a of $\mathbb{Z}_p$, the equality $a^{p-1} = 1$ holds.

Suppose p is a prime number.

1. Show that for r an integer satisfying $1 \leq r \leq p - 1$, the prime p divides $\binom{p}{r}$. (Hint: Use the formula that we have just obtained for binomial coefficients.)

2. The binomial theorem is a statement about polynomials in two variables with integer coefficients. It can be used to obtain a parallel but simpler statement about polynomials in two variables with coefficients in the ring $\mathbb{Z}_p$. Just as we can reduce an integer modulo p, so, too, can we reduce a polynomial in two variables with integer coefficients modulo p, by reducing all the integer coefficients to obtain coefficients in $\mathbb{Z}_p$. This process will be discussed for one-variable polynomials in Section 11.4, but the same process works equally well for two-variable polynomials. For now, without going into further detail, let us assume that it makes sense. Apply the process of reduction modulo p to the polynomials on both sides of the binomial theorem. Deduce that for polynomials in x and y with coefficients in $\mathbb{Z}_p$ the following equality holds:

$$(x + y)^p = x^p + y^p.$$

3. Substitute elements $[a]$ and $[b]$ from $\mathbb{Z}_p$ for the variables x and y in the equality above to deduce that in $\mathbb{Z}_p$,

$$([a] + [b])^p = [a]^p + [b]^p.$$

4. Apply this formula inductively to $([a] + [1])^p$, as a runs from 0 to $p - 1$, to deduce that for every $[a]$ in $\mathbb{Z}_p$,

$$[a]^p = [a].$$

5. If $[a] \neq 0$, you can divide both sides of the equality $[a]^p = [a]$ by $[a]$ to obtain

$$[a]^{p-1} = [1].$$

You have proved Fermat's theorem.

Part II

Polynomials

9

Polynomials and Roots

9.1 Polynomial Equations

The fundamental problem of algebra is the solution of polynomial equations. These are equations of the form

$$a_n x^n + a_{n-1} x^{n-1} + a_{n-2} x^{n-2} + \cdots + a_2 x^2 + a_1 x + a_0 = 0,$$

where the coefficients a_i are specific numbers from some ring such as $\mathbb{Z}$, $\mathbb{Q}$, $\mathbb{R}$, or $\mathbb{C}$. To solve such an equation, we must first determine what numbers we are going to allow as solutions. Typically, we will have a specific set of numbers in mind as the allowable choices for coefficients and for the type of solutions that we are looking for. For instance, we may restrict these values to integers, to rational numbers, to real numbers, or to complex numbers. If we are interested only in integer solutions of the polynomial equation $4x^2 - 1 = 0$, we simply observe that four times the square of an integer cannot equal 1 and declare that the polynomial equation has no solutions. However, if our ring of interest is the rational numbers $\mathbb{Q}$, then we see that $4 \cdot \left(\pm \frac{1}{2} \right)^2 - 1 = 0$, and so the equation has the two solutions $x = \pm \frac{1}{2}$. We do not have to restrict ourselves to the familiar rings mentioned above. We could decide that the coefficients a_i should belong to the finite ring $\mathbb{Z}_m$, for some integer $m > 1$, and look for solutions to polynomial equations from within $\mathbb{Z}_m$. Taking the same polynomial equation that we considered above, but now investigating coefficients and solutions in the ring $\mathbb{Z}_5$, we find that the polynomial equation

$$[4]x^2 - [1] = [0]$$

has the two solutions $x = [2]$ and $x = [3]$.

In solving polynomial equations, we have to perform the usual arithmetic operations of addition, subtraction, multiplication, and division on the coefficients, and as we have just seen, how those operations work depends on the ring in which we are working. That is why we must assume that the coefficients of the polynomial under investigation lie in some fixed ring. In a ring like $\mathbb{Z}_m$ we have not defined division or subtraction, but we can say that we are "dividing" by a nonzero element r when we multiply by its multiplicative inverse r^{-1}, and that we are "subtracting" an element s when we add its additive inverse $-s$. Therefore, we can "divide" by an element r only if r has a multiplicative inverse. If we want to be able to divide freely by the nonzero elements of the ring from which we have chosen the coefficients of our polynomial, we therefore must require that every nonzero element of the ring be a unit. These considerations suggest that the ring of numbers we choose for studying a polynomial equation should be one in which every nonzero element is a unit. Recall that such a ring is called a *field*.

We have seen that the ring $\mathbb{Q}$ of rational numbers is a field, as are the ring $\mathbb{R}$ of real numbers and the ring $\mathbb{C}$ of complex numbers. We have also seen (Theorem 6.2 and Exercise 6.21) that the ring $\mathbb{Z}_m$ is a field if and only if m is a prime number. We will frequently be working with the fields of the form $\mathbb{Z}_p$ for p a prime, so let us introduce a special notation for them. For a prime number p we will sometimes write $\mathbb{F}_p$ for $\mathbb{Z}_p$. This special notation helps to remind us that we are working with a field. Think of $\mathbb{F}_p$ as the finite Field with a *prime* number of elements.

Some of the results about polynomials that we will study hold only for polynomials whose coefficients lie in a specific field, such as one of the fields in our list above. Other results will hold in general, no matter what field contains the coefficients. It does not make sense to prove results of this general type repeatedly, once for $\mathbb{R}$, once for $\mathbb{C}$, once again for $\mathbb{F}_p$, and so on. Instead, we will save a lot of time if we work over an arbitrary field, without paying attention to which field it is. This is far more efficient.

9.2 Rings of Polynomials

Let K, then, be an arbitrary field, and consider polynomials whose coefficients are elements of K. The typical polynomial of this type has the form

$$a_n x^n + a_{n-1} x^{n-1} + a_{n-2} x^{n-2} + \cdots + a_2 x^2 + a_1 x + a_0,$$

with each of the coefficients a_i being an element of K. For economy of notation, we will use expressions such as $f(x)$ to denote polynomials. The set of all polynomials with coefficients in K will be written as $K[x]$. Notice that the elements of K can themselves be regarded as polynomials, namely, as polynomials a_0 that consist of only a constant term. They are called *constant* polynomials. We can add and multiply two polynomials $f(x)$ and $g(x)$ in $K[x]$ in the usual way.

Exercise 9.1. Let K be a field. Verify that the set $K[x]$ of all polynomials with coefficients in K, under the usual addition and multiplication operations, forms a ring. In particular, you must verify that there are additive and multiplicative identities, that addition and multiplication are commutative and associative, that every polynomial has an additive inverse, and that the distributive law holds. Do not write out detailed proofs of all this; just specify what the additive identity is, what the additive inverse of a polynomial is, and what the multiplicative identity is.

A ring R is said to contain *zero-divisors* if it contains one or more nonzero elements that when multiplied together give the result 0. For example, in the ring $\mathbb{Z}_6$, the elements 2 and 3 are zero-divisors, since $2 \times 3 = 0$. The ring $\mathbb{Z}_4$ has the unique zero-divisor 2, since in this ring $2 \times 2 = 0$. The reason for the name "zero-divisor" is that such an element in fact divides (is a factor of) zero. Let us give a precise definition: In a ring R, a *zero-divisor* is a nonzero element r of R for which there is a corresponding nonzero element s of R satisfying $rs = 0$. If R is a ring with no zero-divisors, then R is called an *integral domain*. You can think of "integral" as meaning that in an integral domain, multiplication maintains its *integrity*. For example, $\mathbb{Z}$ is an integral domain, but $\mathbb{Z}_4$ is not. A polynomial $a_n x^n + \cdots + a_0$ in $K[x]$ with $a_n \neq 0$ is said to have *degree* n. The polynomial 0 is assigned the special degree $-\infty$. Notice that the polynomials of degree 0 are precisely the nonzero constant polynomials. We also adopt the convention, in talking about degrees of polynomials, that $-\infty + m = -\infty$. Theorem 9.1 below contains the fundamental facts about degrees and zero-divisors in a polynomial ring.

Theorem 9.1. *Let K be a field.*

1. K *is an integral domain.*
2. *For two polynomials $f(x)$ of degree m and $g(x)$ of degree n in $K[x]$, the product $f(x)g(x)$ has degree $m + n$.*
3. $K[x]$ *is an integral domain.*
4. *A polynomial $f(x)$ in $K[x]$ is a unit in $K[x]$ if and only if $f(x)$ has degree 0.*

Exercise 9.2. Prove Theorem 9.1. (Hint: Use part 1 in proving part 2. For parts 3 and 4, suppose that two polynomials have 0 or 1 as their product. What can their degrees be?)

Assume that K is a field. A polynomial $g(x)$ in $K[x]$ *divides* a polynomial $f(x)$ in $K[x]$ if there is a polynomial $h(x)$ in $K[x]$ such that $f(x) = g(x)h(x)$. One also says that $g(x)$ is a *divisor* of $f(x)$. More generally, in a ring R, an element s *divides* an element r if there is an element t such that $r = st$, in which case s is called a *divisor* of r.

Divisibility and factorization problems are of special interest in integral domains, because we need not worry about the presence of zero-divisors. A factorization $r = ab$ of an element r of R, where r is nonzero and is not a unit, is called *trivial* if either a or b is a unit and *nontrivial* otherwise. An element r of R is *irreducible* if it is nonzero and not a unit in R and every factorization of r in R is trivial.

Exercise 9.3. Let K be a field.

1. Suppose that $a(x)$ and $b(x)$ are polynomials in $K[x]$, and each of them is divisible by a polynomial $g(x)$. Show that $a(x) + b(x)$ and $a(x) - b(x)$ are divisible by $g(x)$. More generally, with $a(x)$ and $b(x)$ as above, prove that for every two polynomials $r(x)$ and $s(x)$ in $K[x]$, $a(x)r(x) + b(x)s(x)$ is also divisible by $g(x)$.
2. Observe that the polynomials in $K[x]$ that are not zero and not units are precisely the polynomials of positive degree.
3. If $f(x)$ is a polynomial of positive degree that has a factorization $f(x) = g(x)h(x)$ for polynomials $g(x)$ and $h(x)$ in $K[x]$, show that this factorization is trivial if and only if either $g(x)$ or $h(x)$ is a constant.
4. Conclude that a factorization of a polynomial $f(x)$ of positive degree as $g(x)h(x)$ is nontrivial if and only if the factors $g(x)$ and $h(x)$ have degrees strictly less than the degree of $f(x)$.
5. Deduce that a polynomial $f(x)$ in $K[x]$ is irreducible if and only if it has positive degree and it cannot be factored as a product of two polynomials of lower degree.
6. Suppose $f(x)$ and $g(x)$ are polynomials in $K[x]$ of the same degree and that $g(x)$ divides $f(x)$. Show that there is a constant c such that $f(x) = cg(x)$.

The criterion given in Exercise 9.3 for a factorization of a polynomial to be nontrivial is reminiscent of the criterion for a factorization of an integer n as ab to be nontrivial. If n, a, and b are all positive integers, then the factorization $n = ab$ is nontrivial if both a and b are less than n; more generally, the factorization is nontrivial if the absolute values $|a|$ and $|b|$ are less than $|n|$. Thus in some way the degree of a nonzero polynomial is a measure of its size, just as the absolute value of a nonzero integer is a measure of the integer's size. We will make use of this analogy frequently as we discover ways in which $K[x]$ behaves like $\mathbb{Z}$.

9.3 Factoring a Polynomial

A fundamental fact about integers is that any integer n greater than 1 is either prime or a product of prime numbers. The analogous factorization result for polynomials is the theorem below.

Theorem 9.2. *Let K be a field. A polynomial $f(x)$ of positive degree in $K[x]$ either is irreducible in $K[x]$ or factors as a product of irreducible polynomials in $K[x]$.*

Let us recall our proof of the integer factorization result. We use the principle of generalized induction. Since 2 is prime, the result holds for 2. Next we consider a positive integer $n > 2$ and assume that the result holds for every integer k from 2 to $n - 1$: Every such k is either prime or a product of prime numbers. If n is prime, we are done. If not, then n must have a nontrivial factorization. Thus there exist positive integers a and b that are not units such that $ab = n$. Since a and b are not units, they are greater than 1, and therefore, they must also be less than n. By the inductive hypothesis about integers between 2 and $n - 1$, we can conclude that each of a and b is itself prime or a product of prime numbers. From this it follows that n, as the product ab, is itself a product of prime numbers.

In this proof, it is helpful to understand that the induction is being done not with respect to the integer n itself, but with respect to the size of the integer. The induction step consists in assuming that the theorem is true for integers of smaller size. Since the size of a positive integer is in fact the integer itself, the two distinct roles played by an integer—(1) an element of the ring $\mathbb{Z}$ and (2) a measure of size of an element of the ring $\mathbb{Z}$—are obscured. In contrast, the size of a polynomial is measured by its degree, and there is no danger of mistaking the degree of a polynomial for the polynomial itself.

Exercise 9.4. Let K be a field and suppose that $f(x)$ is a polynomial of degree $n > 0$ in $K[x]$. Recall from Exercise 9.3 that if $f(x)$ is not irreducible, then $f(x)$ factors as the product of two polynomials of lower degree. Prove Theorem 9.2 by induction on the degree n of $f(x)$. First show that for $n = 1$ the degree-n polynomial $f(x)$ is irreducible. Then perform the inductive step.

We can test the primality of an integer $n > 1$ by checking to see whether each of the integers from 2 to $\sqrt{n}$ divides n. If none of these does, then n is prime. The polynomial analogue takes the following form:

Theorem 9.3. *Let K be a field and let n be a positive integer. Suppose $f(x)$ is a polynomial in $K[x]$ of degree n. If $f(x)$ is not irreducible, then $f(x)$ has a divisor of degree less than or equal to $n/2$.*

Exercise 9.5. Prove Theorem 9.3.

We have now obtained polynomial analogues of two of our initial theorems on prime numbers. A third prime number result is Euclid's theorem: There are infinitely many prime numbers. Once again, there is a polynomial analogue. The analogue is not the statement that the ring $K[x]$ has infinitely many irreducible polynomials. This statement is true, but typically it is of no interest. For example, in $\mathbb{R}[x]$, as r runs through the nonzero real numbers, we obtain infinitely many irreducible polynomials rx. These are all essentially the same, differing from each other by multiplication by constants.

In general, when we are studying factorization questions in an integral domain, we do not bother to distinguish between unit multiples of two elements. In $\mathbb{Z}$ for instance, the integers 2 and -2 are essentially the same; that is, they have the same multiplicative properties. More generally, for every integer r, the integers r and $-r$ are essentially the same. The sense in which they are the same is that if one of the pair r and $-r$ occurs in some factorization of n, then the other occurs in a factorization as well. A factorization of n can be replaced by a new one with an even number of terms in the factorization having their signs changed. For example, 30 factors in $\mathbb{Z}$ as $2 \times 3 \times 5$ and as $(-2) \times 3 \times (-5)$. There is little value in distinguishing between these factorizations.

One way to avoid the clutter of different factorizations of an integer in $\mathbb{Z}$ is to choose from each pair $\{n, -n\}$ of nonzero integers a single one that we regard as *canonical*, that is, as a special or distinguished representative of its class. We have implicitly done so in taking the positive integer from each pair to be the canonical choice. Any other choice would do, but this choice seems the most natural.

The same issue emerges in studying factorization questions in a ring $K[x]$. If a nonzero polynomial $f(x)$ occurs in a factorization of $g(x)$, then so does $cf(x)$, for every nonzero constant c. All the nonzero constant multiples of $f(x)$ are essentially the same, just as the integers n and $-n$ are essentially the same. In studying factorization, we may wish to choose from each family of nonzero constant multiples of a nonzero polynomial a canonical representative. Again, any choice would do, but there is a natural one.

A nonzero polynomial $f(x)$ in $K[x]$ of degree n is called *monic* (from the Greek *monos*, "single") if the coefficient of its term of degree n is 1. A nonzero polynomial $f(x)$ in $K[x]$ can be written as $cp(x)$ for a nonzero constant c in K and a monic polynomial $p(x)$ in $K[x]$. If

$$f(x) = a_n x^n + \cdots + a_1 x + a_0$$

is a nonzero polynomial in $K[x]$, then the monic polynomial in its class of equivalent polynomials is

$$a_n^{-1} f(x) = a_n^{-1} \left(a_n x^n + \cdots + a_1 x + a_0 \right),$$

which equals

$$x^n + \cdots + a_n^{-1} a_1 x + a_n^{-1} a_0.$$

From all the nonzero constant multiples of a given nonzero polynomial $f(x)$, the unique monic polynomial is the natural one to regard as canonical. For example, for the polynomial

$$3x^3 + \frac{1}{7}x - \frac{7}{8}$$

in $\mathbb{Q}[x]$, the monic polynomial

$$x^3 + \frac{1}{21}x - \frac{7}{24}$$

is the canonical polynomial we select from among all the nonzero constant multiples.

The polynomial analogue of Euclid's theorem for integers states not that there are infinitely many irreducible polynomials in $K[x]$ but that there are infinitely many that are not constant multiples of each other. Using monic polynomials, we obtain the following concise formulation:

Theorem 9.4. *The ring of polynomials $K[x]$ with coefficients in a field K contains infinitely many irreducible monic polynomials.*

Exercise 9.6. Prove Theorem 9.4. You can proceed according to the following outline:

1. Show that $K[x]$ has at least one irreducible monic polynomial. You can do so by exhibiting a specific one.
2. We are going to suppose that $K[x]$ has only finitely many irreducible monic polynomials and arrive at a contradiction. Suppose that the distinct polynomials $p_1(x), \ldots, p_k(x)$ make up the entire set of irreducible monic polynomials in $K[x]$.
3. Show that none of $p_1(x), \ldots, p_k(x)$ can divide the polynomial

$$\bigl(p_1(x) \cdots p_k(x)\bigr) + 1.$$

4. Observe that $\bigl(p_1(x) \cdots p_k(x)\bigr) + 1$ has some monic irreducible polynomial $p(x)$ in $K[x]$ as a divisor, and that therefore there is a monic irreducible polynomial in $K[x]$ distinct from $p_1(x), \ldots, p_k(x)$.
5. Deduce that the set $p_1(x), \ldots, p_k(x)$ could not have been the entire collection of irreducible monic polynomials in $K[x]$, and that therefore there must be infinitely many irreducible monic polynomials in $K[x]$.

9.4 The Roots of a Polynomial

A *polynomial equation with coefficients in a field K* is an equation

$$a_n x^n + a_{n-1} x^{n-1} + \cdots + a_1 x + a_0 = 0$$

in which the a_i's are elements of K. A *solution* of the equation is an element γ of K satisfying

$$a_n \gamma^n + a_{n-1} \gamma^{n-1} + \cdots + a_1 \gamma + a_0 = 0.$$

We also say that γ is a *root* of the polynomial $a_n x^n + a_{n-1} x^{n-1} + \cdots + a_1 x + a_0$. In other words, for $f(x)$ a polynomial in $K[x]$, the element γ is a root of $f(x)$, or, equivalently, a solution of the equation $f(x) = 0$, if we obtain 0 when we replace x with γ in the expression for $f(x)$.

We wish to prove that an element γ of K is a root of $f(x)$ if and only if the polynomial $x - \gamma$ divides $f(x)$ in $K[x]$. This relates roots of $f(x)$ to factorization questions about $f(x)$ as an element of the ring $K[x]$. The tool we will use to prove this is a division theorem for polynomials. Recall that the division theorem for integers states that for positive integers a and b there exist unique nonnegative integers q and r with $r < a$ such that $b = aq + r$. In everyday language, the result of dividing b by a is the quotient q with remainder r. The polynomial version is similar:

Theorem 9.5. *Let K be a field. If $a(x)$ and $b(x)$ are nonzero polynomials in $K[x]$, then there exist unique polynomials $q(x)$ and $r(x)$, with $r(x)$ having lower degree than $a(x)$, such that*

$$b(x) = a(x)q(x) + r(x).$$

We call $q(x)$ the *quotient* and $r(x)$ the *remainder* obtained on dividing $b(x)$ by $a(x)$.

Exercise 9.7. In $\mathbb{Q}[x]$, divide $x^4 - 1$ by $x^3 - 2x^2 + x - 2$ using long division. Determine the quotient and the remainder.

Exercise 9.8. Prove the division theorem. Proceed as follows.

1. Begin by reviewing how we proved in the integer case that q and r exist. We did an induction on the size of b, with a fixed. If $b = 1$, the desired result is easy to verify. Assume, then, that $b > 1$ and that the division theorem for integers holds for $b - 1$. Then we obtain nonnegative integers q' and r' with $b - 1 = aq' + r'$ and $r' < a$. Adding 1 to both sides yields $b = aq' + (r' + 1)$. If $r' + 1 < a$, we are done. Otherwise, $r' + 1 = a$, in which case $b = a(q' + 1) + 0$, and again we are done.

2. In the polynomial case, we want to mimic this approach as best we can. The size of a polynomial is measured in terms of its degree. Thus we can try to do an induction on the degree of $b(x)$, with $a(x)$ held fixed. It may be best to do this in three stages. First, deal with the case in which $b(x)$ has degree less than the degree of $a(x)$. This should be easy. Then deal with the case in which $b(x)$ and $a(x)$ have the same degree. This is a little trickier, but it is still entirely elementary. You are now ready for the general induction step.

3. Assume that $b(x)$ has degree n and that n is larger than the degree of $a(x)$. Make the appropriate induction assumption about polynomials of degree $n - 1$. Taking a hint from the integer case, in which we wrote b as the sum of the smaller integer $b - 1$ and 1, we want to write $b(x)$ in terms of a polynomial of degree $n - 1$ and a polynomial of degree 1. Can we write $b(x)$ as $xg(x)$ for some polynomial $g(x)$ of degree $n - 1$? Not necessarily, but we can come close. Observe that you can rewrite $b(x)$ in the form $xg(x) + c$ for some constant c. Use the induction assumption to rewrite $g(x)$ as $a(x)q'(x) + r'(x)$ for suitable polynomials $q'(x)$ and $r'(x)$. What do you know about the degree of $r'(x)$? Plug the expression for $g(x)$ back into $xg(x) + c$ and look at what you have. The argument at this point is reminiscent of the argument for the division theorem for the integers. You will have two cases, depending on the degree of $r'(x)$. In one case, it will be obvious what $q(x)$ and $r(x)$ should be; in the other, some more work will need to be done.

4. To prove the uniqueness portion of the theorem, suppose there is another pair of polynomials $s(x)$ and $t(x)$ with

$$b(x) = a(x)s(x) + t(x)$$

and with the degree of $t(x)$ less than the degree of $a(x)$. Use the degree formula of Theorem 9.1 to show that $r(x) - t(x) = 0 = q(x) - s(x)$.

For our immediate purposes we do not need the full division theorem, but the special case in which $a(x)$ is a polynomial of degree 1. Let us state this case explicitly:

Theorem 9.6. *Let K be a field, $b(x)$ a nonzero polynomial in $K[x]$ of degree n, and γ an element of K. Then there is a unique polynomial $q(x)$ in $K[x]$ of degree less than n and a unique element r of K such that*

$$b(x) = (x - \gamma)q(x) + r.$$

We are ready to relate roots of a polynomial to divisibility questions. Suppose K is a field, γ is an element of K, and $x - \gamma$ is a divisor of $f(x)$. For $x - \gamma$ to be a divisor of $f(x)$ means that there is a polynomial $g(x)$ such that

$$f(x) = (x - \gamma)g(x).$$

What happens if we substitute γ for x in the equation above? On the right side, we get $(\gamma - \gamma)g(\gamma)$, which must be 0. Therefore, $f(\gamma) = 0$, and this means that γ is a root of $f(x)$. In other words, if $x - \gamma$ divides $f(x)$, then γ is a root of $f(x)$. The converse holds as well; it is one of the fundamental results about polynomial equations:

Theorem 9.7. *Let K be a field, let $f(x)$ be a polynomial in $K[x]$, and suppose that γ in K is a root of $f(x)$. Then $x - \gamma$ divides $f(x)$ in $K[x]$.*

Exercise 9.9. Use Theorem 9.6 to prove Theorem 9.7.

We wish to obtain a stronger version of Theorem 9.7. For this we need the next result, which shows that the polynomials in $K[x]$ of degree 1 share with prime numbers an important divisibility property.

Theorem 9.8. *Let K be a field and let γ be an element of K. Suppose $f(x)$ and $g(x)$ are two polynomials in $K[x]$. If $x - \gamma$ divides the product $f(x)g(x)$, then $x - \gamma$ divides at least one of $f(x)$ and $g(x)$.*

Exercise 9.10. Prove Theorem 9.8.

1. Use Theorem 9.6 to rewrite $f(x)$ and $g(x)$ in terms of $x - \gamma$ and multiply the two expressions together to express $f(x)g(x)$ in terms of $x - \gamma$ and a constant.
2. Use the fact that $x - \gamma$ divides this product to show that it divides the constant.
3. Deduce that the constant is 0, and from this conclude that $x - \gamma$ divides $f(x)$ or $g(x)$.

We can use Theorem 9.7 and Theorem 9.8 to give an inductive proof of the following strengthening of Theorem 9.7.

Theorem 9.9. *Let K be a field, let $f(x)$ be a polynomial in $K[x]$, and suppose that $\gamma_1, \ldots, \gamma_m$ are distinct roots of $f(x)$ in K. Then*

$$(x - \gamma_1)(x - \gamma_2) \cdots (x - \gamma_m)$$

divides $f(x)$ in $K[x]$.

Exercise 9.11. Prove Theorem 9.9.

Theorem 9.9 has the following important consequence:

Theorem 9.10. *Let K be a field and let $f(x)$ be a nonzero polynomial of degree n. Then $f(x)$ has at most n distinct roots in K.*

Exercise 9.12. Prove Theorem 9.10. (Hint: If $f(x)$ has $n + 1$ distinct roots, what can you say about the degree of $f(x)$?)

Theorem 9.7 tells us that finding a root in K of a polynomial $f(x)$ in $K[x]$ is equivalent to finding degree-one divisors of $f(x)$ in $K[x]$. In particular, if $f(x)$ is an irreducible polynomial in $K[x]$ of degree greater than one, then $f(x)$ cannot have any roots in K. This links the problem of factoring in $K[x]$ with the problem of finding roots. More generally, if $f(x)$ factors as a product of irreducible polynomials all of which have degrees greater than 1, then again $f(x)$ has no root in K. In contrast, if $f(x)$ has degree-one factors, these factors correspond to roots. Thus, factoring a polynomial $f(x)$ as a product of irreducible polynomials will allow us to find all the roots of $f(x)$.

Exercise 9.13. Let K be a field.

1. Show that a polynomial $f(x)$ in $K[x]$ of degree 2 either has a root in K or is irreducible in $K[x]$.
2. Show that a polynomial $f(x)$ in $K[x]$ of degree 3 either has a root in K or is irreducible in $K[x]$.
3. Give an example of a polynomial $f(x)$ in $\mathbb{R}[x]$ of degree 4 that has no roots in $\mathbb{R}$ yet is not irreducible in $\mathbb{R}[x]$.

9.5 Minimal Polynomials

Suppose that K is a field and $f(x)$ a polynomial in $K[x]$. Then the polynomial $f(x)$ may have no roots in K yet have roots in a larger field L. For example, the polynomial $x^2 - 2$ in $\mathbb{Q}[x]$ has no roots in $\mathbb{Q}$, but it does have the roots $\sqrt{2}$ and $-\sqrt{2}$ in the larger field $\mathbb{R}$. As another example, $x^2 + 1$ has no roots in $\mathbb{R}$, but it does have the roots i and $-i$ in the larger field $\mathbb{C}$. In general, the study of polynomials $f(x)$ with coefficients in a field K leads naturally to the study of larger fields that contain K, these larger fields being the source of missing roots of the polynomials.

A *field extension* is a pair of fields K and L such that K lies inside L and such that the rules of addition and multiplication for K and L are compatible; that is, for two elements of K, their sum and product as members of K coincide with their sum and product as members of L. We often use the notation $K \subseteq L$ to describe a field extension consisting of the fields K and L with K inside L. The symbol $\subseteq$ indicates that K is a part of L. Also, we call L a *field extension of K*. Examples of field extensions are $\mathbb{Q} \subseteq \mathbb{R}$ and $\mathbb{R} \subseteq \mathbb{C}$. For a field extension $K \subseteq L$ and a polynomial $f(x)$ in $K[x]$, a *root of $f(x)$ in L* is an element γ of L such that $f(\gamma) = 0$. We also say the equation $f(x) = 0$ has a solution in L. As we have seen, a polynomial $f(x)$ may have no roots in K but may have roots in a suitable field extension L. This motivates the construction of field extensions as a means of finding roots of polynomials that otherwise have no roots. In Sections 13.4 and 14.5, we will construct many field extensions for this purpose.

Alternatively, one may start with a field extension $K \subseteq L$, in which case elements s of L lead in a natural way to polynomials in $K[x]$, the polynomials that have these elements as roots. For example, starting with the field extension $\mathbb{Q} \subseteq \mathbb{R}$ and the irrational number $\sqrt{2}$, we are led to polynomials in $\mathbb{Q}[x]$ such as $x^2 - 2$. We wish to study this phenomenon systematically.

For a field extension $K \subseteq L$, an element γ of L is *algebraic over K* if there is a nonzero polynomial $f(x)$ in $K[x]$ such that $f(\gamma) = 0$. Thus, γ is algebraic if there are elements $a_0, a_1, \ldots, a_n$ of K, with $n > 0$ and $a_n \neq 0$, such that

$$a_n \gamma^n + a_{n-1} \gamma^{n-1} + \cdots + a_1 \gamma + a_0 = 0.$$

For example, the real number $\sqrt{2}$ is algebraic over $\mathbb{Q}$, and the complex number i is algebraic over $\mathbb{R}$. Furthermore, every element γ of K is algebraic over K, since γ is a root of the polynomial $x - \gamma$, which is in $K[x]$. We wish to associate with γ in a natural way a monic irreducible polynomial in $K[x]$.

Let $K \subseteq L$ be a field extension, and suppose that γ is an element of L and that $f(x)$ is a polynomial in $K[x]$ that has γ as a root; that is, $f(\gamma) = 0$. If $g(x)$ is an arbitrary polynomial in $K[x]$, the product $f(x)g(x)$ will also have γ as a root: $f(\gamma)g(\gamma) = 0 \cdot g(\gamma) = 0$. Thus there is a large collection of polynomials of $K[x]$ that have γ as a root, not just the given one $f(x)$. Despite the large size of this collection, it is possible to give a succinct description of the polynomials in it: It turns out that all the polynomials in $K[x]$ with γ as a root are multiples of a polynomial of *least degree* with γ as a root, which we will call the *minimal polynomial* of γ over K.

To see this, first observe that all the polynomials in $K[x]$ with γ as a root have a positive integer degree, and that in any collection of positive integers there must be a smallest one. Therefore, among all polynomials in $K[x]$ there is a polynomial of *least degree* that has γ as a root. Let n be that least degree. We call n the *degree of γ over K*. Suppose, for example, that γ is an element of K. Then as we saw above, γ is a root of the degree-one polynomial $x - \gamma$

in $K[x]$, so γ has degree 1 over K. As an another example, we have seen that $\sqrt{2}$ has degree 2 over $\mathbb{Q}$: It is a root of the degree-two polynomial $x^2 - 2$ of $\mathbb{Q}[x]$, but it cannot be the root of any degree-one polynomial in $\mathbb{Q}[x]$, since it is not in $\mathbb{Q}$.

Theorem 9.11. *Suppose $K \subseteq L$ is a field extension and γ is an element of L that is algebraic over K of degree n. Let $f(x)$ be a degree-n polynomial in $K[x]$ such that $f(\gamma) = 0$ and let $g(x)$ be an arbitrary polynomial in $K[x]$ such that $g(\gamma) = 0$. Then $f(x)$ divides $g(x)$.*

Exercise 9.14. Prove Theorem 9.11. (Hint: Divide $g(x)$ by $f(x)$ and use the division theorem to write the result in terms of a quotient $q(x)$ and a remainder $r(x)$. Show that $r(\gamma) = 0$. By the assumption on $f(x)$ and the degree bound on $r(x)$, deduce that $r(x) = 0$. Conclude that $f(x)$ divides $g(x)$.)

Theorem 9.11 has the following refinement. The proof is provided. It should be read closely.

Theorem 9.12. *Suppose $K \subseteq L$ is a field extension and γ is an element of L that is algebraic over K of degree n.*

1. *There is a unique monic polynomial $p_\gamma(x)$ of degree n in $K[x]$ having γ as a root.*
2. *The polynomials of $K[x]$ that have γ as a root are precisely the polynomials divisible by $p_\gamma(x)$.*
3. *The polynomial $p_\gamma(x)$ is irreducible in $K[x]$.*

Proof. By definition, there is a polynomial of degree n in $K[x]$ with γ as a root. We can divide this polynomial if necessary by its highest-degree coefficient in order to obtain a monic polynomial of degree n in $K[x]$ with γ as a root. Suppose there are two such polynomials, $f(x)$ and $g(x)$. By Theorem 9.11, $g(x)$ divides $f(x)$, so that $f(x) = g(x)h(x)$ for some polynomial $h(x)$ in $K[x]$. But $f(x)$ and $g(x)$ have the same degree, so $h(x)$ must be an element c of K, by Theorem 9.1. Since $f(x)$ and $g(x)$ are both monic, c must be 1. This proves the uniqueness result. The second part of Theorem 9.12 follows immediately from Theorem 9.11.

For the third part, suppose $p_\gamma(x)$ factors in $K[x]$ as $f(x)g(x)$. To prove that $p_\gamma(x)$ is irreducible, we must show that this factorization is trivial. Substituting γ for x, we find that $0 = p_\gamma(\gamma) = f(\gamma)g(\gamma)$. Thus $f(\gamma)$ and $g(\gamma)$ are elements of K whose product is 0. This can occur only if $f(\gamma)$ or $g(\gamma)$ is 0. (Why? In what ring are the values $f(\gamma)$ and $g(\gamma)$ located?) If the factorization $p_\gamma(x) = f(x)g(x)$ were not trivial, then both $f(x)$ and $g(x)$ would have positive degree less than n (why?), and one of $f(x)$ and $g(x)$ would have γ as a root, contrary to the assumption that γ has degree n. Therefore, the factorization must be trivial, which is what we wished to prove.

The polynomial $p_\gamma(x)$ of Theorem 9.12 is called the *minimal polynomial* of γ over K. Notice that although $p_\gamma(x)$ is irreducible in $K[x]$, it will not be irreducible when regarded as a polynomial in $L[x]$. After all, it has γ as a root, so it has $x - \gamma$ as a divisor. Theorem 9.12 is valuable as a source of irreducible polynomials in $K[x]$.

Polynomials with Real Coefficients

10.1 Quadratic Polynomials

The easiest polynomial equations to solve are those of the lowest degrees. We will consider low-degree polynomials with real coefficients in this chapter. Let us work momentarily with an arbitrary field K before imposing the assumption that K is $\mathbb{R}$.

We begin with first-degree polynomials. Suppose $f(x) = ax + b$ with a and b in the field K and $a \neq 0$. To solve $ax + b = 0$, we simply add $-b$ to both sides to obtain $ax = -b$ and then multiply both sides by the multiplicative inverse a^{-1} of a to obtain $x = -a^{-1}b$. Since K is a field, a^{-1} exists. Since an nth-degree polynomial can have at most n roots, a first-degree polynomial can have at most one root. Therefore, the root of $ax + b$ that we have found is the only one.

Second-degree polynomials are more interesting. Consider a second-degree polynomial $f(x)$ in $K[x]$ of the form $ax^2 + bx + c$. The coefficient a is nonzero, since the degree of $f(x)$ is 2. In solving $f(x) = 0$, we can always divide through by a first, or multiply by its inverse a^{-1}. Thus we can always assume that $a = 1$, that is, that $f(x)$ is monic. Let us do so and start over again, taking $f(x)$ to be

$$x^2 + bx + c$$

for some elements b and c of K.

If $c = 0$, then we are solving the equation $x^2 + bx = 0$. We can factor $x^2 + bx$ as $x(x + b)$, from which we see that the solutions are $x = 0$ and $x = -b$. The next simplest case is that in which $b = 0$, so that we are solving $x^2 + c = 0$. If $c = 0$, then the equation is $x^2 = 0$, and the only solution is 0. (Why?) Assume that $c \neq 0$ and let d be its additive inverse $-c$. Then we are solving the equation $x^2 = d$. Whether or not $x^2 = d$ has a solution amounts to a question about the field K: Are there any elements in K whose squares equal d? In other words, we are asking whether d has square roots in K. As we know from our experience with $\mathbb{R}$, there is no general answer. Some elements of K may have square roots, while others may not. If d has no square root, then $x^2 = d$ has no solutions in K, and therefore $x^2 + c$ does not factor in $K[x]$ as the product of degree-one polynomials.

Let us assume now that K is the field $\mathbb{R}$ of real numbers. We know in this case that if a real number d is positive, it has two real square roots, each the additive inverse of the other; if $d = 0$, it has only one square root, 0; and if d is negative, then it has no real square roots. Thus we have the following result.

Proposition 10.1 *Suppose d is a real number.*

1. *If $d < 0$, then the equation $x^2 - d = 0$ has no solutions in $\mathbb{R}$.*
2. *If $d = 0$, then $x^2 - d = 0$ has one solution in $\mathbb{R}$, the solution $x = 0$.*
3. *If $d > 0$, then $x^2 - d = 0$ has two distinct solutions in $\mathbb{R}$, each the additive inverse of the other.*

Polynomial equations of degree two are called *quadratic* equations (from the Latin *quadrum*, a square, the basic geometric figure of degree two). These equations were studied in many early civilizations, from ancient Babylon and Greece to India and China. In all these civilizations, scholars were able to come up with approaches that led to what we now call the quadratic formula and learn in high school. Recall that for real numbers b and c, this formula describes the roots of $x^2 + bx + c$ as follows:

$$x = -\frac{b}{2} \pm \frac{\sqrt{b^2 - 4c}}{2},$$

where the symbol $\sqrt{\alpha}$ represents the positive square root of the positive real number α.

Exercise 10.1. Let us review how the quadratic formula is obtained.

1. For a real number a, verify that

$$(x + a)^2 = x^2 + 2ax + a^2.$$

2. For a real number b, conclude that the polynomial

$$x^2 + bx + \frac{b^2}{4}$$

is the square of a degree-one polynomial.
3. For real numbers b and c, rewrite

$$x^2 + bx + c$$

by adding and subtracting $\frac{b^2}{4}$ and find that solving the equation $x^2 + bx + c = 0$ is equivalent to solving an equation of the form $\left(x + \frac{b}{2}\right)^2 = \frac{d}{4}$ for a suitable real number d. Write out the number d explicitly in terms of b and c. This number is called the *discriminant* of the polynomial $x^2 + bx + c$. It will be important in what follows.

4. Deduce that if $d = 0$, then $x^2 + bx + c$ factors as $\left(x + \frac{b}{2}\right)^2$, and the one solution to $x^2 + bx + c = 0$ is $x = -\frac{b}{2}$.
5. Deduce that if d is negative, then there is no solution in $\mathbb{R}$ to the equation $x^2 + bx + c = 0$, and $x^2 + bx + c$ is irreducible in $\mathbb{R}[x]$.
6. Deduce that if d is positive, then there are two real solutions to $x^2 + bx + c = 0$. Write out explicitly what these solutions are in terms of b and c.

We can also obtain the quadratic formula by a change of variable. Let us do so, since this will motivate one of the steps we discuss in Sections 10.2 and 10.4 for solving cubic and quartic equations. Starting with the equation $x^2 + bx + c = 0$, we introduce a new variable y and set $x = y - \frac{b}{2}$. Substituting $y - \frac{b}{2}$ for x in the given equation, we obtain

$$\left(y - \frac{b}{2}\right)^2 + b\left(y - \frac{b}{2}\right) + c = 0.$$

When we expand this equation, you can check that we get

$$y^2 - by + by + \frac{b^2}{4} - \frac{b^2}{2} + c = 0.$$

The degree-one terms cancel, leaving us with

$$y^2 - \frac{b^2}{4} + c = 0,$$

or

$$y^2 - \frac{b^2 - 4c}{4} = 0.$$

Our change of variable has allowed us to replace our original quadratic equation with one that has no degree-one term. This represents a great simplification, and to solve the new equation, we must simply calculate the square root of $b^2 - 4c$. We find that

$$y = \pm\frac{\sqrt{b^2 - 4c}}{2}.$$

We can now return to the original variable x:

$$x = y - \frac{b}{2} = -\frac{b}{2} \pm \frac{\sqrt{b^2 - 4c}}{2}.$$

We have obtained the quadratic formula again.

When one says that the quadratic formula provides a solution to the equation

$$x^2 + bx + c = 0,$$

what does that mean? It means that the values for x given by the formula satisfy the equation. Suppose, for the sake of simplification, that $b = 0$. Then the equation has the simpler form

$$x^2 + c = 0,$$

and the quadratic formula tells us that the solutions to the equation are $x = \pm\sqrt{-c}$. In other words, the quadratic formula tells us nothing more than what we knew already. This is what we should expect. The quadratic formula does not tell us how to compute square roots. Rather, the quadratic formula shows how to use algebra to reduce the problem of solving an arbitrary quadratic equation to the problem of computing square roots.

If the discriminant $b^2 - 4c$ of the polynomial $x^2 + bx + c$ is negative, then $b^2 - 4c$ has no real square root, and the equation $x^2 + bx + c = 0$ has no real solutions. This corresponds to the fact that $x^2 + bx + c$ is an irreducible polynomial in $\mathbb{R}[x]$. It cannot be factored as a product of two degree-one polynomials, only trivially as a product of a constant and a degree-two polynomial. If $b^2 - 4c$ is 0, we obtain one solution, $x = -\frac{b}{2}$; the polynomial $x^2 + bx + c$ factors as $(x + \frac{b}{2})^2$. If $b^2 - 4c$ is positive, then we obtain two distinct real solutions r_1 and r_2, and the corresponding result that $x^2 + bx + c$ factors as $(x - r_1)(x - r_2)$. Thus the three possible behaviors of solutions to $x^2 + bx + c = 0$ are mirrored by three possible ways in which $x^2 + bx + c$ can factor in $\mathbb{R}[x]$: It is irreducible; it factors as the square of a degree-one polynomial; or it factors as a product of distinct degree-one polynomials.

If a quadratic equation $x^2 + bx + c = 0$ has a discriminant $b^2 - 4c$ that is negative, then by working with complex numbers we can proceed further. Even though $b^2 - 4c$ has no real square roots, it has two complex square roots, $\left(\sqrt{-b^2 + 4c}\right)i$ and $-\left(\sqrt{-b^2 + 4c}\right)i$. (Note that if $b^2 - 4c$ is negative, then $-b^2 + 4c$ is positive and has a real square root, and so the expression $\sqrt{-b^2 + 4c}$ makes sense.) Working in $\mathbb{C}$ and using these square roots, we can apply the quadratic formula to obtain two complex solutions to $x^2 + bx + c = 0$. Recall that for real numbers m and n, the complex conjugate of $m + ni$ is defined to be the complex number $m - ni$. We see that the two nonreal complex solutions of $x^2 + bx + c = 0$ are complex conjugates of each other. Even though $x^2 + bx + c$ does not factor nontrivially in $\mathbb{R}[x]$, it does factor nontrivially in $\mathbb{C}[x]$, in the form $(x - r_1)(x - r_2)$ with $r_2 = \bar{r}_1$ (where $\bar{z}$ denotes the complex conjugate of a complex number z). We can summarize these facts in the theorem below.

Theorem 10.2. *Let b and c be real numbers. Exactly one of the following three possibilities occurs:*

1. $x^2 + bx + c$ *has two distinct real roots r_1 and r_2, and factors as*

$$(x - r_1)(x - r_2).$$

2. $x^2 + bx + c$ *has only one real root r, and factors as $(x - r)^2$.*
3. $x^2 + bx + c$ *has two distinct nonreal complex roots r and $\bar{r}$, and factors as $(x - r)(x - \bar{r})$.*

Moreover, the first possibility occurs if the discriminant $b^2 - 4c$ is greater than zero, the second possibility occurs if the discriminant $b^2 - 4c$ equals zero, and the third possibility occurs if the discriminant $b^2 - 4c$ is less than zero.

Let us explore the quadratic polynomial $x^2 + bx + c$ a little further in order to obtain an alternative interpretation of the discriminant.

Exercise 10.2. Suppose b and c are real numbers. Let r_1 and r_2 be the two real or complex roots of the polynomial $x^2 + bx + c$.[1]

1. Observe that if r_1 and r_2 are real and distinct, then $(r_1 - r_2)^2 > 0$.
2. Observe that if $r_1 = r_2$, then $(r_1 - r_2)^2 = 0$.
3. The only remaining possibility is that r_1 and r_2 are nonreal, complex numbers that are complex conjugates of each other. In this case, we can write r_1 as $s + ti$ and r_2 as $s - ti$ for some real numbers s and t with $t \neq 0$. (Why can we assume that $t \neq 0$?) Calculate $(r_1 - r_2)^2$ and show that it is a negative real number.
4. Write Δ for $(r_1 - r_2)^2$. We have found that Δ is positive, zero, or negative depending on whether r_1 and r_2 are real and distinct, identical, or nonreal. Using these facts, show that the converse holds as well:
   (a) If $\Delta > 0$, there are two distinct real roots;
   (b) if $\Delta = 0$, there is a real root with multiplicity 2;
   (c) if $\Delta < 0$, there are two distinct complex conjugate roots.

We saw in Theorem 10.2 that the nature of the roots of a quadratic polynomial $x^2 + bx + c$ is determined by the sign of the quantity $b^2 - 4c$. Exercise 10.2 shows that the nature of the roots is determined by the sign of the quantity $(r_1 - r_2)^2$, the square of the difference of the roots, which we are denoting by Δ. Let us compare the two quantities.

Exercise 10.3. Continue with the notation of Exercise 10.2, with the quadratic polynomial $x^2 + bx + c$ factoring as $(x - r_1)(x - r_2)$. Use this to express b and c in terms of r_1 and r_2. Deduce that $(r_1 - r_2)^2 = b^2 - 4c$. Thus Δ and $b^2 - 4c$ coincide.

We have found that the discriminant of $x^2 + bx + c$ has two descriptions. It is the square of the difference of the roots, and it is the quantity $b^2 - 4c$. We will regard $(r_1 - r_2)^2$ as the fundamental quantity, the one that we will take as the definition of the discriminant. The result that $(r_1 - r_2)^2$ equals $b^2 - 4c$ then provides a means of computing the discriminant of $x^2 + bx + c$ in terms of the coefficients b and c alone, without explicit knowledge of the roots.

Using the discriminant, we can determine from b and c alone whether the roots of $x^2 + bx + c$ are real or nonreal. When the roots are real, we can make

---

[1] If there is only one real root, so that $r_1 = r_2$, then we will use the convention that allows us to say that the polynomial has two real roots that happen to be equal. In the next section we will formalize this convention by defining the notion of *multiplicity* of a root.

further use of b and c to determine their signs. We will work out the details in an exercise.

Exercise 10.4. Consider a quadratic polynomial $x^2 + bx + c$, with b and c real and with nonnegative discriminant, so that the roots r_1 and r_2 are real.

1. Recall that $c = r_1 r_2$ and $b = -(r_1 + r_2)$.
2. If $c = 0$, then the nature of the roots is easy to determine. Explain why.
3. Assume that $c \neq 0$. Show that if the roots r_1 and r_2 have the same sign, then $c > 0$, and if the roots have opposite sign, then $c < 0$.
4. Conclude that there is an odd number of positive roots when c is negative and an even number when c is positive.
5. Assume that c is positive. Show that the two roots are both positive precisely when $b < 0$ and both negative precisely when $b > 0$.
6. Conclude that you can use b and c to determine the signs of the roots. Describe exactly how you would do so.

10.2 Cubic Polynomials

A polynomial $f(x)$ with coefficients in a field K has an element r of K as a root if and only if $x - r$ divides $f(x)$ in $K[x]$. Suppose $(x - r)^d$ divides $f(x)$ but $(x - r)^{d+1}$ does not. Then we say that r is a root of $f(x)$ of *multiplicity d*. It is a consequence of the quadratic formula that every quadratic polynomial $f(x)$ with real coefficients has either two distinct real roots, one real root of multiplicity two, or a pair of distinct complex roots that are complex conjugates of each other. We could also say that every quadratic polynomial in $\mathbb{R}[x]$ has exactly two real roots or two complex roots, if we count the roots with multiplicity. Let us obtain a similar statement for cubic polynomials with real coefficients. Standard facts of elementary calculus will guarantee that the polynomial has at least one real root.

Exercise 10.5. Let $f(x)$ be a cubic polynomial of the form $x^3 + ax^2 + bx + c$ with real coefficients.

1. Review an argument from calculus that allows you to deduce that $f(x) \to \infty$ as $x \to \infty$. (This use of ∞ is a shorthand for saying that for every positive number N, the inequality $f(x) > N$ holds once x is sufficiently large.) Similarly, $f(x) \to -\infty$ as $x \to -\infty$.
2. Read a discussion of the intermediate value theorem in a calculus book. This is the foundational result about real numbers, on which all of calculus depends. Without it, calculus would not work. It basically says that if the graph of a continuous function starts at some height and ends at another, then the graph goes through every height in between.
3. Using the intermediate value theorem, deduce for our cubic polynomial $f(x)$ that there is a real number r such that $f(r) = 0$.

4. Conclude that $f(x)$ can be factored as $(x - r)g(x)$ for some quadratic polynomial $g(x)$.

5. Deduce that either $f(x)$ factors in $\mathbb{R}[x]$ as the product of three degree-one polynomials, or $f(x)$ factors in $\mathbb{R}[x]$ as the product of a degree-one polynomial and an irreducible degree-two polynomial.

6. Deduce that either $f(x)$ has three real roots (counting multiplicities) or $f(x)$ has one real root and two nonreal (complex) roots that are complex conjugates of each other.

Now that we know the possibilities for the roots of a cubic polynomial equation, we would like to find those roots; that is, we would like a formula for cubic polynomials analogous to that for quadratic polynomials. To that end, we would like first to achieve a simplification along the lines of what we did for quadratic equations, when we obtained an equivalent equation without the degree-one term.

In the case of a cubic equation

$$x^3 + ax^2 + bx + c = 0,$$

where a, b, and c are real numbers, we will achieve the desired simplification by changing variables to obtain an equation with no degree-two term. Set $x = y - \frac{b}{3}$ and substitute to obtain the equation

$$\left(y - \frac{b}{3}\right)^3 + a\left(y - \frac{b}{3}\right)^2 + b\left(y - \frac{b}{3}\right) + c = 0.$$

If you expand this out and cancel terms, you will find that the y^2 terms cancel, leading to a new cubic equation of the form

$$y^3 + py + q = 0,$$

where

$$p = b - \frac{a^2}{3}$$

and

$$q = c - \frac{ab}{3} + \frac{2a^3}{27}.$$

If we can solve this new equation for y, we can then set $x = y - \frac{b}{3}$ and obtain the solution to the original cubic equation. In this sense, we have reduced the problem of solving the original cubic equation to the problem of solving the simpler equation

$$y^3 + py + q = 0.$$

An equation of the form $y^3 + py + q = 0$ is called a *reduced* cubic equation, and the polynomial $y^3 + py + q$ is called a *reduced* cubic polynomial.

Exercise 10.6. What is the reduced cubic equation that you must solve in order to solve the cubic equation

$$x^3 - 3x^2 - 4x + 12 = 0?$$

We have seen that every cubic polynomial has at least one root and that by a change of variable we can reduce the problem of finding a root of a given cubic polynomial $x^3 + bx^2 + cx + d$ to the problem of finding a root of a cubic polynomial of the form $y^3 + py + q$. But how do we solve this reduced problem?

In Section 10.1 we discussed the idea that the quadratic formula does not allow us to solve quadratic equations directly. Rather, it is a procedure that allows us to reduce the problem of solving a quadratic equation to the problem of calculating square roots of real numbers. Similarly, we should expect that in any attempt to solve cubic equations, we will be content if we can use algebra to reduce the problem to one of cube root calculations. The calculation of cube roots, like the calculation of square roots, is not a problem we should expect algebra to solve. But if we can use algebra to reduce the solution of cubic equations to the problem of calculating real cube roots, we will view this as a success.

To find a root of the reduced cubic polynomial $y^3 + py + q$, we can use a formula known as Cardano's formula. This was discovered by Scipione del Ferro and Niccolò Tartaglia in the early 1500s and publicized by Girolamo Cardano in his book *Ars Magna*, printed in 1545. The formula states that a solution of $y^3 + py + q = 0$ is given by

$$y = \sqrt[3]{-\frac{q}{2} + \sqrt{R}} + \sqrt[3]{-\frac{q}{2} - \sqrt{R}},$$

where R is the quantity

$$R = \left(\frac{p}{3}\right)^3 + \left(\frac{q}{2}\right)^2.$$

Notice that the solution is expressed in terms of the coefficients p and q, addition, multiplication, division by some constants, and the taking of square and cube roots. Thus, just as the quadratic formula reduces the solution of quadratic equations to square root calculations, Cardano's formula reduces the solution of cubic equations to square root and cube root calculations.

After using the formula to find one real root r of $y^3 + py + q$, we can divide $y^3 + py + q$ by $y - r$ to obtain a quadratic polynomial and use the quadratic formula to find the other two roots. (Note: We have implied that Cardano's formula yields a real root. We will have to prove this, and we shall do so later.)

The use of Cardano's formula involves some subtleties, which we will discuss. Before considering examples of its use, let us see how the formula can be derived. There are many ways to do so. In the next exercise, we will work through a derivation described by François Viète in 1591.

Exercise 10.7. We begin with the cubic polynomial $y^3 + py + q$. We can assume that p is nonzero, for if $p = 0$, the equation is $y^3 = -q$, and the solution is easily obtained as the cube root of $-q$. Viète's idea is to introduce a new variable z satisfying

$$y = z - \frac{p}{3z}. \qquad (*)$$

1. Substitute $z - \frac{p}{3z}$ for y in the equation $y^3 + py + q = 0$, expand the cubed term, simplify, and obtain the following equation in z:

$$z^3 - \frac{p^3}{27z^3} + q = 0.$$

2. Multiply by z^3 to clear the variable in the denominator and obtain

$$z^6 + qz^3 - \frac{p^3}{27} = 0.$$

3. Observe that this is a quadratic equation in z^3. Use the quadratic formula to obtain

$$z^3 = -\frac{q}{2} \pm \sqrt{\frac{q^2 + \frac{4p^3}{27}}{4}}.$$

4. Introduce R as an abbreviation for $\left(\frac{p}{3}\right)^3 + \left(\frac{q}{2}\right)^2$ and rewrite the last equality as

$$z^3 = -\frac{q}{2} \pm \sqrt{R}.$$

5. There are two possible values for z^3, namely, $-\frac{q}{2} + \sqrt{R}$ and $-\frac{q}{2} - \sqrt{R}$. Multiply these two values together and simplify. Show that you get

$$\left(-\frac{q}{2} + \sqrt{R}\right)\left(-\frac{q}{2} - \sqrt{R}\right) = \left(-\frac{p}{3}\right)^3.$$

6. Take cube roots of both sides above and deduce that the two values of z have a product satisfying

$$\sqrt[3]{-\frac{q}{2} + \sqrt{R}} \times \sqrt[3]{-\frac{q}{2} - \sqrt{R}} = -\frac{p}{3}.$$

7. Observe that this means that if you choose z to be the cube root of $\frac{q}{2} + \sqrt{R}$, then $-\frac{p}{3z}$ is the cube root of $-\frac{q}{2} - \sqrt{R}$.
8. Recall that z was introduced to satisfy $y = z - \frac{p}{3z}$. You have shown that the two terms on the right of this equation, z and $-\frac{p}{3z}$, are the cube roots of $-\frac{q}{2} + \sqrt{R}$ and $-\frac{q}{2} - \sqrt{R}$, respectively.
9. Conclude that y is the sum of these two cube roots:

$$y = \sqrt[3]{-\frac{q}{2} + \sqrt{R}} + \sqrt[3]{-\frac{q}{2} - \sqrt{R}}.$$

Let us use Cardano's formula to solve three cubic equations.

Exercise 10.8. Solve $y^3 - 3y + 2 = 0$. Use Cardano's formula to find one solution r. This should be easy. Then use this solution to factor $y^3 - 3y + 2$ and find the other two solutions.

Exercise 10.9. Solve $y^3 - 2y + 4 = 0$.

1. Use Cardano's formula to find one solution. You should get

$$y = \sqrt[3]{-2 + \frac{10}{9}\sqrt{3}} + \sqrt[3]{-2 - \frac{10}{9}\sqrt{3}}.$$

   Check this.

2. The resulting value of y is rather complicated, but this is an inevitable result of using Cardano's formula. The formula reduces the problem of solving a cubic equation to the problem of computing cube roots, but cube root calculations can be hard. Depending on our needs, we can leave the answer in this form, use a calculator to get an approximate decimal answer, or try to determine the cube roots in simpler form.

3. Notice that by direct substitution, $y^3 - 2y + 4 = 0$ has the solution $y = -2$. This looks a lot simpler than the solution we found. Given that there is such a simple solution, why has Cardano's formula failed to produce it? In fact, Cardano's formula has produced $y = -2$, but in a highly disguised form. To see this, check by direct calculation that

$$\left(-1 + \frac{\sqrt{3}}{3}\right)^3 = -2 + \frac{10}{9}\sqrt{3}$$

   and

$$\left(-1 - \frac{\sqrt{3}}{3}\right)^3 = -2 - \frac{10}{9}\sqrt{3}.$$

4. Conclude that the solution given by Cardano's formula is

$$y = \sqrt[3]{-2 + \frac{10}{9}\sqrt{3}} + \sqrt[3]{-2 - \frac{10}{9}\sqrt{3}} = \left(-1 + \frac{\sqrt{3}}{3}\right) + \left(-1 - \frac{\sqrt{3}}{3}\right) = -2.$$

   Thus the solution given by Cardano's formula is indeed -2, although this was not apparent.

5. Use this solution to factor $y^3 - 2y + 4$ and find the other two solutions.

   So far, so good, but we are about to encounter one of the greatest surprises in the history of mathematics.

Exercise 10.10. Solve $y^3 - 7y + 6 = 0$.

1. Show that Cardano's formula yields the solution

$$y = \sqrt[3]{-3 + \frac{10}{9}\sqrt{-3}} + \sqrt[3]{-3 - \frac{10}{9}\sqrt{-3}}.$$

Once again, the answer is a complicated sum of cube roots. Worse yet, the formula requires us to compute cube roots of nonreal complex numbers! This suggests that the solutions to $y^3 - 7y + 6 = 0$ must be complicated and that the one we have found may not even be real.

2. Again, the solutions are not complicated; what is complicated is the cube root calculation that Cardano's formula requires. Check that

$$\left(1 + \frac{2}{3}\sqrt{-3}\right)^3 = -3 + \frac{10}{9}\sqrt{-3}$$

and

$$\left(1 - \frac{2}{3}\sqrt{-3}\right)^3 = -3 - \frac{10}{9}\sqrt{-3}.$$

3. Conclude that the solution given by Cardano's formula is simply $y = 2$.
4. Use this solution to factor $y^3 - 7y + 6$ and find the other two solutions, $y = 1$ and $y = -3$.

What happened in our attempt to solve the equation $y^3 - 7y + 6 = 0$ using Cardano's formula deserves a more detailed discussion. The formula did not lead directly to the roots 1, 2, and -3 of $y^3 - 7y + 6$. Instead, it gave us expressions for the roots involving cube roots of $-3 + \frac{10}{9}\sqrt{-3}$ and $-3 - \frac{10}{9}\sqrt{-3}$. In order to obtain $y = 2$, we needed to find explicit expressions for cube roots of $-3 + \frac{10}{9}\sqrt{-3}$ and $-3 - \frac{10}{9}\sqrt{-3}$. In solving $y^3 - 2y + 4 = 0$, we had to compute cube roots of complicated numbers, but at least those were real. What is surprising in working with $y^3 - 7y + 6$ is that even though all three of its roots are real, to find them we have to compute cube roots of *nonreal* complex numbers. We will see in Section 10.3 that this phenomenon occurs whenever Cardano's formula is applied to a cubic polynomial $y^3 + py + q$ that has three distinct real roots.

In our derivation of Cardano's formula in Exercise 10.7, we ignored some subtleties that need to be addressed. We will do so in the next two exercises.

Exercise 10.11. Write ω for the complex number

$$-\frac{1}{2} + \frac{\sqrt{-3}}{2}.$$

As you may know, ω is the lowercase letter *omega* in the Greek alphabet and is the final letter in the alphabet. In Greek, one goes from alpha to omega rather than from A to Z.

1. Show that

$$\omega^2 = -\frac{1}{2} - \frac{\sqrt{-3}}{2}.$$

2. Show that
$$\omega^3 = 1.$$

Deduce that the three cube roots of 1 are 1, ω, and ω^2.
3. Verify that $1 + \omega + \omega^2 = 0$.
4. Suppose that s is a real number and r is its real cube root, so that $r^3 = s$. Verify that $(\omega r)^3 = s$ and $(\omega^2 r)^3 = s$. Deduce that s has three cube roots in the complex numbers: r, ωr, and $\omega^2 r$.
5. More generally, suppose that s is a complex number and r is a cube root of s. Show that ωr and $\omega^2 r$ are also cube roots of s.

Exercise 10.12. Suppose p and q are real numbers. Write R for the quantity

$$\left(\frac{p}{3}\right)^3 + \left(\frac{q}{2}\right)^2.$$

Continue to write ω for $-\frac{1}{2} + \frac{\sqrt{-3}}{2}$. Cardano's formula for the root of the cubic polynomial $y^3 + py + q$ takes the form

$$y = \sqrt[3]{-\frac{q}{2} + \sqrt{R}} + \sqrt[3]{-\frac{q}{2} - \sqrt{R}}.$$

1. Observe that since every nonzero number in $\mathbb{C}$ has three distinct cube roots, each summand on the right side of Cardano's formula has three possible values.
2. Check once again, as you did in Exercise 10.7, that the product of $-\frac{q}{2} + \sqrt{R}$ and $-\frac{q}{2} - \sqrt{R}$ is $-\frac{p^3}{27}$.
3. Choose one cube root of $-\frac{q}{2} + \sqrt{R}$ and call it A; choose one cube root of $-\frac{q}{2} - \sqrt{R}$ and call it B. Notice that there are three choices for A and three choices for B, so that there are nine choices altogether for the pair of numbers A and B. Observe that regardless of which of the nine choices is made, $A^3 B^3 = -\frac{p^3}{27}$. Deduce that there are three possibilities for the product AB, the numbers $-\frac{p}{3}$, $-\frac{p\omega}{3}$, and $-\frac{p\omega^2}{3}$.
4. From the last observation, deduce that whatever choices we make for A and B, one of the three products AB, ωAB, and $\omega^2 AB$ will equal $-\frac{p}{3}$. Observe also that since B is one of the three cube roots of $-\frac{q}{2} - \sqrt{R}$, the numbers ωB and $\omega^2 B$ are the other two cube roots of $-\frac{q}{2} - \sqrt{R}$.
5. Fix A as one of the three cube roots of $-\frac{q}{2} + \sqrt{R}$. We have chosen B randomly as one of the three cube roots of $-\frac{q}{2} - \sqrt{R}$, with a resulting uncertainty about whether AB equals $-\frac{p}{3}$, $-\frac{p\omega}{3}$, or $-\frac{p\omega^2}{3}$. Using the last part, observe that with the choice of A fixed, we can choose B among the three cube roots of $-\frac{q}{2} - \sqrt{R}$ so that $AB = -\frac{p}{3}$. Let us make this choice.
6. Show that with A and B chosen in this way, the numbers ωA and $\omega^2 B$ are respectively cube roots of $-\frac{q}{2} + \sqrt{R}$ and $-\frac{q}{2} - \sqrt{R}$ with the property that their product equals $-\frac{p}{3}$. Show also that the numbers $\omega^2 A$ and ωB

are respectively cube roots of $-\frac{q}{2} + \sqrt{R}$ and $-\frac{q}{2} - \sqrt{R}$ with the property that their product equals $-\frac{p}{3}$.

7. Conclude that if A and B are chosen as cube roots of $-\frac{q}{2} + \sqrt{R}$ and $-\frac{q}{2} - \sqrt{R}$ satisfying $AB = -\frac{p}{3}$, then the three roots of $y^3 + py + q$ have the form

$$r_1 = A + B; \quad r_2 = wA + w^2B; \quad r_3 = w^2A + wB.$$

Exercise 10.12 removes the imprecision that was present in our derivation of Cardano's formula earlier. In Exercise 10.7, we had not specified sufficiently how to choose cube roots in the formula. Now we have.

Exercise 10.13. Use Cardano's formula, as clarified in Exercise 10.12, to obtain all three solutions to the cubic equation

$$y^3 - 7y + 6 = 0.$$

1. Write down the solution given by the formula as a sum of cube roots. Observe that it involves the cube roots of

$$-3 + \frac{10}{9}\sqrt{-3}$$

and

$$-3 - \frac{10}{9}\sqrt{-3}.$$

2. Using the numbers w and w^2 and the earlier determination of one cube root of $-3 + \frac{10}{9}\sqrt{-3}$, write expressions for the three complex numbers that are cube roots of $-3 + \frac{10}{9}\sqrt{-3}$. Also write down the three complex numbers that are cube roots of $-3 - \frac{10}{9}\sqrt{-3}$.

3. Pair the cube roots of $-3 + \frac{10}{9}\sqrt{-3}$ and $-3 - \frac{10}{9}\sqrt{-3}$ as specified in Exercise 10.12 to get three pairs such that the product of the complex numbers in each pair equals $\frac{7}{3}$.

4. Add together the complex numbers in each pair to obtain all three solutions of $y^3 - 7y + 6 = 0$. Note that all three solutions are real.

10.3 The Discriminant of a Cubic Polynomial

For a quadratic polynomial $x^2 + bx + c$ with real coefficients whose real or complex roots are r_1 and r_2, we saw in Exercise 10.3 that the squared difference $(r_1 - r_2)^2$ of the roots is expressible in terms of the coefficients:

$$(r_1 - r_2)^2 = b^2 - 4c.$$

This allows us to determine from the coefficients alone some partial information on the roots. In particular, if $b^2 - 4c$ is positive, the roots are real and distinct; if $b^2 - 4c$ is zero, the roots are real and identical; and if $b^2 - 4c$ is negative, then the roots are a pair of nonreal, complex conjugate numbers.

We would like similarly to be able to obtain information about the roots of a cubic polynomial in terms of its coefficients. Recall from the discussion that followed Exercise 10.3 that the two quantities $b^2 - 4c$ and $(r_1 - r_2)^2$ associated with the quadratic polynomial $x^2 + bx + c$ can both be regarded as its discriminant, but that the more fundamental of these quantities is $(r_1 - r_2)^2$, and we took this to be the definition of the discriminant. Consider a cubic polynomial $x^3 + bx^2 + cx + d$, and let us write r_1, r_2, and r_3 for its roots. We define the *discriminant* of $x^3 + bx^2 + cx + d$ to be the analogous quantity

$$(r_1 - r_2)^2 (r_1 - r_3)^2 (r_2 - r_3)^2,$$

and we denote this quantity by Δ. We will show that Δ determines the nature of the roots and that we can compute Δ in terms of the coefficients of the polynomial. This will allow us, as in the quadratic case, to determine the nature of the roots from the coefficients alone, without having advance knowledge of the roots themselves.

Exercise 10.14. Let $f(x) = x^3 + bx^2 + cx + d$. We will relate the sign of the discriminant Δ of $f(x)$ to the nature of the roots of $f(x)$.

1. Suppose that the three roots are real and distinct. Show that $\Delta > 0$.
2. Suppose $f(x)$ has a multiple root (and hence all roots are real; why?). Show that $\Delta = 0$.
3. Suppose that one root is real and the other two are nonreal complex conjugates of each other. Show that $\Delta < 0$. (Hint: Suppose the roots are r, $a + bi$, and $a - bi$, with r, a, and b all real numbers, $b \neq 0$. Using this notation, calculate the product of the root differences first, before squaring.)
4. We have found that Δ is positive, zero, or negative depending on whether the roots are real and distinct, there is a multiple root, or only one root is real. Using these facts, show that the converse holds as well:
   (a) If $\Delta > 0$, then $f(x)$ has three distinct real roots;
   (b) if $\Delta = 0$, then $f(x)$ has a multiple root and all its roots are real;
   (c) if $\Delta < 0$, then $f(x)$ has one real root and two nonreal complex conjugate roots.

What makes the discriminant Δ of a quadratic polynomial $x^2 + bx + c$ useful is that Δ can be computed in terms of b and c. Thus one can determine the nature of the roots from the coefficient data alone, without actually computing the roots. Similarly, we would like to be able to compute the discriminant of a cubic polynomial in terms of its coefficients, so that we can determine the nature of the roots without actually computing the roots. Let us first consider the case of a reduced cubic polynomial.

Exercise 10.15. Suppose p and q are real numbers and consider the polynomial $y^3 + py + q$. We will use the notation and results of Exercise 10.12. Accordingly, we choose A to be one of the three cube roots of $-\frac{q}{2} + \sqrt{R}$ and then choose B to be the unique cube root of $-\frac{q}{2} - \sqrt{R}$ satisfying $AB = -\frac{p}{3}$. Recall that with these choices, the three roots of $y^3 + py + q$ are given by the expressions

$$r_1 = A + B; \quad r_2 = \omega A + \omega^2 B; \quad r_3 = \omega^2 A + \omega B.$$

Recall also that $1 + \omega + \omega^2 = 0$.

1. Verify that
$$r_1 - r_2 = (1 - \omega)(A - \omega^2 B),$$
that
$$r_1 - r_3 = -\omega^2(1 - \omega)(A - \omega B),$$
and that
$$r_2 - r_3 = \omega(1 - \omega)(A - B).$$

2. Verify that
$$(1 - \omega)^3 = 3(\omega^2 - \omega) = -3\sqrt{3}i$$
and deduce that
$$(r_1 - r_2)(r_1 - r_3)(r_2 - r_3) = -(1 - \omega)^3 \left(A^3 - B^3\right) = 3\sqrt{3}i \left(A^3 - B^3\right).$$

3. Recall that by definition, $A^3 = -\frac{q}{2} + \sqrt{R}$ and $B^3 = -\frac{q}{2} - \sqrt{R}$. Deduce that
$$A^3 - B^3 = 2\sqrt{R}$$
and that
$$(r_1 - r_2)(r_1 - r_3)(r_2 - r_3) = 6\sqrt{3}i\sqrt{R}.$$

4. Deduce that the discriminant Δ of $y^3 + py + q$ is given by the formula
$$\Delta = -108R.$$

5. Finally, recall the definition of R and obtain the following formula:
$$\Delta = -4p^3 - 27q^2.$$

We have succeeded in expressing the discriminant of a reduced cubic polynomial $y^3 + py + q$ in terms of its coefficients p and q. We have also found that the discriminant is simply a constant multiple of the quantity R that played such a prominent role in Cardano's formula. Thus we can rewrite Cardano's formula in terms of Δ rather than R. This makes the similarity between the quadratic formula and Cardano's formula even clearer: Both are expressions involving the discriminant of the given polynomial.

Exercise 10.16. Using the formula for the discriminant of a reduced cubic polynomial, calculate the discriminants of the cubic polynomials below. State for each polynomial whether it has a multiple root (and therefore all its roots are real), a real root and two nonreal roots, or three distinct real roots.

1. $y^3 - 3y + 2$.
2. $y^3 + 5y + 1$.
3. $y^3 - 5y + 1$.

Next we will obtain a formula for the discriminant of a general cubic polynomial in terms of its coefficients.

Exercise 10.17. Consider the cubic polynomial $x^3 + bx^2 + cx + d$, with real coefficients b, c, and d.

1. Recall from Section 10.2 the change of variable

$$x = y - \frac{b}{3}$$

that transforms $x^3 + bx^2 + cx + d$ into a polynomial of the form $y^3 + py + q$. Review this substitution and recall also the explicit formulas for p and q in terms of b, c, and d.
2. Write the roots of $y^3 + py + q$ as r_1, r_2, and r_3, as we did above. What are the roots of $x^3 + bx^2 + cx + d$ in terms of r_1, r_2, and r_3?
3. Observe that although the roots of $x^3 + bx^2 + cx + d$ and $y^3 + px + q$ are different, the three root differences are the same. Deduce that the discriminant of $x^3 + bx^2 + cx + d$ and the discriminant of $y^3 + px + q$ must be equal.
4. Using the formula for the discriminant of $y^3 + px + q$ and the formulas for p and q in terms of b, c, and d, obtain the formula

$$\Delta = 18bcd - 4b^3d + b^2c^2 - 4c^3 - 27d^2$$

for the discriminant of $x^3 + bx^2 + cx + d$ in terms of b, c, and d.
5. Conclude that you can determine the nature of the roots of $x^3 + bx^2 + cx + d$ in terms of b, c, and d, without explicit knowledge of the roots.

Exercise 10.18. Calculate the discriminant of each cubic polynomial below and state whether it has a multiple root (and hence all roots are real), a real root and two nonreal roots, or three distinct real roots.

1. $x^3 + 2x^2 - 4$.
2. $x^3 + 3x^2 - 5x + 1$.

The discriminant Δ of a cubic polynomial $x^3 + bx^2 + cx + d$ can be used to decide whether the polynomial has a multiple root, three distinct real roots, or one real root and two complex conjugate roots. In turn, we can express Δ in terms of the coefficients b, c, and d, and this allows us to determine the real or nonreal character of the roots from the coefficients alone. We will next see how to use the coefficients to determine how many of the real roots are positive and how many are negative. This is interesting in its own right, but it will also be useful to us in our study of quartic (fourth-degree) polynomials.

Exercise 10.19. Consider a cubic polynomial $x^3 + bx^2 + cx + d$ with real coefficients and suppose that its three roots are r_1, r_2, and r_3.

1. Check that

$$b = -(r_1 + r_2 + r_3); \quad c = r_1 r_2 + r_1 r_3 + r_2 r_3; \quad d = -r_1 r_2 r_3.$$

   (Hint: Use the factorization of $x^3 + bx^2 + cx + d$ as $(x - r_1)(x - r_2)(x - r_3)$.)
2. If $d = 0$, then the nature of the roots is easy to determine. Explain why.
3. Assume that $d \neq 0$. Recall that if $\Delta < 0$, then there is exactly one real root, occurring with multiplicity one. Show that in this case, the sign of the root is determined by the sign of d. (Hint: Use the formula above for d.)
4. Assume for the remainder of the exercise that $\Delta \geq 0$. Recall that in this case all three roots are real.
5. Show that if b, c, and d are all positive, then all the roots are negative. (Hint: What happens in this case if you substitute a positive real number for x in $x^3 + bx^2 + cx + d$?)
6. Conversely, show that if all the roots are negative, then b, c, and d are all positive. (Hint: Examine the expressions in the first part of the exercise for b, c, and d in terms of the roots.)
7. Show that if b is negative, c is positive, and d is negative, then all the roots are positive. (Hint: What happens in this case if you substitute a negative real number for x in $x^3 + bx^2 + cx + d$?)
8. Conversely, show that if all the roots are positive, then b is negative, c is positive, and d is negative.
9. Conclude that if there are three positive roots or three negative roots, this can be detected from the data of b, c, and d.
10. It remains to discover how to use the coefficients to decide whether there is exactly one positive root or exactly two. Show that $d > 0$ if there are zero or two positive roots; show that $d < 0$ if there are one or three positive roots.
11. Use the last part to explain how you can use the coefficients b, c, and d to determine whether exactly one real root is positive or exactly two are positive.
12. Conclude with a detailed review of how you would use the coefficients b, c, and d to determine how many positive real roots $x^3 + bx^2 + cx + d$ has and how many negative real roots it has.

Exercise 10.20. For each cubic polynomial below, use Exercise 10.19 to determine from the coefficients how many real roots there are and how many of the real roots are positive or negative.

1. $x^3 - 6x^2 + 11x - 6$.
2. $x^3 - 5x^2 + 9x - 5$.
3. $x^3 + 6x^2 + 3x + 18$.
4. $x^3 - 3x^2 - 10x + 24$.

Recall that when we used Cardano's formula in Exercise 10.10 to solve the equation $y^3 - 7y + 6 = 0$, we had to work with nonreal complex numbers, even though all three solutions are real. As was noted after that exercise, this surprising phenomenon occurs often, and we can use our discriminant formula to determine when it does.

Exercise 10.21. Consider the reduced cubic polynomial $y^3 + py + q$.

1. Observe from our calculation of the discriminant Δ in terms of R that Δ and R have opposite signs.
2. Review from Exercise 10.14 the relation between the sign of Δ and the nature of the roots. Rephrase this as a relation between the sign of R and the nature of the roots.
3. Deduce that if $y^3 + py + q$ has one real root and two nonreal roots, then Cardano's formula expresses the real root as the sum of cube roots of real numbers. Deduce as well that if $y^3 + py + q$ has a multiple root, so that all the roots are real, then Cardano's formula expresses these roots as sums of cube roots of real numbers.
4. Deduce that in contrast, if $y^3 + py + q$ has three distinct real roots, then Cardano's formula expresses all three of them as sums of cube roots of nonreal complex numbers.

We see that for a reduced cubic polynomial $y^3 + py + q$, there are two fundamentally different cases, depending on the sign of R, or of Δ. Cardano called the case in which $\Delta > 0$ and $R < 0$ the *irreducible case*. In this case, the expression R will be negative, the quantities $-(\frac{q}{2}) \pm \sqrt{R}$ will be nonreal complex numbers that are conjugate to each other, and the three roots of $y^3 + py + q$ will be expressed by Cardano's formula as sums of pairs of nonreal complex numbers. Thus what we found in our determination of the roots of $y^3 - 7y + 6$ turns out to have been not unusual but inevitable. As in that example, the numbers in the pair given by Cardano's formula are conjugate, the sum is a real number, but the initial expression for the root necessarily involves nonreal complex numbers.

This aspect of Cardano's formula is a great surprise. It was especially so in the 1500s, when the concept of complex number did not even exist! To use the formula when a cubic polynomial has real coefficients and three real roots, we must temporarily step outside the realm of the real numbers in order to make cube root calculations with complex numbers before we can return to the real numbers. This puzzling fact was recognized by Cardano, who published the first use of complex numbers in his treatment of cubic equations in *Ars Magna*.

10.4 Quartic Polynomials

Thus far, we have seen that a quadratic polynomial $f(x)$ in $\mathbb{R}[x]$ either has no roots in $\mathbb{R}$ and is irreducible in $\mathbb{R}[x]$, or else has two roots counting multiplicity (that is, two distinct roots or one root of multiplicity two) and factors completely in $\mathbb{R}[x]$ into linear (first-degree) factors. A cubic polynomial, in contrast, always has at least one real root, and so always factors nontrivially in $\mathbb{R}[x]$, either—if $f(x)$ has only one real root (counting multiplicity)—as the product of a linear factor and an irreducible (in $\mathbb{R}[x]$) quadratic factor or—if $f(x)$ has three real roots (counting multiplicity)—into three linear factors.

In each case, we see that a quadratic or cubic polynomial has a nontrivial factorization in $\mathbb{R}[x]$ if and only if it has at least one real root. In contrast, a quartic (fourth-degree) polynomial in $\mathbb{R}[x]$ can factor nontrivially yet have no real roots. It is easy to produce examples. For instance, $x^4 + 2x^2 + 1$ factors in $\mathbb{R}[x]$ as $\left(x^2 + 1\right)\left(x^2 + 1\right) = \left(x^2 + 1\right)^2$, where each factor $\left(x^2 + 1\right)$ is irreducible in $\mathbb{R}[x]$, while in $\mathbb{C}[x]$ it factors completely into linear factors as $(x+i)^2(x-i)^2 = (x-i)(x-i)(x+i)(x+i)$. We see that its only roots are i and $-i$, and so there are indeed no real roots.

If a quartic polynomial $f(x)$ with real coefficients has a real root r, then it factors as $(x-r)g(x)$ for some cubic polynomial $g(x)$ with real coefficients. The factor $g(x)$ is a cubic polynomial with real coefficients, and so has at least one real root. Thus $f(x)$ has at least two real roots. In other words, if $f(x)$ has a real root, it has at least two real roots, where we are now always counting roots with multiplicity. Similarly, if $f(x)$ has three real roots r, s, and t (which may or may not all be distinct), it must have a fourth. This leads to the following possibilities, where we note again that the number of roots mentioned is always with multiplicity:

1. $f(x)$ has four real roots, in which case it factors in $\mathbb{R}[x]$ as the product of four degree-one polynomials.
2. $f(x)$ has only two real roots, in which case there are real numbers r and s and a quadratic polynomial $h(x)$ in $\mathbb{R}[x]$ such that $f(x)$ factors in $\mathbb{R}[x]$ as $(x-r)(x-s)h(x)$. In this case, since there are only two real roots, $h(x)$ must be irreducible in $\mathbb{R}[x]$, with two nonreal complex roots that are complex conjugates of each other. That is, in this case $f(x)$ has two real roots and a pair of nonreal, complex conjugate roots.
3. $f(x)$ has no real roots.

What happens in this last situation? Since $f(x)$ has no real roots, it has no degree-one factors in $\mathbb{R}[x]$. Thus either it is irreducible in $\mathbb{R}[x]$, or it factors in $\mathbb{R}[x]$ as the product of two quadratic polynomials. Are both of these options available to fourth-degree polynomials in $\mathbb{R}[x]$ with no real roots? One might guess that some of these polynomials factor nontrivially and others remain irreducible. However, such a guess would be wrong: Every fourth-degree polynomial can be factored nontrivially in $\mathbb{R}[x]$, which is the content of the following theorem.

Theorem 10.3. *A quartic polynomial $f(x)$ in $\mathbb{R}[x]$ cannot be irreducible in $\mathbb{R}[x]$. One of the following occurs:*

1. $f(x)$ *is the product of four degree-one polynomials and has four real roots.*
2. $f(x)$ *is the product of two degree-one polynomials and an irreducible degree-two polynomial; it has two real roots and two nonreal conjugate complex roots.*
3. $f(x)$ *is the product of two irreducible degree-two polynomials. It has four nonreal complex roots, occurring in two complex conjugate pairs.*

To prove Theorem 10.3, we will show that every quartic polynomial in $\mathbb{R}[x]$ factors as a product of two quadratic polynomials in $\mathbb{R}[x]$. Once we know this, we are led immediately to three possibilities: The two quadratic polynomials both factor as products of degree-one polynomials in $\mathbb{R}[x]$, one factors as a product of degree-one polynomials in $\mathbb{R}[x]$ but the other is irreducible in $\mathbb{R}[x]$, or both are irreducible in $\mathbb{R}[x]$. These are the three possibilities described in Theorem 10.3.

The result that every quartic polynomial in $\mathbb{R}[x]$ factors as a product of quadratic polynomials in $\mathbb{R}[x]$ was essentially proved in the 1500s by Cardano and Luigi Ferrari. Here we shall sketch a later approach of René Descartes, with some details omitted. We begin with the quartic polynomial

$$x^4 + ax^3 + bx^2 + cx + d.$$

Our first step, as with cubics, is to make a change of variable to eliminate the second-highest power.

Exercise 10.22. Let

$$f(x) = x^4 + bx^3 + cx^2 + dx + e,$$

for real numbers b, c, d, and e. Let us see what happens under the change of variable $x = z + a$ for a real number a.

1. Substitute $z + a$ for x and obtain a new polynomial $g(z)$ in the new variable z. Write it as
$$z^4 + Bz^3 + Cz^2 + Dz + E$$
   and obtain explicit formulas for the coefficients B, C, D, and E in terms of a and the old coefficients b, c, d, and e. To do this, you should verify and use the identities
$$(A + B)^3 = A^3 + 3A^2B + 3AB^2 + B^3$$
   and
$$(A + B)^4 = A^4 + 4A^3B + 6A^2B^2 + 4AB^3 + B^4,$$
   which hold for every pair of real numbers A and B.

2. Observe that there is a particular choice of a for which $B = 0$. Thus for this choice of a, changing variables provides a new polynomial $g(z)$ of the form

$$z^4 + Cz^2 + Dz + E.$$

3. What is the relation between a root of $f(x)$ and a root of $g(z)$?
4. Conclude that you will be able to solve the equation $f(x) = 0$ if you can solve $g(z) = 0$.

An equation of the form

$$z^4 + qz^2 + rz + s = 0$$

is called a *reduced* quartic equation, and $z^4 + qz^2 + rz + s$ is called a *reduced* quartic polynomial. You have shown that by making the substitution $x = z - \frac{b}{4}$, you can pass from a problem of solving an arbitrary quartic equation to an equivalent problem of solving a reduced quartic equation.

Exercise 10.23. What is the reduced quartic equation that you must solve in order to solve the quartic equation

$$x^4 + 4x^3 - 2x + 4 = 0?$$

We wish to prove that every quartic polynomial in $\mathbb{R}[x]$ factors in $\mathbb{R}[x]$ as a product of two quadratic polynomials. It suffices to prove that every *reduced* quartic polynomial in $\mathbb{R}[x]$ factors as such. After all, if this is the case, we can pass from a quartic polynomial $x^4 + bx^3 + cx^2 + dx + e$ to a reduced quartic polynomial $z^4 + qz^2 + rz + s$ by the change of variable $x = z - \frac{b}{4}$, factor $z^4 + qz^2 + rz + s$ as a product of quadratic polynomials, then perform the reverse substitution $z = x + \frac{b}{4}$ to obtain a factorization of the original quartic polynomial as a product of quadratic polynomials.

Let us therefore focus on a reduced quartic polynomial $z^4 + qz^2 + rz + s$ in $\mathbb{R}[x]$ and try to factor it as a product of quadratic polynomials.

Exercise 10.24. We can immediately dispose of one special case, that in which $r = 0$. Consider the quartic polynomial $z^4 + qz^2 + s$.

1. Factor $z^4 + qz^2 + s$ (which is a quadratic polynomial in z^2) in the form $(z^2 - r_1)(z^2 - r_2)$ for two real or possibly complex numbers r_1 and r_2.
2. If r_1 and r_2 are real, you have obtained the desired factorization of $z^4 + qz^2 + s$ in $\mathbb{R}[x]$.
3. Alternatively, if r_1 and r_2 are nonreal, observe that they are complex conjugates of each other. Using their square roots, factor $z^4 + qz^2 + s$ as a product of degree-one polynomials in $\mathbb{C}[x]$. Show that the degree-one terms can be regrouped and combined in pairs to obtain a factorization of $z^4 + qz^2 + s$ as a product of quadratic polynomials in $\mathbb{R}[x]$.

We wish to factor $z^4 + qz^2 + rz + s$ in $\mathbb{R}[x]$ as a product of quadratic polynomials. By Exercise 10.24, we can assume that r is nonzero. If the desired factorization exists, then it must look like

$$z^4 + qz^2 + rz + s = \left(z^2 + kz + l\right)\left(z^2 + hz + m\right),$$

for real numbers k, l, h, m. Since the coefficient of z^3 is zero, we see that we must have $h = -k$ (why?), and so we can write the assumed factorization in the form

$$z^4 + qz^2 + rz + s = \left(z^2 + kz + l\right)\left(z^2 - kz + m\right), \qquad (**)$$

for some real numbers k, l, and m. Furthermore, since we have assumed that the coefficient of z is nonzero, we must have $k \neq 0$ (why?).

If we multiply out the right-hand side of equation $(**)$ and equate coefficients, we will have found the desired factorization if we can express as real numbers the unknown quantities k, l, and m in terms of the given real coefficients. We therefore regard the quantities k, l, and m as unknowns, in contrast to q, r, and s, which are constants given to us as the coefficients of the quartic polynomial.

So let us take pencil in hand and multiply out $\left(z^2 + kz + l\right)\left(z^2 - kz + m\right)$, combine terms, and compare the resulting coefficients with the coefficients of $z^4 + qz^2 + rz + s$. We thereby obtain three equations for the unknowns k, l, and m:

$$l + m - k^2 = q; \qquad k(m - l) = r; \qquad lm = s. \qquad (\lozenge)$$

Let us rewrite the first equation $(\lozenge)$ as $m + l = q + k^2$. Since we have established that $k \neq 0$, in solving these equations we are free to divide by k. Doing so in the second equation $(\lozenge)$, we obtain $m - l = \frac{r}{k}$. Thus we obtain the two equations

$$m + l = q + k^2; \qquad m - l = \frac{r}{k}.$$

If we add these equations together, we obtain

$$2m = q + k^2 + \frac{r}{k},$$

while if we subtract the second equation from the first, we obtain

$$2l = q + k^2 - \frac{r}{k}.$$

This shows that if we find a value for k, then the unknown coefficients l and m will also have been found as well. We now use the last of the three equations $(\lozenge)$, $lm = s$, rewritten in the form $(2l)(2m) = 4s$. We have just obtained expressions for $2l$ and $2m$ in terms of q, r, and k. Substituting these expressions in the equation $(2l)(2m) = 4s$, we obtain an equation involving the single unknown k and the constants q, r, and s. If you work this out, multiplying through at the end by k^2 to clear denominators, you will obtain

$$k^6 + (2q)k^4 + \left(q^2 - 4s\right)k^2 - r^2 = 0.$$

Notice that this equation can be regarded as a cubic polynomial equation in k^2. Let us therefore introduce a new variable for k^2, setting $k^2 = j$. Then the equation becomes

$$j^3 + (2q)j^2 + \left(q^2 - 4s\right)j - r^2 = 0.$$

If we can solve this cubic for j, we can then take square roots to obtain the two values k and $-k$, use these to obtain values for l and m, and then factor $z^4 + qz^2 + rz + s$ as a product of quadratic polynomials.

We have reduced the problem of factoring the quartic polynomial $z^4 + qz^2 + rz + s$ with real coefficients as a product of quadratic polynomials with real coefficients to the problem of solving the new cubic equation in j, and we know how to solve cubic equations. However, one subtlety arises. In our factorization of $z^4 + qz^2 + rz + s$, we must obtain values for k and $-k$ that are real numbers. Therefore, although our cubic equation in j, like all cubic equations, has a real solution, in this case there had better be a solution that is a *positive* real number, since for j negative there is no *real* number k for which $(\pm k)^2 = j$. Thus, we need to know that

$$j^3 + (2q)j^2 + \left(q^2 - 4s\right)j - r^2 = 0$$

has a solution that is both real and positive. One can show this purely algebraically, but let us instead make a short argument using calculus.

Exercise 10.25. Use the intermediate value theorem of elementary calculus to show that the cubic polynomial $j^3 + (2q)j^2 + \left(q^2 - 4s\right)j - r^2$ has a positive real root.

This real and positive solution allows us to conclude that the quartic polynomial $z^4 + qz^2 + rz + s$ factors as a product of quadratic polynomials with real coefficients. Moreover, we have found a procedure for obtaining such a factorization, and thus we can solve every quartic equation.

Let us review the procedure. We pass from a quartic polynomial

$$x^4 + bx^3 + cx^2 + dx + e$$

to a reduced quartic polynomial

$$z^4 + qz^2 + rz + s$$

by changing variables. We handle the easy case of $r = 0$ with the quadratic formula. In the difficult $r \neq 0$ case, we write down the cubic polynomial

$$j^3 + (2q)j^2 + \left(q^2 - 4s\right)j - r^2$$

associated with the reduced quartic polynomial $z^4 + qz^2 + rz + s$. This cubic polynomial in j is called the *cubic resolvent* of the quartic. We find a positive real root of the cubic resolvent, take its square roots $\pm k$, and use these to factor the reduced quartic polynomial as a product of quadratic polynomials. We can use the quadratic formula to obtain the roots of these quadratic polynomials, thereby obtaining the roots of the reduced quartic polynomial. Finally, we pass from these to the roots of the original quartic polynomial. This can be a long and complicated process, but in principle it works.

Exercise 10.26. Find the four solutions of each quartic equation below. These equations are chosen so that once you write down the cubic resolvent polynomial, you can find a positive real root of the cubic resolvent by guessing rather than by using Cardano's formula. After you have determined a positive real root of the cubic resolvent, use it to factor the quartic polynomial as a product of two quadratic polynomials and find their roots.

1. $z^4 - 3z^2 + 6z - 2 = 0$.
2. $z^4 - 10z^2 - 4z + 8 = 0$.

10.5 A Closer Look at Quartic Polynomials

We have obtained a procedure for solving a reduced quartic equation. With a little more work, we can produce an explicit and simple formula for the solutions in terms of the solutions of the cubic resolvent equation. Let us see how.

Consider once again the reduced quartic equation

$$z^4 + qz^2 + rz + s = 0$$

and its cubic resolvent equation

$$j^3 + 2qj^2 + \left(q^2 - 4s\right)j - r^2 = 0.$$

Suppose the three roots of the cubic resolvent equation are j_1, j_2, and j_3. Our goal is to describe the roots of $z^4 + qz^2 + rz + s$ in terms of j_1, j_2, and j_3. We have already shown that at least one of the cubic resolvent's roots is positive and real. Let us assume that j_1 is such a root, and let k_1 and $-k_1$ be its two square roots. Because j_1 is positive, the square roots k_1 and $-k_1$ are real. The other two roots of the cubic resolvent, j_2 and j_3, may be real with any sign or may be complex conjugates of each other. Whether j_2 and j_3 are real or not, we have seen that the two square roots of each are additive inverses of each other. Thus we can write k_2 and $-k_2$ for the square roots of j_2, and k_3 and $-k_3$ for the square roots of j_3. The numbers k_i are determined only up to sign. Do not be fooled by the notation. Seeing the pair k_i and $-k_i$, you should not assume that k_i is positive or that $-k_i$ is negative. In fact, they may not even be real. The notation should convey only that whatever they are, each is the additive inverse of the other.

Exercise 10.27. Use the notation just above.

1. Explain why there are eight possible choices for the triple of numbers (k_1, k_2, k_3), depending on choices of signs.
2. Use the first part of Exercise 10.19 to obtain the equalities

$$j_1 + j_2 + j_3 = -2q$$

and

$$j_1 j_2 j_3 = r^2.$$

3. Deduce from the second equation above that $k_1 k_2 k_3$ equals either r or $-r$, depending on which square roots of the j_i's are chosen as the k_i's. More precisely, deduce that four of the eight choices of the triple (k_1, k_2, k_3) result in the equality $k_1 k_2 k_3 = r$ and that the other four choices of triple result in the equality $k_1 k_2 k_3 = -r$.
4. For the remainder of the exercise, fix the triple of square roots (k_1, k_2, k_3) to be one of the four choices satisfying $k_1 k_2 k_3 = -r$.
5. With our fixed choice of k_1 as one of the square roots of the positive real number j_1, we obtain a factorization

$$z^4 + qz^2 + rz + s = \left(z^2 + k_1 z + l_1\right)\left(z^2 - k_1 z + m_1\right),$$

where l_1 and m_1 are expressed in terms of k_1 by the formulas we obtained in Section 10.4. Using these expressions explicitly, show that

$$z^4 + qz^2 + rz + s$$
$$= \left(z^2 + k_1 z + \frac{1}{2}\left(q + k_1^2 - \frac{r}{k_1}\right)\right)\left(z^2 - k_1 z + \frac{1}{2}\left(q + k_1^2 + \frac{r}{k_1}\right)\right).$$

6. Use the quadratic formula to obtain expressions for the roots of

$$z^2 + k_1 z + \frac{1}{2}\left(q + k_1^2 - \frac{r}{k_1}\right).$$

Show that you get

$$z = -\frac{k_1}{2} \pm \frac{1}{2}\sqrt{k_1^2 - 2\left(q + k_1^2 - \frac{r}{k_1}\right)}.$$

7. Use the formula we stated earlier in this exercise for $-2q$ and our choice of the k_i's to satisfy $k_1 k_2 k_3 = -r$ in order to rewrite this expression for z as

$$z = -\frac{k_1}{2} \pm \frac{1}{2}\sqrt{(k_2 - k_3)^2}.$$

It should be emphasized that this equality depends on our having chosen the triple (k_1, k_2, k_3) to satisfy $k_1 k_2 k_3 = -r$.

8. Conclude that the roots of this quadratic factor are

$$z = \frac{1}{2}(-k_1 + k_2 - k_3); \quad z = \frac{1}{2}(-k_1 - k_2 + k_3).$$

9. Follow the same procedure to find the roots of

$$z^2 - k_1 z + \frac{1}{2}\left(q + k_1^2 + \frac{r}{k_1}\right)$$

and show that you get

$$z = \frac{1}{2}(k_1 + k_2 + k_3); \quad z = \frac{1}{2}(k_1 - k_2 - k_3).$$

For the reduced quartic polynomial $z^4 + qz^2 + rz + s$, we have shown in Exercise 10.27 that once the three roots j_1, j_2, j_3 of the cubic resolvent polynomial are found, the roots of $z^4 + qz^2 + rz + s$ can be expressed in terms of square roots of j_1, j_2, and j_3. Specifically, the four roots of $z^4 + qz^2 + rz + s$ are four of the eight numbers

$$\frac{1}{2}\left(\pm\sqrt{j_1} \pm \sqrt{j_2} \pm \sqrt{j_3}\right).$$

The four choices of sign that produce the roots are those for which the product of the three square roots $\sqrt{j_1}$, $\sqrt{j_2}$, and $\sqrt{j_3}$ is $-r$.

Let us solve two quartic equations using this approach. The equations are rigged so that the roots of the cubic resolvent polynomials are easily found by trial and error.

Exercise 10.28. Find all four solutions of the quartic equation

$$z^4 - 3z^2 + \sqrt{6}z - \frac{1}{2} = 0$$

and of the quartic equation

$$z^4 - 3z^2 - \sqrt{6}z - \frac{1}{2} = 0.$$

In doing the previous exercise, you should have found that the two quartic equations have the same cubic resolvent equation. This is a special case of a general phenomenon.

Exercise 10.29. Explain why the quartic equations

$$z^4 + qz^2 + rz + s = 0$$

and

$$z^4 + qz^2 - rz + s = 0$$

have the same cubic resolvent equation. Describe how the four solutions of the first equation are related to the four solutions of the second equation.

10.6 The Discriminant of a Quartic Polynomial

Using our refined analysis in Section 10.5 of the roots of a reduced quartic polynomial, we can obtain a formula for the discriminant of any quartic
polynomial. Recall that the discriminant of a quadratic or a cubic polynomial is defined as the square of the product of differences of the roots. The
discriminant of a quartic polynomial

$$x^4 + bx^3 + cx^2 + dx + e$$

is defined in the same way. Suppose its roots are r_1, r_2, r_3, and r_4. Then its
discriminant Δ is the product

$$(r_1 - r_2)^2(r_1 - r_3)^2(r_1 - r_4)^2(r_2 - r_3)^2(r_2 - r_4)^2(r_3 - r_4)^2.$$

Notice that because the root differences are squared, we get the same result
regardless of how the roots are ordered.

For quadratic and cubic polynomials, the discriminant is important for two
reasons. First, its sign gives us information on the nature of the roots: how
many are real numbers and how many are nonreal complex numbers. Second,
there is a formula expressing the discriminant in terms of the coefficients of
the polynomial, allowing us to calculate the discriminant from the data of
the coefficients alone. We can then obtain information about the roots of a
quadratic or a cubic polynomial without knowing the roots explicitly. For a
quartic polynomial, the discriminant again gives information on the nature
of the roots, although the information is not as complete as in the quadratic
and cubic cases, and again the discriminant can be calculated from the data
of the coefficients alone.

Exercise 10.30. Let us see what the discriminant of a quartic polynomial
tells us about the nature of its roots. Let $f(x)$ be such a polynomial, with
discriminant Δ, and let r_1, r_2, r_3, and r_4 be its four roots.

1. Check that $\Delta = 0$ precisely when at least two of the roots coincide.
2. Assume in the remainder of the exercise that the four roots are distinct,
   so that $\Delta \neq 0$. Since $f(x)$ factors as the product of two quadratic polynomials, there are three possibilities for the roots: All are real, two are real
   and two are nonreal complex conjugates, or none is real and there are two
   pairs of nonreal complex conjugates.
3. Show that if all four roots are real (and distinct), then $\Delta > 0$.
4. Suppose two roots are real (and distinct) and two are complex conjugates.
   Show that $\Delta < 0$. (Hint: Recall that the product of a nonzero complex
   number and its conjugate is a positive real number. Name the four roots,
   for example r, s, $a+bi$, and $a-bi$, so that you can work with them explicitly
   as real and complex numbers. There are six root differences. You should
   observe that four of the six differences occur as pairs of conjugate complex
   numbers, so that their products are real. You should also find that one of

the remaining root differences is real and that the other is the product of a real number and i. From this you can conclude that Δ is negative.)

5. Suppose finally that the four distinct roots occur in two pairs of nonreal complex conjugates. Show that $\Delta > 0$. (Hint: You can again pair four of the differences so that they are conjugates and their products are real. The other two root differences should be products of real numbers and i.)

6. Conclude that the sign of Δ gives some information on the nature of the roots, but not complete information. In particular, if $\Delta \geq 0$, there is some ambiguity.

We wish to calculate the discriminant Δ of a quartic polynomial in terms of its coefficients. We will handle the reduced case first.

Exercise 10.31. Let us relate the discriminant Δ of a reduced quartic polynomial

$$z^4 + qz^2 + rz + s$$

to the discriminant of its cubic resolvent polynomial.

1. Using the notation we employed in Section 10.5, write the four roots r_i of $z^4 + qz^2 + rz + s$ as halves of sums of $\pm k_i$. Using these expressions, show that the six differences of roots are the quantities

$$k_1 \pm k_2; \quad k_1 \pm k_3; \quad k_2 \pm k_3.$$

2. Deduce that

$$\Delta = \left(k_1^2 - k_2^2\right)^2 \left(k_1^2 - k_3^2\right)^2 \left(k_2^2 - k_3^2\right)^2.$$

3. Recall that the k_i's are the square roots of the solutions j_1, j_2, j_3 of the cubic resolvent equation. Using the j_i's, rewrite the equation for Δ above as

$$\Delta = (j_1 - j_2)^2 (j_1 - j_3)^2 (j_2 - j_3)^2.$$

4. Observe that since the j_i's are by definition the roots of the cubic resolvent polynomial, the product on the right side of the last equation is in fact the discriminant of the cubic resolvent.

5. Conclude that the discriminant of $z^4 + qz^2 + rz + s$ coincides with the discriminant of its cubic resolvent $j^3 + 2qj^2 + \left(q^2 - 4s\right) j - r^2$.

Exercise 10.32. Since the discriminant of a reduced quartic polynomial $z^4 + qz^2 + rz + s$ coincides with the discriminant of its cubic resolvent, we can obtain a formula for the discriminant of $z^4 + qz^2 + rz + s$ using our results on discriminants of cubics. Let us do so.

1. Recall from Exercise 10.17 that the discriminant of a cubic polynomial $x^3 + bx^2 + cx + d$ is given by the formula

$$18bcd - 4b^3d + b^2c^2 - 4c^3 - 27d^2.$$

2. The cubic resolvent polynomial of $z^4 + qz^2 + rz + s$ has coefficients $2q$, $q^2 - 4s$, and $-r^2$. Substitute these expressions for b, c, and d in the discriminant formula above (without expanding and simplifying) and obtain an expression for the discriminant of the cubic resolvent in terms of q, r, and s. Conclude that this complicated expression in q, r, and s is also the discriminant of the reduced quartic polynomial $z^4 + qz^2 + rz + s$ with which we began.

3. Expand the formula and simplify in order to obtain the formula

$$\Delta = 144qr^2s - 128q^2s^2 - 4q^3r^2 + 16q^4s - 27r^4 + 256s^3$$

for the discriminant of $z^4 + qz^2 + rz + s$.

Exercise 10.33. Compute the discriminants of the quartic polynomials below and say what you can about the number of real roots of each.

1. $z^4 - 3z^2 + 6z - 2$.
2. $z^4 - 2z^2 - 8z - 3$.
3. $z^4 - 3z^2 + \sqrt{6}z - \frac{1}{2}$.

We saw in Exercise 10.30 that if a quartic polynomial $z^4 + qz^2 + rz + s$ has positive discriminant, then there are two possibilities for its roots: They are all real, or they are all nonreal, forming two pairs of complex conjugates. We wish to be able to determine from the coefficients q, r, and s alone which is the case. In case $r = 0$, this is a straightforward matter. We can regard $z^4 + qz^2 + s$ as a quadratic polynomial in z^2, use the sign of $q^2 - 4s$ to determine whether the two values of z^2 occurring as roots are real or nonreal, and in the real case use Exercise 10.19 to determine the signs of these values of z^2 in terms of q and s. From this, we can determine whether the four values of z are real or nonreal. We will therefore restrict ourselves in the exercise below to the case in which r is nonzero.

Exercise 10.34. Consider the quartic polynomial

$$z^4 + qz^2 + rz + s.$$

Assume that the discriminant Δ is positive and that r is nonzero.

1. Recall that Δ is also the discriminant of the cubic resolvent polynomial

$$j^3 + 2qj^2 + (q^2 - 4s)j - r^2$$

and that since $\Delta > 0$, the roots j_1, j_2, and j_3 of the cubic resolvent are all real.

2. Observe that the constant term of the cubic resolvent is negative. Review Exercise 10.19 and conclude that either all of the cubic resolvent's roots j_i are positive or one j_i is positive and the other two are negative.

3. Observe that if all three roots j_i are positive, then their square roots $\pm k_i$ are all real numbers. Conclude that the four roots of the original quartic $z^4 + qz^2 + rz + s$ are all real.

4. Alternatively, suppose j_1 is positive and j_2 and j_3 are negative. Conclude that the two square roots $\pm k_1$ are real while $\pm k_2$ and $\pm k_3$ are pure imaginary (that is, of the form ri for some real number r). Show that in this case, the four roots of $z^4 + qz^2 + rz + s$ are all nonreal complex numbers.

5. Conclude that if $\Delta > 0$, the question of whether the four roots of $z^4 + qz^2 + rz + s$ are all real or all nonreal reduces to the question of whether the cubic resolvent polynomial has three positive real roots or one positive real root and two negative real roots.

6. Using Exercise 10.19 again, conclude that if $q < 0$ and $q^2 - 4s > 0$, then the roots of the cubic resolvent are all positive, and the roots of the given quartic are all real; alternatively, if the inequalities $q < 0$ and $q^2 - 4s > 0$ do not both hold, then only one root of the cubic resolvent is positive and no root of $z^4 + qz^2 + rz + s$ is real.

7. Conclude that you can decide whether all the roots of $z^4 + qz^2 + rz + s$ are real or all are nonreal by examining the signs of q and $q^2 - 4s$. If the first is negative and the second is positive, all the roots of $z^4 + qz^2 + rz + s$ are real. Otherwise, no root of $z^4 + qz^2 + rz + s$ is real.

Exercise 10.35. For each of the quartic polynomials below, use the discriminant and the coefficients to decide how many roots are real.

1. $z^4 + z + 1$.
2. $z^4 + z - 1$.
3. $z^4 + z + s$. (Your answer will depend on s.)
4. $z^4 - 3z^2 + 6z - 2$.
5. $z^4 - 2z^2 - 8z - 3$.
6. $z^4 - 3z^2 + \sqrt{6}z - \frac{1}{2}$.

Exercise 10.34 shows how to use the coefficients of a reduced quartic polynomial with positive discriminant to determine whether the roots are all real or all nonreal. For a reduced quartic polynomial with zero discriminant, there are also several possibilities for the nature of the roots, and again the coefficients of the polynomial can be used to determine which possibility occurs. The analysis proceeds along similar lines: The associated cubic resolvent polynomial will also have zero discriminant, so it will have only real roots. Using Exercise 10.19 again, we can sort out how many of these roots are positive and how many are negative. It is then a simple matter to decide how many roots of the reduced quartic polynomial are real and how many are nonreal.

We have been focused in the last few exercises on reduced quartic polynomials. Let us return to arbitrary monic quartic polynomials and use some good old-fashioned algebra to obtain a general formula for their discriminants.

Exercise 10.36. Consider the quartic polynomial

$$x^4 + bx^3 + cx^2 + dx + e.$$

1. Observe, as you did for cubic polynomials in Exercise 10.17, that although the roots of a general quartic polynomial and the roots of its associated reduced quartic polynomial are not the same, the differences of the roots are. Deduce that the discriminant of the quartic polynomial is the same as the discriminant of the associated reduced quartic polynomial.

2. Conclude that in principle you can obtain a formula for the discriminant of $x^4 + bx^3 + cx^2 + dx + e$ in terms of b, c, d, and e. To do so, you can change variables to obtain a reduced quartic polynomial $z^4 + qz^2 + rz + s$. Each of the new coefficients q, r, and s can then be written explicitly in terms of b, c, d, and e. You can then substitute the expressions for q, r, and s in terms of b, c, d, and e in the formula for the discriminant of $z^4 + qz^2 + rz + s$ and obtain a formula for the discriminant of $x^4 + bx^3 + cx^2 + dx + e$ in terms of b, c, d, and e. This is a lengthy calculation, but it is entirely elementary algebra.

3. Carry out this calculation and obtain the formula below for the discriminant Δ of $x^4 + bx^3 + cx^2 + dx + e$:

$$\begin{aligned}
\Delta = {} & 18bcd^3 + 18b^3cde - 80bc^2de - 6b^2d^2e + 144cd^2e \\
& + 144b^2ce^2 - 128c^2e^2 - 192bde^2 + b^2c^2d^2 - 4b^3d^3 \\
& - 4c^3d^2 - 4b^2c^3e + 16c^4e - 27d^4 - 27b^4e^2 + 256e^3.
\end{aligned}$$

In Exercise 10.34, we saw how to use the coefficients of a reduced quartic polynomial with positive discriminant to determine whether its roots are all real or all nonreal. For an arbitrary quartic polynomial, the roots are all real exactly when the roots of its associated reduced quartic polynomial are all real, and they are all nonreal when the roots of the associated reduced quartic polynomial are all nonreal. This allows us to obtain a version of the result in Exercise 10.34 for quartic polynomials with positive discriminant. The interested reader can work out the details.

10.7 The Fundamental Theorem of Algebra

Finding the roots of a polynomial in $\mathbb{R}[x]$ can be a difficult matter, even if one knows that they exist. We have seen that one can use the quadratic formula for quadratic polynomials, Cardano's formula for cubic polynomials, and a combination of them for quartic polynomials. The work of various Italian mathematicians in the sixteenth century that culminated in the formulas for roots of cubic and quartic polynomials was the most important mathematical achievement in centuries. But there the matter stood for over two more centuries.

The argument given in Exercise 10.5 to show that a cubic polynomial in $\mathbb{R}[x]$ has a real root applies equally well to a polynomial of any odd degree in $\mathbb{R}[x]$, allowing one to prove that a polynomial in $\mathbb{R}[x]$ of odd degree has a factor of degree one and therefore cannot be irreducible. In particular, a polynomial of degree five—a *quintic* polynomial—has at least one real root. Naturally, one would like a formula for this root, analogous to the formulas for roots of polynomials of lower degree. No one was able to find a general procedure that would determine this root, or the roots of polynomials of any higher degree, in terms of the coefficients of the polynomial using the four basic arithmetic operations and the extraction of roots.

When two centuries of attempts to find such a formula failed, speculation arose that perhaps no such formula existed. Finally, in the 1820s, the young Norwegian mathematician Niels Henrik Abel (1802–1829) proved that indeed there could be no such formula. A few years later, the even younger French mathematician Évariste Galois (1811–1832) gave a proof of the more general fact that for every integer n greater than four there cannot be a formula for the roots of a general degree-n polynomial in terms of the polynomial's coefficients and extraction of suitable mth roots. Thus the quadratic formula, the cubic formula, and the quartic formula are the only such formulas possible. Galois's proof introduced a number of revolutionary ideas involving the relation between field extensions (like those we studied in Chapter 9) and abstract algebraic structures called *groups*. The entire subject of modern abstract algebra can be said to stem from Galois's results. His is one of the greatest of all contributions in the history of mathematics.

Even though we cannot obtain formulas for the roots of polynomials in $\mathbb{R}[x]$ of an arbitrary positive degree, we can still study factorization questions in $\mathbb{R}[x]$. You may have noticed in our study of linear, quadratic, cubic, and quartic polynomials that in thinking about the roots of the polynomial, it is perhaps more natural to think of the polynomial as living in $\mathbb{C}[x]$ than in $\mathbb{R}[x]$. Why? Because while it is rather complicated to list all the possibilities for the number of roots in $\mathbb{R}$, the statement for the number of roots in $\mathbb{C}$ is extremely simple:

Theorem 10.4. *Let $f(x)$ be a polynomial in $\mathbb{R}$ of degree $n = 1$, 2, 3, or 4. Then $f(x)$ has exactly n roots in $\mathbb{C}$ (counting multiplicity). Equivalently, $f(x)$ factors completely in $\mathbb{C}[x]$ into n linear factors.*

It turns out that the statement of Theorem 10.4 is valid not only for polynomials of degrees one, two, three, and four, but also for polynomials of all higher degrees in $\mathbb{R}[x]$. And that takes us to the topic of the *fundamental theorem of algebra*.

For a field K, a polynomial in $K[x]$ of positive degree factors as a product of irreducible polynomials. This leads us to focus on the question of what the irreducible polynomials of $K[x]$ are for a particular choice of K. These irreducible polynomials play a fundamental role in the ring $K[x]$, just as prime numbers do in $\mathbb{Z}$. We have found that a polynomial in $\mathbb{R}[x]$ of odd degree has a real root and therefore cannot be irreducible. We have also found that a polynomial in $\mathbb{R}[x]$ of degree 4 cannot be irreducible. What about polynomials in $\mathbb{R}[x]$ of even degree greater than 4? Can they be irreducible? The answer is no, and in fact, the following result is equivalent to the fundamental theorem of algebra:

Theorem 10.5 (Gauss). *Every irreducible polynomial in $\mathbb{R}[x]$ has degree either one or two.*

There are many proofs of the fundamental theorem of algebra. The first correct proof was given in 1799 by Gauss in his doctoral thesis. Every proof requires the use of some results from mathematical analysis (calculus). The amount of analysis required can be reduced to just some elementary calculus, but the weaker the results of analysis that a proof uses, the more intricate the proof tends to be. No proof will be provided here. However, given the profound importance of the fundamental theorem of algebra, one should read through a proof of it at least one time in one's life, if only to get a sense of what is involved in proving it.

Even though we will not prove the fundamental theorem, we will illustrate its power by deducing from it the following result.

Theorem 10.6. *Let $f(x)$ be a polynomial in $\mathbb{R}[x]$ of positive degree n.*

1. *The polynomial $f(x)$ factors in $\mathbb{R}[x]$ as the product of polynomials of degree 1 or 2.*
2. *The polynomial $f(x)$ has n roots in $\mathbb{C}$ (counting multiplicity). In particular, there are nonnegative integers r and s satisfying $r + 2s = n$ such that $f(x)$ has r real roots and s pairs of nonreal conjugate complex numbers as roots.*
3. *The polynomial $f(x)$ factors in $\mathbb{C}[x]$ as the product of n degree-one polynomials.*

Exercise 10.37. Prove Theorem 10.6.

The field of complex numbers is created from the field of real numbers by adjoining a single number i. This is the number needed in order for the polynomial $x^2 + 1$ to have a root, and adjoining it to $\mathbb{R}$ allows us to factor $x^2 + 1$ as the product $(x + i)(x - i)$ of two first-degree polynomials. Theorem 10.6 shows that by adjoining i to the real numbers we obtain a sufficiently rich set of numbers so that every polynomial $f(x)$ in $\mathbb{R}[x]$ of positive degree has a complete set of roots, permitting the factorization of $f(x)$ into a product of first-degree polynomials.

Theorems 10.5 and 10.6 tell us that every irreducible polynomial in $\mathbb{R}[x]$ has degree either one or two and that every polynomial in $\mathbb{R}[x]$ of positive degree, whether it has a root in $\mathbb{R}$ or not, has a root in $\mathbb{C}$. What about polynomials of positive degree in $\mathbb{C}[x]$? If we now allow our polynomials to have complex coefficients, do they also have roots in $\mathbb{C}$? The positive answer to this question is what is usually called the fundamental theorem of algebra:

Theorem 10.7. *Every nonconstant polynomial in $\mathbb{C}[x]$ has a complex root. Equivalently, the only irreducible polynomials in $\mathbb{C}[x]$ are the polynomials of degree one, and every polynomial in $\mathbb{C}[x]$ of positive degree factors as a product of degree-one polynomials.*

It is not hard to prove Theorem 10.7 as a consequence of Theorem 10.5. In order to do so, we must study *conjugation* of polynomials in $\mathbb{C}[x]$. First, let us obtain a formula for the coefficients of the product of two polynomials. We have been assuming that polynomials have coefficients in a field, but a ring will do just as well. For instance, we can consider the set $\mathbb{Z}[x]$ of polynomials with integer coefficients. We will work in this level of generality.

Exercise 10.38. Let K be a field, or ring, and let $g(x)$ and $h(x)$ be polynomials in $K[x]$ of degrees m and n. Write $g(x)$ as

$$g(x) = b_m x^m + b_{m-1} x^{m-1} + \cdots + b_1 x + b_0$$

and $h(x)$ as

$$h(x) = c_n x^n + c_{n-1} x^{n-1} + \cdots + c_1 x + c_0.$$

Let $f(x)$ be the product $g(x)h(x)$, and write $f(x)$ as

$$a_{m+n} x^{m+n} + \cdots + a_1 x + a_0.$$

1. Write a formula for a_0 in terms of the b_i's and c_j's.
2. Do the same for a_1, for a_2, and for a_3.
3. Let k be a nonnegative integer. With the previous parts as a guide, write a formula for a_k in terms of the b_i's and c_j's.

If a is a complex number of the form $r + si$, with r and s real, its conjugate $\bar{a}$ is the complex number $r - si$. We can extend the process of conjugation from complex numbers to polynomials $f(x)$ with complex numbers as coefficients. To do so, we simply conjugate each of the coefficients of $f(x)$, leaving the powers of x alone. For instance, the conjugate of the polynomial

$$3x^3 + (4 - i)x^2 + (1 + 5i)x - 5i$$

is the polynomial

$$3x^3 + (4 + i)x^2 + (1 - 5i)x + 5i.$$

Let us write $\overline{f}(x)$ for the conjugate of the polynomial $f(x)$. We will need the following result.

Theorem 10.8. *Let $f(x)$ be a polynomial in $\mathbb{C}[x]$.*

1. *Suppose that $f(x)$ factors as*

$$f(x) = g(x)h(x)$$

for polynomials $g(x)$ and $h(x)$ in $\mathbb{C}[x]$. Then

$$\overline{f}(x) = \overline{g}(x)\overline{h}(x).$$

2. *Suppose the complex number r is a root of $f(x)$. Then the conjugate $\bar{r}$ of r is a root of the conjugate polynomial $\bar{f}(x)$.*

Exercise 10.39. Verify the following facts.

1. For complex numbers a and b,

$$\overline{a+b} = \bar{a} + \bar{b}; \qquad \overline{ab} = \bar{a}\bar{b}.$$

2. For complex numbers $a_0, \ldots, a_k$,

$$\overline{a_0 + \cdots + a_k} = \overline{a_0} + \cdots + \overline{a_k}; \qquad \overline{a_0 \cdots a_k} = \overline{a_0} \cdots \overline{a_k}.$$

3. For complex numbers $a_0, \ldots, a_k$ and $b_0, \ldots, b_k$,

$$\overline{a_0 b_k + a_1 b_{k-1} + \cdots + a_{k-1} b_1 + a_k b_0} = \overline{a_0}\overline{b_k} + \cdots + \overline{a_k}\overline{b_0}.$$

Using this last fact, prove the first part of Theorem 10.8 by comparing the coefficients of $\bar{f}(x)$ and $\bar{g}(x)\bar{h}(x)$. Deduce the second part by conjugating $f(r)$.

Using Theorem 10.5, we have proved that the nonreal complex roots of a polynomial with real coefficients occur in conjugate pairs. This can be proved directly, without reference to the fundamental theorem of algebra, using Theorem 10.8. A slightly stronger statement can be proved as well:

Theorem 10.9. *Suppose that $f(x)$ is a polynomial of positive degree in $\mathbb{R}[x]$ and that r is a root of $f(x)$ in $\mathbb{C}$.*

1. *The conjugate $\bar{r}$ of r is also a root of $f(x)$.*
2. *The polynomial $(x - r)(x - \bar{r})$ lies in $\mathbb{R}[x]$.*
3. *If r is not real, then $(x - r)(x - \bar{r})$ is a divisor of $f(x)$ in $\mathbb{R}[x]$.*

Exercise 10.40. Prove Theorem 10.9.

1. For the first part, show that $\bar{f}(x) = f(x)$.
2. For the second part, check that the coefficients of $(x - r)(x - \bar{r})$ are real, so that $(x - r)(x - \bar{r})$ lies in $\mathbb{R}[x]$.
3. Deduce that if r is a nonreal complex number and $x - r$ divides $f(x)$ in $\mathbb{C}[x]$, then $(x - r)(x - \bar{r})$ divides $f(x)$ in $\mathbb{C}[x]$.
4. Observe that to prove that $(x-r)(x-\bar{r})$ divides $f(x)$ in $\mathbb{R}[x]$, it suffices to prove the following statement: Suppose $f(x)$ and $g(x)$ are nonzero polynomials in $\mathbb{R}[x]$ and $h(x)$ is a polynomial in $\mathbb{C}[x]$ such that $f(x) = g(x)h(x)$. Then $h(x)$ lies in $\mathbb{R}[x]$.
5. Prove this last statement.

Finally, let us deduce Theorem 10.7 from Theorem 10.5.

Exercise 10.41. Suppose $f(x)$ is a nonconstant polynomial in $\mathbb{C}[x]$.

1. Show that $f(x)\bar{f}(x)$ is unchanged by conjugation and deduce that $f(x)\bar{f}(x)$ lies in $\mathbb{R}[x]$.
2. Use Theorem 10.5 to show that $f(x)\bar{f}(x)$ has a root in $\mathbb{C}$.
3. Deduce that at least one of $f(x)$ and $\bar{f}(x)$ has a root in $\mathbb{C}$.
4. Using Theorem 10.8, conclude that $f(x)$ itself has a root in $\mathbb{C}$.

Polynomials with Rational Coefficients

11.1 Polynomials over $\mathbb{Q}$

We have found formulas for roots of polynomials in $\mathbb{R}[x]$ of degree 4 or less and found that there are no such general formulas for polynomials of higher degree. When we are dealing with a particular polynomial of degree greater than 4 we may be lucky in finding a root or in discovering one or more factors, but in the general case, when luck fails us, if we want to find the roots of a higher-degree polynomial we must use numerical approximation techniques such as Newton's method of calculus. Such approximation involves numerical calculation, so one begins by approximating the real coefficients of a polynomial $f(x)$ by rational numbers, replacing $f(x)$ in $\mathbb{R}[x]$ with a polynomial in $\mathbb{Q}[x]$. Therefore, it is important from a calculational point of view to be able to find roots of polynomials with rational coefficients, or to be able to factor polynomials with rational coefficients. Moreover, the theory of polynomials in $\mathbb{Q}[x]$ is of great interest; it is, for example, the basis Galois's proof that there are no general formulas for the roots of polynomials of degree greater than 4.

According to the fundamental theorem of algebra, the irreducible polynomials of $\mathbb{R}[x]$ are the linear (degree-one) polynomials and the quadratic (degree-two) polynomials of negative discriminant, while the only irreducible polynomials of $\mathbb{C}[x]$ are the linear polynomials. In contrast, in $\mathbb{Q}[x]$ there are irreducible polynomials of every positive degree n, as we will prove in this chapter. One source of irreducible polynomials in $\mathbb{Q}[x]$ is polynomials whose roots are real or complex numbers that are algebraic over $\mathbb{Q}$, such as the minimal polynomial that we met in Section 9.5. Let us review the relevant notions from that section.

Let $K \subseteq L$ be a field extension. The elements of L that are roots of polynomials in $K[x]$ are said to be *algebraic* over K. For example, the fields $\mathbb{R}$ and $\mathbb{C}$ are field extensions of $\mathbb{Q}$, and the real or complex numbers that are algebraic over $\mathbb{Q}$ are known more simply as *algebraic numbers*. In other words, a real or complex number is *algebraic* if it is a root of a polynomial in $\mathbb{Q}[x]$. Every rational number α is an algebraic number, since it is the root of the linear polynomial $x - \alpha$ in $\mathbb{Q}[x]$. But that is just the beginning. We can find many other algebraic numbers, such as all complex numbers of the form $\alpha + \beta i$, for α and β in $\mathbb{Q}$.

Exercise 11.1. Prove that all numbers of the form $\alpha + \beta i$, for α and β in $\mathbb{Q}$, are algebraic over $\mathbb{Q}$.

We have also seen that irrational numbers such as $\sqrt{2}$ are algebraic ($\sqrt{2}$ is a root of the polynomial $x^2 - 2$). According to Theorem 9.12, for each algebraic number γ, there is a unique monic irreducible polynomial $p_\gamma(x)$ in $\mathbb{Q}[x]$ having γ as a root. Moreover, the polynomials in $\mathbb{Q}[x]$ that have γ as a root are precisely the polynomials divisible by $p_\gamma(x)$. The degree of $p_\gamma(x)$ is called the *degree* of γ. The study of algebraic numbers essentially coincides with the study of irreducible polynomials in $\mathbb{Q}[x]$.

Many of the numbers we use on a daily basis are algebraic. For example, the nth roots of an integer m are algebraic, since they are roots of $x^n - m$. Are all real and complex numbers algebraic? The irrational numbers that immediately come to mind, namely, those involving radicals, like $\sqrt[3]{47}$ and $\frac{1}{3} - \sqrt{2 + \sqrt[5]{1 - \sqrt[3]{7}}}$, are all algebraic. And there are many more algebraic numbers that cannot be expressed using radicals; that is the content of Galois's theorem.

Exercise 11.2. Find a polynomial with rational coefficients that has

$$\frac{1}{3} - \sqrt{2 + \sqrt[5]{1 - \sqrt[3]{7}}}$$

as a root.

Are there any numbers that are not algebraic? What about some of the important constants of mathematics, like e, the base of the natural logarithm, and π, the ratio of the circumference of a circle to its diameter. Are these numbers algebraic? These questions have had an important role in the history of mathematics.

In fact, there are many real and complex numbers that are not algebraic. A number that is not algebraic is called *transcendental*. The first proof that transcendental numbers exist was given by Joseph Liouville (say lee-you–veal') in 1844. He proved a theorem regarding how well irrational algebraic numbers can be approximated by rational numbers. (Of course, every real number can be approximated arbitrarily well by rational numbers, as we discussed in Section 6.1; what Liouville considered was how well irrational numbers can be approximated by rational numbers *with small denominators*: He proved that algebraic numbers cannot be approximated as well as transcendental numbers.) As a consequence of his theorem, one obtains a recipe for the construction of infinitely many transcendental numbers, such as

$$\mathcal{L} = 0.1001000000010000000000000000000000000001\ldots.$$

This number is the decimal obtained by writing a one, then 2 zeros, then a one, then 3×2 zeros, then a one, then $4 \times 3 \times 2$ zeros, and so on. The trick of employing ever larger strings of zeros produces a number that can be rationally approximated very well; for example, $\mathcal{L}$ can be approximated to 35 decimal places by a rational number with a denominator of only 11 digits, and the further out you go, the better this type of approximation becomes.

Liouville's construction has the disadvantage that the transcendental numbers that it produces do not arise naturally in everyday mathematics. This problem was rectified in 1873, when Charles Hermite proved that the number e is transcendental. A few years later, in 1882, Carl Lindemann proved that π is transcendental. Thus, no polynomial with rational coefficients has e or π as a root.

The transcendentality of π has as a consequence a solution to the famous centuries-old geometric problem of squaring the circle (that is, the problem of constructing a square whose area is the same as that of given circle). It is not particularly difficult to show that all lengths that can be constructed with straightedge and compass (so-called constructible numbers), starting with a reference length 1, are algebraic numbers. If one could construct a square whose area was the same as that of given circle, say with radius 1, then that square would have area π (why?), and one would thus have constructed a square with sides of length $\sqrt{\pi}$. That would make $\sqrt{\pi}$ a constructible number. But if $\sqrt{\pi}$ were constructible, then π would also be constructible (exercise) and therefore algebraic. However, π is transcendental, and thus it cannot be a constructible number. Therefore, the circle cannot be squared.

Following Hermite and Lindemann, mathematicians have proved that many other commonly occurring real and complex numbers are transcendental. Furthermore, in a suitable sense, *most* real and complex numbers are transcendental, as Georg Cantor proved in 1874. Cantor had introduced a way of describing different sizes of infinity for other purposes, but then showed by a famous argument that the rational numbers make up an insignificantly small proportion of the set of real numbers. He then showed that the algebraic numbers form a set that has the same size as the rational numbers, so that "most" real numbers must be transcendental.

The collection of algebraic numbers is extraordinarily complicated, as reflected in the fact that there are infinitely many algebraic numbers of every degree n. Corresponding to this, there are infinitely many irreducible monic polynomials in $\mathbb{Q}[x]$ of every degree n, as we will prove in Section 11.3.

We still have not demonstrated the irreducibility of any specific polynomials in $\mathbb{Q}[x]$. Let us do so now. Every degree-one polynomial in $\mathbb{Q}[x]$ is irreducible, as is every degree-two polynomial whose discriminant has no rational square root. Consider degree 3. We know that no cubic polynomial in $\mathbb{R}[x]$ is irreducible. However, there are many irreducible cubic polynomials in $\mathbb{Q}[x]$. For example, consider $x^3 - 2$. The same sort of argument used to prove that the square roots of 2 are irrational also shows that the real cube root of 2 is irrational. Therefore, $x^3 - 2$ has no rational root and is irreducible as an element of $\mathbb{Q}[x]$. Next consider degree 4 and the quartic polynomial $x^4 - 2$. We can show that $x^4 - 2$ has no real root, but this does not ensure the irreducibility of $x^4 - 2$ in $\mathbb{Q}[x]$. We must also show that $x^4 - 2$ cannot factor in $\mathbb{Q}[x]$ as a product of two quadratic polynomials.

Exercise 11.3. Prove that $x^4 - 2$ is irreducible in $\mathbb{Q}[x]$ by following the outline below.

1. Factor $x^4 - 2$ in $\mathbb{R}[x]$ as the product of an irreducible polynomial of degree two and two polynomials of degree 1.
2. From the above factorization, read off what the four complex roots of $x^4 - 2$ are and observe that none of these roots is a rational number.
3. Deduce that $x^4 - 2$ has no linear factor in $\mathbb{Q}[x]$, so that either $x^4 - 2$ is irreducible in $\mathbb{Q}[x]$, as we wish to prove, or it factors as the product of two polynomials of degree two in $\mathbb{Q}[x]$.
4. Suppose that $x^4 - 2$ does factor as the product of two polynomials of degree two in $\mathbb{Q}[x]$. Write out these factors explicitly, say as $ax^2 + bx + c$ and $dx^2 + ex + f$. Multiply the factors together to obtain five equations that the coefficients a, b, c, d, e, and f must satisfy. We wish to show that these equations cannot be simultaneously satisfied by rational numbers. This would prove that $x^4 - 2$ is irreducible. (Why?)
5. Use two equations involving only a, b, d, and e to obtain the equation $e + bd^2 = 0$.
6. Use two equations involving only b, c, e, and f to obtain $-2b + c^2e = 0$.
7. Combine these two equations to find that $b(c^2d^2 + 2) = 0$ and deduce that $b = e = 0$.
8. Returning to your original five equations, you should find that you are left with three equations involving a, c, d, and f. Multiplying through the one that involves all four coefficients by ac and using the other two, obtain the relation $c^2 = 2a^2$.
9. Observe that since a and c are assumed to be rational, any solution to the equation $c^2 = 2a^2$ would yield the rationality of $\sqrt{2}$, a contradiction. Therefore, the five equations cannot be simultaneously solved by rational numbers.

We wish to extend our analysis of $x^3 - 2$ and $x^4 - 2$ by proving the following theorem.

Theorem 11.1. *For every positive integer n, the polynomial $x^n - 2$ is irreducible in $\mathbb{Q}[x]$. More generally, for every positive integer n and every prime number p, the polynomial $x^n - p$ is irreducible in $\mathbb{Q}[x]$. Hence, there are infinitely many monic irreducible polynomials of every positive degree n in $\mathbb{Q}[x]$.*

It is not hard to see that for n and p as in the statement of the theorem, $x^n - p$ has no real rational root; this tells us $x^n - p$ has no factor of degree one in $\mathbb{Q}[x]$. That $x^n - p$ is irreducible in $\mathbb{Q}[x]$ is a much deeper result. How can we prove it? We certainly do not want to try the kind of approach we used in Exercise 11.3 in proving that $x^4 - 2$ is irreducible. The computations are bound to get too hard as n increases, and to use the argument for a general n could lead to a notational nightmare. A better approach does exist. It is treated in Sections 11.2 and 11.3.

11.2 Gauss's Lemma

In order to prove Theorem 11.1, we will relate the factorization of polynomials in $\mathbb{Q}[x]$ to the factorization of polynomials in $\mathbb{Z}[x]$. To do so, we begin with some elementary observations about the factorization of polynomials in $\mathbb{Z}[x]$.

For every field K, we know that the units of $K[x]$ are the polynomials of degree 0 and that a polynomial $f(x)$ of positive degree is irreducible if and only if it does not factor as the product of lower-degree polynomials. Since $\mathbb{Z}$ is not a field, the situation in the ring $\mathbb{Z}[x]$ is different. To understand it, we must first determine the units in $\mathbb{Z}[x]$.

Exercise 11.4. Prove that the only units in $\mathbb{Z}[x]$ are 1 and -1. To do so, use the usual degree argument to show that every unit in $\mathbb{Z}[x]$ lies in $\mathbb{Z}$, and then deduce that the units are as stated.

Since 1 and -1 are the only units of $\mathbb{Z}[x]$, every factorization of a polynomial in $\mathbb{Z}[x]$ involving a nonzero integer besides 1 and -1 must be regarded as nontrivial. For example, the factorization

$$2x^2 + 2 = 2 \cdot (x^2 + 1)$$

is a nontrivial factorization of $2x^2 + 2$ in $\mathbb{Z}[x]$. Of course, $2x^2 + 2$ cannot be factored in $\mathbb{Z}[x]$ as a product of polynomials of lower degree. Yet $2x^2 + 2$ is not irreducible in $\mathbb{Z}[x]$.

These considerations lead to the notion of primitivity. A polynomial $f(x)$ in $\mathbb{Z}[x]$ is *primitive* if the greatest common divisor of its coefficients is 1. For example, $x^3 - 3x^2 + 9$ is primitive, as is $6x^2 - 10x + 15$, but $2x^2 + 2$ is not primitive, because its coefficients are divisible by 2, and $6x^2 - 9x + 15$ is not primitive, because its coefficients are divisible by 3. A degree-zero polynomial is understood to be primitive only if it is 1 or -1.

Exercise 11.5. This exercise treats basic facts about primitive polynomials.

1. Verify that a polynomial $g(x)$ in $\mathbb{Z}[x]$ is primitive if and only if there is no prime number p that divides every coefficient of $g(x)$.
2. Write the polynomial $3x^7 + 12x^5 - 15x^2 + 21$ as the product of an integer and a primitive polynomial.
3. More generally, show that if $f(x)$ is a polynomial in $\mathbb{Z}[x]$, there are a positive integer m and a primitive polynomial $p(x)$ in $\mathbb{Z}[x]$ such that $f(x) = mp(x)$. Show that the only choice for m is the greatest common divisor of the coefficients of $f(x)$.

4. Suppose m and n are positive integers and $p(x)$ and $q(x)$ are primitive polynomials satisfying $mp(x) = nq(x)$. Prove that $m = n$ and $p(x) = q(x)$.
5. Prove that a primitive polynomial of positive degree that does not factor in $\mathbb{Z}[x]$ as a product of lower-degree polynomials is irreducible in $\mathbb{Z}[x]$.
6. Prove Theorem 11.2 below. (Hint: Show that the only integers that are irreducible in $\mathbb{Z}$ are the prime numbers and their negatives. Deduce that the only integers that are irreducible in $\mathbb{Z}[x]$ are the prime numbers and their negatives. Then treat polynomials of positive degree.)

Theorem 11.2. *The irreducible elements of* $\mathbb{Z}[x]$ *are*

1. *Prime numbers of* $\mathbb{Z}$ *and their negatives;*
2. *Polynomials of positive degree that are primitive and that do not factor as products of lower-degree polynomials.*

In view of Theorem 11.2, we must be cautious in working in $\mathbb{Z}[x]$. A positive-degree polynomial $f(x)$ in $\mathbb{Z}[x]$ that does not factor as product of lower-degree polynomials is not necessarily irreducible in $\mathbb{Z}[x]$, for its coefficients may have a greatest common divisor greater than 1. On the other hand, $f(x)$ will be close to being irreducible. Let d be the greatest common divisor of the coefficients of $f(x)$. Then we can factor d out of each of the coefficients of $f(x)$, writing $f(x)$ as $d \cdot g(x)$ for a primitive polynomial $g(x)$ that will be irreducible in $\mathbb{Z}[x]$. Thus, even though $f(x)$ is not itself irreducible, it will be the product of an integer and a primitive irreducible polynomial.

If we are going to be working with primitive polynomials, we would like to know that the product of primitive polynomials is again primitive. This result is known as Gauss's lemma:

Theorem 11.3. *The product of primitive polynomials is primitive: If* $g(x)$ *and* $h(x)$ *are two primitive polynomials in* $\mathbb{Z}[x]$, *then their product* $g(x)h(x)$ *is also primitive.*

Exercise 11.6. Prove Theorem 11.3 by following the outline below.

1. Choose a prime number p. Since $g(x)$ is primitive, there is at least one coefficient of $g(x)$ not divisible by p. Similarly for $h(x)$. Using this, we will show that there is a coefficient of $g(x)h(x)$ not divisible by p.
2. Write out $g(x)$ and $h(x)$ using the notation of Exercise 10.38. As a warmup, intended to motivate the general argument, suppose that the three coefficients of $g(x)$ of lowest degree (the coefficients of x^0, x^1, and x^2) are divisible by p, but the fourth is not, and suppose that the two coefficients of $h(x)$ of lowest degree (the coefficients of x^0 and x^1) are divisible by p, but the third is not. Write out the first few coefficients of $g(x)h(x)$. Show that they start out being divisible by p, but that you soon reach a coefficient that is not divisible by p.
3. Recall from Exercise 10.38 how to write down the coefficients of $g(x)h(x)$ in terms of the coefficients of $g(x)$ and $h(x)$. Using this expression and

an idea that should emerge from the previous part, show in general that if $g(x)$ has a coefficient not divisible by p, and $h(x)$ has a coefficient not divisible by p, then $g(x)h(x)$ has a coefficient not divisible by p.

A polynomial of positive degree in $\mathbb{Z}[x]$ factors as a product of prime numbers and a primitive polynomial. Suppose we are given two positive-degree polynomials in $\mathbb{Z}[x]$, in the form $p_1 \cdots p_m g(x)$ and $q_1 \cdots q_n h(x)$, where the p_i's and q_j's are prime numbers and $g(x)$ and $h(x)$ are primitive. Then Theorem 11.3 assures us that the primitive polynomial part of

$$(p_1 \cdots p_m q_1 \cdots q_n)g(x)h(x)$$

is $g(x)h(x)$. The product $g(x)h(x)$ contributes no new prime numbers to the product of prime numbers in front.

We wish to use Theorem 11.3 to reduce factorization questions in $\mathbb{Q}[x]$ to factorization questions in $\mathbb{Z}[x]$. Notice that for every polynomial $g(x)$ in $\mathbb{Q}[x]$ we can find a positive integer m such that $m \cdot g(x)$ has integer coefficients. For instance, for the polynomial

$$\frac{1}{2}x^3 - \frac{2}{7}x + 5,$$

we can multiply through by 14 to get

$$7x^3 - 4x + 70.$$

By exercising a little care, we can choose m such that $m \cdot g(x)$ is primitive. This is the case in the example above, but it would not have been had we carelessly cleared denominators by multiplying by 28.

Beginning with a polynomial $g(x)$ with rational coefficients, we can thus obtain a primitive polynomial $m \cdot g(x)$ in $\mathbb{Z}[x]$. If we can factor $m \cdot g(x)$ in $\mathbb{Q}[x]$, we can divide the factorization by m to obtain a factorization of $g(x)$ in $\mathbb{Q}[x]$. Thus in studying the factorization of polynomials with rational coefficients, we can always assume that we are working with polynomials with integer coefficients. Once we clear denominators to get a polynomial $f(x)$ with integer coefficients, we might try to factor $f(x)$ as a product of polynomials that themselves have integer coefficients. This leads to a question: If $f(x)$ does not factor as a product of lower-degree polynomials with integer coefficients, can $f(x)$ nonetheless factor as a product of lower-degree polynomials with rational coefficients? Gauss's lemma on primitive polynomials allows us to prove a result, also known as Gauss's lemma, that says that this cannot happen; namely, if a polynomial $f(x)$ with integer coefficients has a factorization in $\mathbb{Q}[x]$ of lower-degree polynomials, then it also has a factorization in $\mathbb{Z}[x]$ of lower-degree polynomials:

Theorem 11.4. *Let $f(x)$ be a polynomial of positive degree in $\mathbb{Z}[x]$. Suppose*

$$f(x) = g(x)h(x)$$

*for polynomials $g(x)$ and $h(x)$ in $\mathbb{Q}[x]$. Then there exists a rational number r
such that $r \cdot g(x)$ and $\frac{1}{r} \cdot h(x)$ have integer coefficients. Thus $f(x)$ factors in
$\mathbb{Z}[x]$ as the product*

$$\left(r \cdot g(x)\right)\left(\frac{1}{r} \cdot h(x)\right).$$

Proof. Let us prove the theorem under the special assumption that $f(x)$ is
primitive in $\mathbb{Z}[x]$. This will be sufficient for our applications, and it is a small
matter to give the general proof once one treats this special case.

Suppose, then, that we have a factorization of the primitive polynomial
$f(x)$ as $f(x) = g(x)h(x)$, where $g(x)$ and $h(x)$ have coefficients in $\mathbb{Q}$. We
can rewrite $g(x)$ as $s \cdot g_1(x)$ for some positive rational number s and some
primitive polynomial $g_1(x)$. Similarly, write $h(x)$ as $r \cdot h_1(x)$ for some positive
rational number r and some primitive polynomial $h_1(x)$. Thus $g(x)h(x) =
rs \cdot g_1(x)h_1(x)$. Since rs is a rational number, we can write it as $\frac{m}{n}$ for some
integers m and n. From this, we obtain the equation

$$f(x) = \frac{m}{n} \cdot g_1(x)h_1(x)$$

in $\mathbb{Q}[x]$ and the equation

$$nf(x) = m \cdot g_1(x)h_1(x)$$

in $\mathbb{Z}[x]$. By Gauss's lemma (Theorem 11.3) the product $g_1(x)h_1(x)$ is a primitive polynomial.

Using the fourth part of Exercise 11.5, we find that $m = n$ and $f(x) =
g_1(x)h_1(x)$. The equality $m = n$ implies that $\frac{m}{n} = 1$ and that $s = \frac{1}{r}$. Thus we
find that $r \cdot g(x)$ is a primitive polynomial in $\mathbb{Z}[x]$, that $\frac{1}{r} \cdot h(x)$ is a primitive
polynomial in $\mathbb{Z}[x]$, and that $f(x)$ factors in $\mathbb{Z}[x]$ as the product of $r \cdot g(x)$
and $\frac{1}{r} \cdot h(x)$. This proves the theorem.

We will use the contrapositive of Theorem 11.4. Let us state it explicitly.

Corollary 11.5 *Let $f(x)$ be a polynomial of positive degree in $\mathbb{Z}[x]$. If $f(x)$
has no factorization as a product of lower-degree polynomials in $\mathbb{Z}[x]$, then
$f(x)$ is irreducible in $\mathbb{Q}[x]$.*

11.3 Eisenstein's Criterion

Gauss's lemma, in the form of Corollary 11.5, allows us to prove the irreducibility of polynomials in $\mathbb{Q}[x]$ by treating the easier problem of nonfactorizability
in $\mathbb{Z}[x]$ in terms of lower-degree polynomials. Let us use this philosophy to
prove Theorem 11.1, which states that for a positive integer n and a prime
number p, the polynomial $x^n - p$ is irreducible in $\mathbb{Q}[x]$. We will prove this in
the next two exercises, dealing first with $p = 2$ and then with an arbitrary
prime p.

Exercise 11.7. Let n be an integer greater than 1. Prove that $x^n - 2$ is irreducible in $\mathbb{Q}[x]$ by proceeding as follows.

1. Suppose $x^n - 2 = g(x)h(x)$, where $g(x)$ and $h(x)$ are polynomials in $\mathbb{Z}[x]$ of degrees k and l, with $k < n$ and $l < n$. We wish to obtain a contradiction. Write out explicit expressions for $g(x)$ and $h(x)$.

2. Show that 2 divides the constant coefficient of $g(x)$ or the constant coefficient of $h(x)$ but not both. Make a choice; for instance, suppose that 2 divides the constant coefficient of $g(x)$ but not the constant coefficient of $h(x)$.

3. Now look at the degree-one coefficient of $g(x)h(x)$, written in terms of the coefficients of $g(x)$ and $h(x)$, and use this to prove that 2 divides the degree-one coefficient of $g(x)$.

4. Similarly, show that 2 divides the degree-two coefficient of $g(x)$ and the degree-three coefficient of $g(x)$.

5. The last two parts are just a warmup, so you can see what is going on. Now start over again and use the fact that 2 divides the constant coefficient of $g(x)$ along with induction to show that for every i from 0 to k we have that 2 divides the degree-i coefficient of $g(x)$. Conclude from this that in particular, 2 divides the degree-k coefficient of $g(x)$.

6. Show that this implies that 2 divides the degree-n coefficient of the product $g(x)h(x)$. Observe that this is a contradiction, and conclude that $g(x)$ and $h(x)$ as assumed cannot exist.

7. Use Corollary 11.5 to deduce that $x^n - 2$ is irreducible in $\mathbb{Q}[x]$.

Exercise 11.8. Let n be an integer greater than 1 and suppose that m is an odd integer.

1. Review the steps of the argument you made in Exercise 11.7 in proving that $x^n - 2$ does not factor in $\mathbb{Z}[x]$ as a product of lower-degree polynomials. Observe that they apply equally well to prove that $x^n - 2m$ does not factor in $\mathbb{Z}[x]$ as a product of lower-degree polynomials.

2. Conclude that for every positive integer n and every odd integer m, the polynomial $x^n - 2m$ is irreducible in $\mathbb{Q}[x]$. In particular, for every positive integer n there exist infinitely many irreducible monic polynomials of degree n in $\mathbb{Q}[x]$.

3. Contrast this with what is true about irreducibility of polynomials in $\mathbb{R}[x]$ and $\mathbb{C}[x]$.

Exercise 11.9. Let p be a prime number.

1. Review the steps of the argument you made in Exercise 11.7 in proving that $x^n - 2$ does not factor in $\mathbb{Z}[x]$ as a product of lower-degree polynomials. Observe that they apply equally well to prove that $x^n - p$ does not factor in $\mathbb{Z}[x]$ as a product of lower-degree polynomials. In other words, the only property of 2 that you used in your proof above is its primality, and 2 can be replaced in the argument by any prime number p.

2. Conclude that $x^n - p$ is irreducible in $\mathbb{Q}[x]$ for every positive integer n, so that Theorem 11.1 is proved.
3. Review the steps of the argument you made in Exercise 11.8 in proving for m odd that $x^n - 2m$ does not factor in $\mathbb{Z}[x]$ as a product of lower-degree polynomials. Observe that they apply equally well to prove that $x^n - pm$ does not factor in $\mathbb{Z}[x]$ as a product of lower-degree polynomials for m relatively prime to p.

The argument used in Exercises 11.7, 11.8, and 11.9 to prove the irreducibility of polynomials of the form $x^n - pm$ can be applied to certain other polynomials as well.

Exercise 11.10. Using the same kind of arguments you made in Exercises 11.7, 11.8, and 11.9, prove that $x^{14} - 27x^{11} + 15x^3 + 12$ does not factor in $\mathbb{Z}[x]$ as a product of lower-degree polynomials and therefore is irreducible in $\mathbb{Q}[x]$. Use 3 in the role played by 2 in Exercises 11.7 and 11.8 and by p in Exercise 11.9.

The most general result we can prove by the argument used in the last four exercises is the following theorem, known as Eisenstein's criterion for irreducibility. The hypotheses may seem odd, but they are just what is needed to extend our arguments.

Theorem 11.6. *Let $f(x)$ be a polynomial of degree $n > 1$ in $\mathbb{Z}[x]$. Suppose that $f(x) = a_n x^n + \cdots + a_1 x + a_0$ and that there is a prime number p such that the following hold:*

1. *the coefficient a_n is not divisible by p;*
2. *every coefficient a_i with $i < n$ is divisible by p; and*
3. *the constant coefficient a_0 is not divisible by p^2.*

Then $f(x)$ does not factor in $\mathbb{Z}[x]$ as a product of lower-degree polynomials. Hence $f(x)$ is irreducible in $\mathbb{Q}[x]$.

Exercise 11.11. Prove Eisenstein's criterion. To do so, proceed as follows.

1. Suppose $f(x) = g(x)h(x)$ where $g(x)$ and $h(x)$ are polynomials of degrees k and l, with $k < n$ and $l < n$. We wish to obtain a contradiction. Write out expressions for $g(x)$ and $h(x)$.
2. Show that p divides the constant coefficient of $g(x)$ or the constant coefficient of $h(x)$ but not both. Make a choice; for instance, suppose that p divides the constant coefficient of $g(x)$ but not the constant coefficient of $h(x)$.
3. Make an induction argument to show that p divides the degree-i coefficient of $g(x)$ for every i from 0 to k. You have already taken care of the case $i = 0$. Conclude from this that p divides the degree-k coefficient of $g(x)$.
4. Show that this implies that p divides the degree-n coefficient of the product $g(x)h(x)$. Observe that this is a contradiction and conclude that $g(x)$ and $h(x)$ as hypothesized cannot exist.

Exercise 11.12. Use Eisenstein's criterion to show that the following polynomials do not factor in $\mathbb{Z}[x]$ as products of lower-degree polynomials. Deduce that they are irreducible in $\mathbb{Q}[x]$:

1. $x^{22} + 7x^3 + 7$;
2. $x^{35} + 35x^{15} - 90$;
3. $1662x^{384} - 35x^{100} + 625x^{44} + 100x^{10} - 75x + 20$;
4. $6x^{31} + 35x^{21} + 245x^{11} + 175$.

11.4 Polynomials with Coefficients in $\mathbb{F}_p$

We have identified the irreducible polynomials in $\mathbb{R}[x]$ and $\mathbb{C}[x]$, and we have studied the irreducibility of polynomials in $\mathbb{Q}[x]$. After these investigations of polynomials with coefficients in one of the infinite fields $\mathbb{Q}$, $\mathbb{R}$, $\mathbb{C}$, a natural topic is the study of polynomials with coefficients in one of the finite fields $\mathbb{F}_p$. For a prime number p, one can in principle test the irreducibility of a polynomial in $\mathbb{F}_p[x]$, since there are only finitely many polynomials of lower degree. This is analogous to testing the primality of a positive integer. For a positive integer n, the only candidates for prime factors of n besides n itself are the prime numbers in the range 2 to $n - 1$. Since there are only finitely many such prime numbers, one can simply test them all. If none divides n, then n is prime. Similarly, if $f(x)$ is a polynomial of degree n in $\mathbb{F}_p[x]$, every irreducible polynomial dividing $f(x)$, besides $f(x)$ itself or a constant multiple of $f(x)$, must have degree between 1 and $n - 1$. Because $\mathbb{F}_p$ is finite, the collection of polynomials in $\mathbb{F}_p[x]$ of degree less than n is also finite. Thus we can test them all. If none divides $f(x)$, then $f(x)$ is irreducible. Alternatively, if $f(x)$ is not irreducible, we can find a factorization of $f(x)$ as a product of irreducible polynomials, using the finite list of possible factors. Thus in principle, factorization questions for a particular polynomial in $\mathbb{F}_p[x]$ can always be answered.

Exercise 11.13. Let us determine all irreducible polynomials of low degree in $\mathbb{F}_2[x]$ and $\mathbb{F}_3[x]$.

1. Write down all degree-two polynomials in $\mathbb{F}_2[x]$. Decide which ones are irreducible and which ones have roots in $\mathbb{F}_2$. For each degree-two polynomial $f(x)$ that does have roots, describe the roots and the corresponding factorization of $f(x)$ in $\mathbb{F}_2[x]$ as a product of two degree-one polynomials.
2. Write down all degree-three polynomials in $\mathbb{F}_2[x]$. Decide which ones are irreducible and which ones have roots in $\mathbb{F}_2$. For each degree-three polynomial $f(x)$ that does have roots, describe the roots and the corresponding factorization of $f(x)$ in $\mathbb{F}_2[x]$, either as a product of three degree-one polynomials or as a product of a degree-one and an irreducible degree-two polynomial.

3. Write down all degree-two polynomials in $\mathbb{F}_3[x]$. Decide which ones are irreducible and which ones have roots in $\mathbb{F}_3$. For each degree-two polynomial $f(x)$ that does have roots, describe the roots and the corresponding factorization of $f(x)$ in $\mathbb{F}_3[x]$ as a product of two degree-one polynomials.

The irreducible polynomials in $\mathbb{C}[x]$ all have degree 1, and the irreducible polynomials in $\mathbb{R}[x]$ all have degree 1 or 2, while in contrast, there are irreducible polynomials in $\mathbb{Q}[x]$ of every positive degree. The rings $\mathbb{F}_p[x]$ are closer in this regard to $\mathbb{Q}[x]$ than they are to $\mathbb{C}[x]$ or $\mathbb{R}[x]$:

Theorem 11.7. *For every prime number p, the polynomial ring $\mathbb{F}_p[x]$ has irreducible polynomials of arbitrarily high degree; that is, there is no positive integer n such that all the irreducible polynomials of $\mathbb{F}_p[x]$ have degree less than or equal to n.*

One can prove a stronger theorem than Theorem 11.7, showing that there are irreducible polynomials in $\mathbb{F}_p[x]$ of every degree. See Theorem 17.7 and the discussion in Section 17.3. We content ourselves here with the weaker Theorem 11.7, which is much easier to prove.

Exercise 11.14. Prove Theorem 11.7. (Hint: For a positive integer n, how many polynomials are there in $\mathbb{F}_p[x]$ of degree at most n? Use this along with Theorem 9.4.)

The irreducible polynomials in $\mathbb{F}_p[x]$ cannot be easily described. However, as we have already observed, it is always possible in principle to determine the irreducibility of a particular polynomial in a particular $\mathbb{F}_p[x]$.

In this book we are interested in the rings $\mathbb{F}_p[x]$ primarily as a tool in our study of polynomials in $\mathbb{Z}[x]$ and $\mathbb{Q}[x]$. Using arguments involving polynomials in $\mathbb{F}_p[x]$, we can obtain simple proofs of Gauss's lemma and of Eisenstein's criterion for irreducibility. We will see how to do so after studying the relationship between polynomials in $\mathbb{Z}[x]$ and polynomials in $\mathbb{F}_p[x]$.

Fix a prime number p. For an integer a in $\mathbb{Z}$, we have already discussed the process of passing to an element $[a]$, or $[a]_p$, in $\mathbb{F}_p$. Sometimes we have used the bracket notation, and sometimes we have written elements of $\mathbb{F}_p$ using the numbers from 0 to $p - 1$, depending on whether there is danger of confusion and whether we wish to emphasize that the element $[a]_p$ in $\mathbb{F}_p$ represents the congruence class of integers congruent to a modulo p. The process of replacing an integer a by the element of $\mathbb{F}_p$ that it represents is called *reduction modulo* p. For example, the reduction of 25 modulo 7 is $[4]_7$, or $[4]$, or simply 4 if it is understood that by 4 we mean not the integer 4 but the element 4 of $\mathbb{F}_7$. Similarly, the reduction of -20 modulo 11 is $[2]$, or just 2.

We can reduce not just integers modulo p but also polynomials with integer coefficients. The *reduction modulo* p of a polynomial $f(x)$ in $\mathbb{Z}[x]$ is the polynomial $[f](x)$ in $\mathbb{F}_p[x]$ obtained by replacing each integer coefficient of $f(x)$ by the corresponding element of $\mathbb{F}_p$. We could also write $[f]_p(x)$ for the reduction, but this makes the notation a bit cumbersome, so we shall drop the subscript and assume that p is understood. Explicitly, if

$$f(x) = a_n x^n + a_{n-1} x^{n-1} + \cdots + a_1 x + a_0$$

is a polynomial in $\mathbb{Z}[x]$, then

$$[f](x) = [a_n] x^n + [a_{n-1}] x^{n-1} + \cdots + [a_1] x + [a_0]$$

is the associated reduced polynomial (modulo p) in $\mathbb{F}_p[x]$. For example, suppose $p = 7$ and

$$f(x) = 9x^5 - 17x^3 + 3x^2 - 14x + 20.$$

Then

$$[f](x) = [9] x^5 - [17] x^3 + [3] x^2 - [14] x + [20] = [2] x^5 + [4] x^3 + [3] x^2 + [6].$$

If we are writing the elements of $\mathbb{F}_7$ as ordinary symbols $0, 1, \ldots, 6$, we would rewrite this as

$$[f](x) = 2x^5 + 4x^3 + 3x^2 + 6.$$

The coefficients here are understood to be elements of $\mathbb{F}_7$, not ordinary integers.

The process of reducing a polynomial in $\mathbb{Z}[x]$ modulo p shares some properties with the process of conjugating a polynomial in $\mathbb{C}[x]$. Recall that a polynomial in $\mathbb{C}[x]$ is conjugated by conjugating each coefficient (that is, replacing each coefficient by its complex conjugate) and leaving the powers of x alone. The fundamental result about conjugation of polynomials is the compatibility of conjugation with multiplication, in the sense that if $f(x)$ in $\mathbb{C}[x]$ factors as

$$f(x) = g(x)h(x)$$

for polynomials $g(x)$ and $h(x)$ in $\mathbb{C}[x]$, then

$$\overline{f}(x) = \overline{g}(x)\overline{h}(x).$$

A similar result holds for reduction modulo p and is proved in the same way.

Theorem 11.8. *Suppose a polynomial $f(x)$ in $\mathbb{Z}[x]$ factors as*

$$f(x) = g(x)h(x)$$

for polynomials $g(x)$ and $h(x)$ in $\mathbb{Z}[x]$. Then the reductions of these polynomials modulo a prime number p satisfy

$$[f](x) = [g](x)[h](x)$$

in $\mathbb{F}_p[x]$.

Exercise 11.15. Assume that p is a prime number and use brackets to denote reduction modulo p. Verify the following facts:

1. For integers a and b,

$$[a+b] = [a] + [b]; \qquad [ab] = [a][b].$$

(This is true because of the way in which addition and multiplication are defined in $\mathbb{F}_p$.)

2. For integers $a_0, \ldots, a_k$,

$$[a_0 + \cdots + a_k] = [a_0] + \cdots + [a_k]; \qquad [a_0 \cdots a_k] = [a_0] \cdots [a_k].$$

3. For integers $a_0, \ldots, a_k$ and $b_0, \ldots, b_k$,

$$[a_0 b_k + a_1 b_{k-1} + \cdots + a_{k-1} b_1 + a_k b_0] = [a_0][b_k] + \cdots + [a_k][b_0].$$

Using this last part, prove Theorem 11.8 by comparing the coefficients of $[f](x)$ and $[g](x)[h](x)$.

Theorem 11.8 has the following important application:

Theorem 11.9. *Suppose $f(x)$ is a polynomial of positive degree in $\mathbb{Z}[x]$ and p is a prime number that does not divide the highest-degree coefficient of $f(x)$. If the reduction $[f](x)$ of $f(x)$ modulo p is irreducible in $\mathbb{F}_p[x]$, then $f(x)$ does not factor in $\mathbb{Z}[x]$ as a product of lower-degree polynomials.*

Exercise 11.16. Prove Theorem 11.9, using Theorem 11.8.

Theorem 11.9 provides us with a new technique for proving that certain polynomials $f(x)$ in $\mathbb{Z}[x]$ do not factor as a product of lower-degree polynomials: Find a prime number p that does not divide the highest-degree coefficient of $f(x)$ (there are infinitely many from which to choose!) for which the reduction $[f](x)$ is irreducible in $\mathbb{F}_p[x]$. The attraction of this technique resides in the fact that we can always determine, by the brute-force method of testing every possibility, whether a particular polynomial in $\mathbb{F}_p[x]$ is irreducible. In particular, the irreducibility $[f](x)$ in $\mathbb{F}_p[x]$ may be easier to prove than the nonfactorizability of $f(x)$ in $\mathbb{Z}[x]$ as a product of lower-degree polynomials.

Let us consider two examples.

Exercise 11.17. Use reduction modulo p to prove that $x^5 + x^2 + 1$ does not factor in $\mathbb{Z}[x]$ as a product of lower-degree polynomials. (Hint: Explain why it suffices to show simply that there exists a prime p such that the reduction of $x^5 + x^2 + 1$ modulo that prime p is irreducible in $\mathbb{F}_p[x]$. Try $p = 2$, and write the reduced polynomial also as $x^5 + x^2 + 1$. Show that $x^5 + x^2 + 1$ has no root in $\mathbb{F}_2$ and conclude that $x^5 + x^2 + 1$ has no degree-one factor in $\mathbb{F}_2[x]$. Show that $x^2 + x + 1$ does not divide $x^5 + x^2 + 1$ and conclude that $x^5 + x^2 + 1$ has no degree-two factor in $\mathbb{F}_2[x]$. Conclude that $x^5 + x^2 + 1$ is irreducible in $\mathbb{F}_2[x]$.)

Exercise 11.18. Use reduction modulo p to prove that $x^5 + x^4 + 2x^3 + 2x + 2$ does not factor in $\mathbb{Z}[x]$ as a product of lower-degree polynomials. (Hint: Notice that reduction modulo the prime $p = 2$ will not work, since the polynomial *is* reducible modulo 2. Try reduction modulo $p = 3$ instead. Show first that there are no degree-one factors. If there is a degree-two factor, you can assume that it has the form $x^2 + bx + c$. Show that only three of the nine polynomials in $\mathbb{F}_3[x]$ of this form are irreducible, and that none of these is a divisor of $x^5 + x^4 + 2x^3 + 2x + 2$.)

It is not particularly easy to demonstrate directly the nonfactorizability of $x^5 + x^2 + 1$ and $x^5 + x^4 + 2x^3 + 2x + 2$ in $\mathbb{Z}[x]$ as products of polynomials of lower degree. Theorem 11.9 reduces the verification to much easier calculations.

We now have two methods of showing that a polynomial in $\mathbb{Z}[x]$ does not factor in $\mathbb{Z}[x]$ as a product of lower-degree polynomials: reduction modulo a prime p and Eisenstein's criterion. These two methods are great when they work, but there are many irreducible polynomials in $\mathbb{Z}[x]$ and $\mathbb{Q}[x]$ to which they do not apply. One example is $x^4 + 1$: Eisenstein's criterion does not apply, since no prime number divides 1. One can also show that Theorem 11.9 does not apply either, for regardless of the prime number p we choose, the reduction of $x^4 + 1$ modulo p factors nontrivially in $\mathbb{F}_p[x]$. Yet $x^4 + 1$ is irreducible in $\mathbb{Z}[x]$, as we can show by arguments of the sort used in Exercise 11.3.

Let us use reduction modulo p to prove once again the two major theorems of Sections 11.2 and 11.3.

Exercise 11.19. Prove Gauss's lemma, Theorem 11.3, again. (Hint: First show that a polynomial $f(x)$ in $\mathbb{Z}[x]$ is primitive if and only if for every prime number p the reduction $[f](x)$ in $\mathbb{F}_p[x]$ is nonzero. Then consider $g(x)h(x)$ and its reductions modulo primes p.)

Exercise 11.20. Prove Eisenstein's criterion again. (Hint: Suppose the polynomial $f(x)$ in the statement of Eisenstein's criterion factors as the product $g(x)h(x)$ of lower-degree polynomials. Consider reductions of these polynomials modulo p. Show that $[f](x)$ has an especially simple form. Show that $[g](x)$ and $[h](x)$ cannot both have constant term 0. Obtain a contradiction.)

12

Polynomial Rings

12.1 Unique Factorization for Integers Revisited

We will review the fundamental theorem of arithmetic with an eye to proving an analogous theorem for polynomials. The fundamental theorem of arithmetic states that the prime factorization of an integer greater than 1 is unique. Recall that formulating a precise statement of uniqueness requires some care. The basic difficulty is that we do not want to distinguish between two factorizations that differ only in the order of the factors. For instance, 6 factors as both 2×3 and 3×2, and we do not want to regard these as different. The following version of the fundamental theorem is one way to handle this problem.

Theorem 12.1. *Let a be an integer greater than 1. Suppose that $p_1 p_2 \cdots p_m$ and $q_1 q_2 \cdots q_n$ are two prime factorizations of a. Then $m = n$, and the order of the factors in the second factorization can be changed so that $p_i = q_i$ for each index i.*

In studying factorization questions for integers, we have generally ignored negative integers. Let us not do so any longer. An integer is *irreducible* if it is not zero or a unit and if it has no nontrivial factorizations. Since the units in $\mathbb{Z}$ are 1 and -1, the irreducible integers are the integers of absolute value greater than 1 that have no nontrivial factorizations. It is easy to see that these are the prime numbers and the negatives of prime numbers. In fact, a stronger statement was proved in Exercise 11.5.

By allowing negative integers, we can obtain different factorizations of an integer as a product of irreducible integers not only by changing the order of the factors, but also by changing the signs of the factors. For instance, in addition to the factorizations 2×3 and 3×2 of 6, there are the factorizations $(-2) \times (-3)$ and $(-3) \times (-2)$. If we are to allow all integers in factorizations, we must reformulate the statement of the fundamental theorem to take this into account. The following is one such reformulation.

Theorem 12.2. *Let a be an integer of absolute value greater than 1. Suppose $p_1 p_2 \cdots p_m$ and $q_1 q_2 \cdots q_n$ are two factorizations of a as a product of irreducible integers. Then $m = n$, and the order of the factors in the second factorization can be changed so that for each index i either $q_i = p_i$ or $q_i = -p_i$.*

We have seen that every polynomial $f(x)$ of positive degree in $K[x]$, where K is a field, is irreducible or factors as the product of irreducible polynomials. Such a statement is the polynomial analogue of the result for $\mathbb{Z}$ that every integer a greater than 1 is prime or a product of prime numbers. We would like to obtain a polynomial analogue of the fundamental theorem.

In the formulation of the fundamental theorem just stated as Theorem 12.2, we are permitted to change factors by multiplying by -1. The significance of -1 is that it is the only unit in $\mathbb{Z}$ besides 1. The units in $K[x]$ are the nonzero constants. Just as we do not wish to distinguish between two factorizations of an integer as a product of irreducible numbers when the factorizations differ by having the signs of some factors changed, so too, we do not want to distinguish between two different factorizations of a polynomial as a product of irreducible polynomials when the factorizations differ by having some factors altered by constant multiples. For example, we do not want to distinguish between the factorizations of $x^2 - 1$ in $\mathbb{R}[x]$ as $(x-1)(x+1)$ and as $(5x - 5)\left(\frac{x}{5} + \frac{1}{5}\right)$. This suggests the following unique factorization theorem for $K[x]$.

Theorem 12.3. *Let K be a field and let $a(x)$ be a polynomial in $K[x]$ of positive degree. Suppose $p_1(x) \cdots p_m(x)$ and $q_1(x) \cdots q_n(x)$ are two factorizations of $a(x)$ as a product of irreducible polynomials in $K[x]$. Then $m = n$, and the order of the factors in the second factorization can be changed so that for each index i there is a nonzero constant c_i such that $q_i(x) = c_i p_i(x)$.*

The statements of Theorems 12.2 and 12.3 can be made to look more alike if we change the concluding phrases. In the statement of Theorem 12.2, the wording of the concluding phrase can be changed to "...for each index i there is a unit u_i of $\mathbb{Z}$ such that $q_i = u_i p_i$." Similarly, in the statement of Theorem 12.3, the wording of the concluding phrase can be changed to "...for each index i there is a unit u_i of $K[x]$ such that $q_i(x) = u_i p_i(x)$."

Let us review the route we took in proving the fundamental theorem of arithmetic. This will provide us with a map we can try to follow in proving Theorem 12.3. The route begins with the division theorem, our first major result about $\mathbb{Z}$.

Theorem 12.4. *For positive integers a and b, there exist unique nonnegative integers q and r, with $r < a$, such that*

$$b = aq + r.$$

Iterating the division theorem leads to the Euclidean algorithm, a procedure for finding the greatest common divisor of two integers. The Euclidean algorithm also has Bézout's theorem as a consequence. Recall its statement:

Theorem 12.5. *Let d be the greatest common divisor of two integers a and b. Then there exist integers r and s such that*

$$d = ar + bs.$$

The Euclidean algorithm and Bézout's theorem are proved by an induction argument based on the number of steps required for the algorithm to terminate. The following somewhat special result is a consequence of Bézout's theorem. Let us recall its proof.

Theorem 12.6. *Suppose that a and b are relatively prime integers, and suppose that c is an integer such that a divides the product bc. Then a divides c.*

Proof. Since a and b are relatively prime, Bézout's theorem yields the existence of integers r and s satisfying $ar + bs = 1$. Multiplying through by c yields $acr + bcs = c$. Certainly, a divides a, and by assumption a divides bc. Therefore, a divides acr and bcs, and thus their sum $acr + bcs$. But this sum is c, so a divides c, as desired.

We are interested less in Theorem 12.6 than in the following consequence.

Theorem 12.7. *Suppose the prime number p divides the product bc of integers b and c. Then p divides b, or p divides c.*

Proof. If p divides b, we are done. If not, since p is prime and fails to divide b, the integers p and b are relatively prime. Theorem 12.6 applies to this situation, implying that p divides c.

Using Theorem 12.7 and induction, we proved the following more general result.

Corollary 12.8 *If a prime number p divides a product $a_1 a_2 \cdots a_n$ of integers, then p divides one of the factors a_i.*

Corollary 12.8 allowed us to prove the following:

Corollary 12.9 *Suppose $p_1 p_2 \cdots p_m$ and $q_1 q_2 \cdots q_n$ are two factorizations of an integer $a > 1$ as a product of prime numbers. Then there is an index j such that $q_j = p_m$, and $p_1 p_2 \cdots p_{m-1} = q_1 q_2 \cdots q_{j-1} q_{j+1} \cdots q_n$. In particular, we can reorder the second product so that $q_n = p_m$ and $p_1 p_2 \cdots p_{m-1} = q_1 q_2 \cdots q_{n-1}$.*

Corollary 12.9 is the tool needed to give an inductive proof of unique factorization in $\mathbb{Z}$. In the next sections we will prove analogous results for the ring $K[x]$.

12.2 The Euclidean Algorithm

The first step in the proof of the fundamental theorem of arithmetic is the division theorem, which states that if b and a are integers, with $a \neq 0$, then b can be written uniquely as $b = aq + r$, for integers q and r, with $0 \leq r < a$. Similarly, in studying factorization in polynomial rings $K[x]$, we start with the division theorem for polynomials. Recall that this was proved in Section 9.4 as Theorem 9.5. It states that for a field K, a nonzero polynomial $b(x)$ in $K[x]$, and another polynomial $a(x)$ of positive degree in $K[x]$, there exist unique polynomials $q(x)$ and $r(x)$, with $r(x)$ having lower degree than $a(x)$, such that

$$b(x) = a(x)q(x) + r(x).$$

For a field K and polynomials $a(x)$ and $b(x)$ in $K[x]$, a polynomial $g(x)$ is a *common divisor* of $a(x)$ and $b(x)$ if $g(x)$ divides both of them. What should we mean by the *greatest* common divisor of $a(x)$ and $b(x)$? The greatest common divisor of integers a and b is a common divisor that is the largest of all common divisors. For polynomials the role of size is played by degree. It is natural then to define a *greatest common divisor* of polynomials $a(x)$ and $b(x)$ to be a polynomial $d(x)$ that is a common divisor of $a(x)$ and $b(x)$ and that additionally has the largest possible degree; that is, its degree is the maximum of the degrees of all the common divisors of $a(x)$ and $b(x)$. Notice that there must be a maximum degree among the common divisors, since a common divisor of $a(x)$ and $b(x)$ has degree no greater than the smaller of the degrees of $a(x)$ and $b(x)$.

This definition of greatest common divisor provides for the possibility that there may be many greatest common divisors of two polynomials $a(x)$ and $b(x)$. For example, for one greatest common divisor $d(x)$, all the nonzero constant multiples $c \cdot d(x)$ of $d(x)$ are also greatest common divisors. We shall soon prove that these are the only greatest common divisors: Any two greatest common divisors of a pair of polynomials $a(x)$ and $b(x)$ are nonzero constant multiples of each other. As an example, consider the two polynomials $x^2 - 1$ and $x^3 - 1$ in $\mathbb{R}[x]$. Certainly, $x - 1$ is a common divisor. Any common divisor has degree at most 2, but you can check (do so!) that there is no common divisor of degree 2. Therefore, $x - 1$ is a greatest common divisor. For a nonzero real number c, the polynomial $cx - c$ is also a greatest common divisor.

Exercise 12.1. For a field K, prove the following analogues for polynomials in $K[x]$ of results you proved earlier for integers.

1. Suppose $a(x)$ divides $b(x)$. Then $a(x)$ is a greatest common divisor of $a(x)$ and $b(x)$.
2. For polynomials $a(x)$, $b(x)$, and $q(x)$, a polynomial $g(x)$ is a common divisor of $a(x)$ and $b(x)$ if and only if $g(x)$ is a common divisor of $a(x)$ and $b(x) - a(x)q(x)$. A polynomial $g(x)$ is a greatest common divisor of $a(x)$ and $b(x)$ if and only if $g(x)$ is a greatest common divisor of $a(x)$ and $b(x) - a(x)q(x)$.

3. Suppose that $b(x)$ can be written as $a(x)q(x) + r(x)$ for some polynomials $q(x)$ and $r(x)$. Then a polynomial $g(x)$ is a greatest common divisor of $b(x)$ and $a(x)$ if and only if $g(x)$ is a greatest common divisor of $a(x)$ and $r(x)$.

4. Suppose we wish to calculate a greatest common divisor of two polynomials $a(x)$ and $b(x)$, and suppose further that the degree of $b(x)$ is greater than or equal to the degree of $a(x)$. If $a(x)$ divides $b(x)$, then we have seen that $a(x)$ is a greatest common divisor of $a(x)$ and $b(x)$. If not, use the preceding part to show that we can replace the initial problem of finding a greatest common divisor of $a(x)$ and $b(x)$ with the problem of finding a greatest common divisor for a new pair $r(x)$ and $a(x)$, where $r(x)$ has smaller degree than $a(x)$.

5. Use this idea to calculate a greatest common divisor in $\mathbb{Q}[x]$ of the polynomials $x^4 - 1$ and $x^3 - 2x^2 + x - 2$. (Hint: Recall from an earlier exercise that you can write $x^4 - 1$ as $(x^3 - 2x^2 + x - 2)(x + 2) + (3x^2 + 3)$. Use this to replace the original pair of polynomials with a new pair; then show that one polynomial in the pair divides the other, and use this to write down a greatest common divisor.)

Suppose that $a(x)$ and $b(x)$ are polynomials for which we wish to find a greatest common divisor, and assume that $b(x)$ has degree at least as large as the degree of $a(x)$. We can use the division theorem to write

$$b(x) = a(x)q_1(x) + r_1(x),$$

where the degree of the remainder $r_1(x)$ is smaller than the degree of $a(x)$. Exercise 12.1 tells us that instead of computing a greatest common divisor of $b(x)$ and $a(x)$, we can compute a greatest common divisor of $a(x)$ and $r_1(x)$. If $r_1(x)$ divides $a(x)$, we are done: $r_1(x)$ is a greatest common divisor. If not, we can use the division theorem again to obtain

$$a(x) = r_1(x)q_2(x) + r_2(x)$$

for some polynomial $r_2(x)$ of degree less than the degree of $r_1(x)$. The pair $a(x)$ and $r_1(x)$ has the same greatest common divisors as the pair $r_1(x)$ and $r_2(x)$, so we can pass to this new pair. If $r_2(x)$ divides $r_1(x)$, we are done. If not, we continue, obtaining

$$r_1(x) = r_2(x)q_3(x) + r_3(x)$$

for some polynomial $r_3(x)$ of degree less than the degree of $r_2(x)$. Now we replace the pair $r_1(x)$ and $r_2(x)$ with the pair $r_2(x)$ and $r_3(x)$. If $r_3(x)$ divides $r_2(x)$, we are done. Otherwise, we keep going.

This should look familiar. If we drop the x's and pretend that the polynomials are integers, it is the Euclidean algorithm for $\mathbb{Z}$. Since at each step along the way the degree of $r_{i+1}(x)$ is lower than the degree of $r_i(x)$, the process cannot go on forever. It must terminate, which it does when we reach a remainder $r_{i+1}(x)$ that divides the previous remainder $r_i(x)$. This final remainder $r_{i+1}(x)$ is what we are after. It is a greatest common divisor of $a(x)$ and $b(x)$.

The procedure we have described, using the division theorem repeatedly until we obtain a remainder that divides the preceding one, is the *Euclidean algorithm* for pairs of polynomials in $K[x]$. To show rigorously that the final nonzero remainder is a greatest common divisor, we must make an inductive argument, as we did in the case of the integers.

Exercise 12.2. For polynomials $a(x)$ and $b(x)$, prove that the last nonzero remainder obtained by the Euclidean algorithm applied to $a(x)$ and $b(x)$ is a greatest common divisor of $a(x)$ and $b(x)$. (Hint: Do so by induction on the number of steps required until the Euclidean algorithm terminates.)

Exercise 12.3. Use the Euclidean algorithm to find greatest common divisors of the following pairs of polynomials:

1. $x^2 + 1$ and $x^5 + 1$ in $\mathbb{Q}[x]$.
2. $x^2 + 2x + 1$ and $x^3 + 2x^2 + 2$ in $\mathbb{F}_3[x]$.

12.3 Bézout's Theorem

The Euclidean algorithm for polynomials yields a polynomial version of Bézout's theorem, just as the original Euclidean algorithm yields Bézout's theorem for integers.

Theorem 12.10. *For a field K, let $a(x)$ and $b(x)$ be polynomials in $K[x]$ and let $d(x)$ be the greatest common divisor of $a(x)$ and $b(x)$ produced by the Euclidean algorithm. There exist polynomials $r(x)$ and $s(x)$ in $K[x]$ such that*

$$d(x) = a(x)r(x) + b(x)s(x).$$

Exercise 12.4. Prove Bézout's theorem. This is most easily done by induction on the number of steps in the Euclidean algorithm for $a(x)$ and $b(x)$. If the algorithm terminates right away, which means that $a(x)$ divides $b(x)$, the result is easily proved. Next assume that you know the result for pairs of polynomials for which the algorithm terminates after k steps and deduce the result for pairs of polynomials for which the algorithm terminates after $k+1$ steps.

The proof of Bézout's theorem shows not only that $r(x)$ and $s(x)$ exist, but also that we can compute them by performing the Euclidean algorithm. Let us try some calculations of this sort.

Exercise 12.5. For the pair of polynomials $a(x)$ and $b(x)$ below, use the Euclidean algorithm to find polynomials $r(x)$ and $s(x)$ such that $a(x)r(x) + b(x)s(x)$ equals a greatest common divisor of $a(x)$ and $b(x)$:

1. $a(x) = x^2 + 1$ and $b(x) = x^5 + 1$ in $\mathbb{Q}[x]$.
2. $a(x) = x^2 + 2x + 1$ and $b(x) = x^3 + 2x^2 + 2$ in $\mathbb{F}_3[x]$.

The polynomial version of Bézout's theorem allows us to characterize all the greatest common divisors of a pair of polynomials.

Corollary 12.11 *For a field K, let $a(x)$ and $b(x)$ be polynomials in $K[x]$ and let $d(x)$ be the greatest common divisor of $a(x)$ and $b(x)$ produced by the Euclidean algorithm. Every common divisor of $a(x)$ and $b(x)$ divides $d(x)$. In particular, every greatest common divisor of $a(x)$ and $b(x)$ is a nonzero constant multiple of $d(x)$.*

Exercise 12.6. Prove Corollary 12.11. (Hint: For the first statement, use Bézout's theorem and basic divisibility facts. For the second statement, use the fact that if one polynomial divides another of equal degree, then they are constant multiples of each other.)

Since the greatest common divisors of two polynomials $a(x)$ and $b(x)$ are all constant multiples of each other, there is a natural way to single one out to be called *the* greatest common divisor of $a(x)$ and $b(x)$. To do so, select the unique monic greatest common divisor. For instance, suppose the Euclidean algorithm produces $\pi x^2 - \sqrt{37}$ as a greatest common divisor of a particular pair of polynomials $a(x)$ and $b(x)$ in $\mathbb{R}[x]$. Then we can replace $\pi x^2 - \sqrt{37}$ with $x^2 - \frac{\sqrt{37}}{\pi}$ and call this *the* greatest common divisor of $a(x)$ and $b(x)$. If the greatest common divisor of two polynomials $a(x)$ and $b(x)$ is 1, we say that $a(x)$ and $b(x)$ are *relatively prime*. Equivalently, $a(x)$ and $b(x)$ are relatively prime if they have no common divisor other than constants.

The polynomial version of Bézout's theorem yields the following polynomial analogue of Theorem 12.6.

Theorem 12.12. *Let K be a field. Suppose that $a(x)$ and $b(x)$ are relatively prime polynomials in $K[x]$, and suppose that $c(x)$ is a polynomial in $K[x]$ such that $a(x)$ divides the product $b(x)c(x)$ in $K[x]$. Then $a(x)$ divides $c(x)$ in $K[x]$.*

Exercise 12.7. Prove Theorem 12.12, mimicking the proof of Theorem 12.6.

12.4 Unique Factorization for Polynomials

Using the general theorems we have obtained for polynomial rings, we can prove Theorem 12.3, the uniqueness theorem for factorizations of polynomials as products of irreducible polynomials. We will imitate the proof of the fundamental theorem of arithmetic, as reviewed in Section 12.1. Our first step is to use Theorem 12.12 to prove Theorem 12.13.

Theorem 12.13. *Let K be a field. Let $p(x)$ be an irreducible polynomial in $K[x]$, and suppose $p(x)$ divides the product $b(x)c(x)$ of polynomials $b(x)$ and $c(x)$ in $K[x]$. Then $p(x)$ divides $b(x)$, or $p(x)$ divides $c(x)$.*

Exercise 12.8. Prove Theorem 12.13 by mimicking the proof of Theorem 12.7. Then prove Corollary 12.14 below by induction.

Corollary 12.14 *Let K be a field. Let $p(x)$ be an irreducible polynomial in $K[x]$ that divides a product $a_1(x)a_2(x)\cdots a_n(x)$ of polynomials in $K[x]$. Then $p(x)$ divides one of the factors $a_i(x)$.*

These last two results are the polynomial analogues of Theorem 12.7 and Corollary 12.8. Corollary 12.14 can be used to prove Corollary 12.15, the polynomial analogue of Corollary 12.9.

Corollary 12.15 *Let K be a field. Suppose*

$$p_1(x)p_2(x)\cdots p_m(x) \quad and \quad q_1(x)q_2(x)\cdots q_n(x)$$

are two factorizations of a polynomial $a(x)$ of positive degree as a product of irreducible polynomials. Then there are an index j and a nonzero constant c of K such that $q_j(x) = cp_m(x)$ and

$$p_1(x)\cdots p_{m-1}(x) = cq_1(x)\cdots q_{j-1}(x)q_{j+1}(x)\cdots q_n(x).$$

In particular, we can reorder the second product so that $q_n(x) = cp_m(x)$ and

$$p_1(x)\cdots p_{m-1}(x) = cq_1(x)\cdots q_{n-1}(x).$$

Exercise 12.9. Use Corollary 12.14 to prove Corollary 12.15.

We are in position to prove Theorem 12.3, the unique factorization theorem for polynomials. In the situation of Corollary 12.15, we can keep going, reordering the factors in the last equality if necessary to deduce that $q_{n-1}(x)$ is a constant multiple of $p_{m-1}(x)$, and so on. Eventually, we should find that $m = n$ and that for a suitable ordering, each $q_i(x)$ is a constant multiple of $p_i(x)$. In order to make this rigorous, we can make an induction argument on the number of factors m in one of the factorizations of $a(x)$.

Exercise 12.10. Prove Theorem 12.3. To do so, use induction on the number m of factors in one of the irreducible factorizations of $a(x)$.

1. Deal with the case $m = 1$. Observe that this means that $a(x)$ is irreducible.
2. Perform the inductive step. For each integer $k \geq 1$, show that if the result we want to prove is true for polynomials that can be factored as a product of k irreducible polynomials, then it is true for polynomials that can be factored as a product of $k + 1$ irreducible polynomials.

13

Quadratic Polynomials

13.1 Square Roots

In our study of roots of polynomials, we have focused on polynomials with coefficients in specific fields, such as $\mathbb{R}$ and $\mathbb{Q}$. We wish to study roots of polynomials with coefficients in an arbitrary field. As a first step, we will study quadratic polynomials. In order to do so, we need some elementary information about square roots of elements in arbitrary fields.

From our experience with the real numbers, we are accustomed to the idea that if a nonzero number c has a square root b, then it has two square roots, the other square root being the additive inverse $-b$ of b. This is not true for square roots in arbitrary fields, but it almost is. There are some subtleties that need to be addressed. First we must obtain some elementary results on additive inverses in fields, and while we are at it, we may as well consider the more general case of additive inverses in rings.

Suppose R is a ring. As usual, let us write 0 for its additive identity and 1 for its multiplicative identity. Our experience with familiar rings suggests that the first two parts of the result below are true. However, to *know* that they are true, and that they are true in general, for any ring, no matter how weird, we must supply proofs.

Proposition 13.1 *Let R be a ring and let r be an arbitrary element of R.*

1. *Suppose s and t are additive inverses of r; that is, $r + s = r + t = 0$. Then $s = t$. Thus each element of R has a unique additive inverse.*
2. *Suppose c is an element of R such that $r + c = c$. Then $r = 0$. Therefore, R possesses a unique additive identity.*
3. $r \cdot 0 = 0$.
4. *Suppose a is the unique additive inverse of 1 in R, so that $a + 1 = 0$. Then ar is an additive inverse of r.*

Proof. Our proof of the first statement is an exercise in the laws of commutativity and associativity that hold in every ring:

$$s = s + 0 \qquad \qquad \text{(additive identity)}$$
$$= s + (r + t) \qquad \text{(substitution)}$$
$$= (s + r) + t \qquad \text{(associativity)}$$
$$= (r + s) + t \qquad \text{(commutativity)}$$
$$= 0 + t \qquad \qquad \text{(substitution)}$$
$$= t \qquad \qquad \quad \text{(additive identity)}$$

For the second part, suppose that d is the additive inverse of c. Adding d to both sides of the equaltiy $r + c = c$ yields

$$r + c + d = c + d.$$

But $c + d = 0$, so the left side of the equality above becomes $r + 0$, or r, and the right side becomes 0. We have proved that $r = 0$.

The third statement relates multiplication to the additive identity, and so our proof will have to use the distributive law, which relates multiplication to addition. Let c be any element of r. Then $r \cdot c + r \cdot 0 = r \cdot (c + 0) = r \cdot c$. But $r \cdot c + r \cdot 0 = r \cdot c$ says that $r \cdot 0$ acts like an additive identity, and statement 2, which we just proved, says that the additive identity is unique. Therefore, we must have $r \cdot 0 = 0$.

For the fourth part, we must show that $ar + r = 0$. By the distributive law, $ar + r = (a + 1)r$. Since $a + 1 = 0$, we obtain

$$ar + r = (a + 1)r = 0 \cdot r = 0.$$

It is natural to adopt the notational convention that -1 is the additive inverse of 1 in R, so that -1 satisfies $1 + (-1) = 0$. With this notation, the fourth part of Proposition 13.1 states that for every r in R, the product $(-1) \cdot r$ is the additive inverse of r. In other words, $r + (-1) \cdot r = 0$. It is natural also to adopt the convention that the additive inverse of b is written as $-b$. The fourth part of Proposition 13.1 can be restated as follows:

Proposition 13.2 *Let R be a ring and let -1 denote the additive inverse of 1 in R. For each element r of R, let $-r$ denote its additive inverse. Then*

$$(-1) \cdot r = -r.$$

Let us review the meaning of Proposition 13.2. In a ring R, we know that every element r must have a unique additive inverse, which we are choosing to write as $-r$. In particular, the multiplicative identity 1 has an additive inverse, which we write as -1. We now compare the additive inverse $-r$ of r to the product of r and -1, the additive inverse of 1. Does $-r$ equal $-1 \cdot r$?

We know that this is the case in the rings we have grown up using, such as $\mathbb{Z}$ and $\mathbb{R}$. This does not imply that $-r$ and $-1 \cdot r$ must be equal in every ring, but they are, and we have just proved it. As a special case, suppose r is -1. Then

$$(-1) \cdot (-1) = -(-1).$$

But $-(-1)$ is the additive inverse of -1, which is 1. Therefore, we find that the additive inverse -1 of 1 satisfies

$$(-1) \cdot (-1) = 1.$$

We have seen that although a ring is defined with two apparently unrelated operations, addition and multiplication, the distributive law links them together in a way strongly reminiscent of the relationship between these two operations in the familiar ring of the integers. Now, finally, we can prove a result about square roots in a ring.

Proposition 13.3 *Let R be a ring. Suppose b and c are elements of R such that $b^2 = c$. Then $(-b)^2 = c$. Thus if b is a square root of an element c of R, then so is $-b$.*

Proof. We have proved that $-b = (-1) \cdot b$ and that $(-1)^2 = 1$. Combining these results with the associative law, we obtain

$$(-b)^2 = (-b) \cdot (-b) = (-1) \cdot b \cdot (-1) \cdot b = (-1) \cdot (-1) \cdot b \cdot b$$
$$= (-1)^2 \cdot b^2 = 1 \cdot b^2 = b^2.$$

Proposition 13.3 tells us that if an element b of R is a square root of c, then the additive inverse $-b$ of b is also a square root of c.

For our further discussion of additive inverses, let us work in fields. Let K, then, be a field. Then every element b of K has an additive inverse $-b$. Assume that b is not the zero element of the field. Then 0 cannot be the additive inverse of b (why?), so $-b$ must also be nonzero. We might expect that b and $-b$ are two different elements of K. Need they be? No. For example, consider the field $\mathbb{F}_2$, which has only two elements, $[1]$ and $[0]$. In this field the element $[1]$ is the multiplicative identity, and $[1] + [1] = [0]$. Thus, the additive inverse of $[1]$ is itself.

Suppose K is a field with additive identity 0 and multiplicative identity 1, and suppose there is a nonzero element b that is its own additive inverse; that is, $b + b = 0$. We now engage in a bit of algebraic manipulation involving multiplicative inverses and the distributive law. Since K is a field and b is nonzero, b has a multiplicative inverse b^{-1}. We have

$$0 = b^{-1}(0) = b^{-1}(b + b) = b^{-1}b + b^{-1}b = 1 + 1.$$

Thus we obtain $1 + 1 = 0$, or $-1 = 1$. We have shown that if some nonzero element of K coincides with its additive inverse, then in particular the multiplicative identity 1 coincides with its additive inverse, and this then implies that $c + c = 0$ for every element c in K. In other words, if one nonzero element of a field K is its own additive inverse, then *every* element of K is its own additive inverse. This condition has the strange consequence that addition and subtraction in K coincide. After all, $b = -b$ for every b in K, so adding b and subtracting b are the same operation. We have seen two examples of fields satisfying this odd condition that addition and subtraction coincide. One is $\mathbb{F}_2$. The second is the last fruit ring in Section 6.3, as an examination of its addition table shows.

In general, for a field K with additive identity written as 0 and multiplicative identity written as 1, let us write 2 for the element $1 + 1$. We see that it may be the case in K that 2 coincides with 0, and that when this happens, addition and subtraction coincide. In our analysis of polynomial equations of degree 2, we will need to take special care to consider whether we are working with such a field or not. The need for this distinction will become clear in Section 13.2.

13.2 The Quadratic Formula

In Section 9.1 we reviewed and proved the quadratic formula for quadratic equations with real coefficients. Is the formula applicable more generally? Let us try it out in a couple of examples, for which we choose $\mathbb{F}_5$ as our coefficient field. We will write elements of $\mathbb{F}_5$ without brackets, for simplicity, but keep in mind that the symbols we write are representatives of congruence classes modulo 5. Also keep in mind that we have no way yet of knowing whether the quadratic formula works as a means of solving quadratic equations with coefficients in $\mathbb{F}_5$. We are simply going to try it out and see what happens.

First, let us try to solve the equation

$$x^2 + 2x + 2 = 0.$$

According to the quadratic formula, the solutions should have the form

$$x = -\frac{2}{2} \pm \frac{\sqrt{2^2 - 4 \cdot 2}}{2}.$$

But since we are working modulo 5, dividing by 2 is the same as multiplying by its multiplicative inverse 3. This allows us to rewrite our putative solutions as

$$x = -2 \cdot 3 \pm 3 \cdot \sqrt{-4}.$$

Since -4 is simply 1, its square root is 1, and we obtain

$$x = -6 \pm 3.$$

Replacing the two values -3 and -9 of x by the congruent values 2 and 1, we find that the quadratic formula yields $x = 1$ and $x = 2$ as solutions to the equation $x^2 + 2x + 2 = 0$. Substituting, we see that this is in fact correct.

Let us also try to solve the equation

$$x^2 + 2x + 3 = 0$$

using the quadratic formula. We get

$$x = -\frac{2}{2} \pm \frac{\sqrt{2^2 - 4 \cdot 3}}{2},$$

which upon simplification becomes

$$x = -6 \pm 3 \cdot \sqrt{2}.$$

What is the square root of 2 in $\mathbb{F}_5$? An examination of the possibilities shows that there is no such square root. If the quadratic formula works in this setting, then we are led to conclude that $x^2 + 2x + 3 = 0$ has no solutions in $\mathbb{F}_5$. Indeed, we can check directly by trying all five elements of $\mathbb{F}_5$ that $x^2 + 2x + 3 = 0$ has no solution in $\mathbb{F}_5$. This is analogous to the situation that occurs when we study a real quadratic equation with a negative discriminant. As we have learned, there is no real solution to such an equation, but there are two complex solutions.

These examples suggest that the quadratic formula may apply to quadratic equations with coefficients chosen from an array of different fields. In order to find out what happens in general, let us study quadratic equations with coefficients in an arbitrary field K. We begin with the simplest quadratic equation, one of the form $x^2 + c = 0$, or equivalently, $x^2 = d$, with d an element of K. Solving this amounts to finding square roots of d in K.

If d has no square root, then $x^2 = d$ has no solutions, and therefore $x^2 + c$ does not factor in $K[x]$ as the product of first-degree polynomials. Suppose d does have a square root in K, and that it is the element γ: We thus have $\gamma^2 = d$. By Proposition 13.3, the additive inverse $-\gamma$ also satisfies $(-\gamma)^2 = d$. However, as we saw in the discussion following Proposition 13.3, depending on the field K, the elements γ and $-\gamma$ may or may not be distinct. If K is a field in which $2 \neq 0$, then γ and $-\gamma$ are distinct. By the theorems of Chapter 9, both γ and $-\gamma$ are solutions of the equation $x^2 - d = 0$, and these must be all the solutions, since a polynomial equation of degree n can have at most n distinct solutions. In other words, if d has a square root γ in K and $2 \neq 0$ in K, then d has two distinct square roots in K, and these are the two solutions of $x^2 - d = 0$. This is exactly what we expect from our experience working over $\mathbb{R}$. We can record the result as follows.

Proposition 13.4 *Suppose K is a field in which $1 + 1 \neq 0$, and let d be a nonzero element of K. Then the equation*

$$x^2 - d = 0$$

has either no solutions in K or two distinct solutions in K, each the additive inverse of the other.

What happens if K is a field in which $2 = 0$? Can a nonzero element d still have two square roots? This is treated in the following exercise.

Exercise 13.1. Let K be a field with additive identity 0 and multiplicative identity 1, and suppose that $2 = 1 + 1 = 0$ in K.

1. Verify that for elements a and b of K,

$$(a + b)^2 = a^2 + b^2.$$

2. Verify that for an element a in K,

$$(x + a)^2 = x^2 + a^2$$

in $K[x]$.
3. Consider the equation $x^2 + a^2 = 0$. Show that $x = a$ is a solution. Suppose that $x = b$ is also a solution, so that $b^2 + a^2 = 0$. Deduce that $(a+b)^2 = 0$, so that $a + b = 0$. Conclude that $b = a$, so that a is the only solution of $x^2 + a^2 = 0$.
4. Deduce that if d is an element of K that has a square root γ in K, then it has only one square root, and $x^2 + d$ factors as $(x + \gamma)^2$.

For the special fields K in which every element is its own additive inverse, the last exercise shows that if d is an element of K with a square root γ, then γ is the only square root of d in K. In contrast, for other fields, $-\gamma$ is an additional square root. Notice that both situations are covered by the statement that if d has square root γ, then the only square roots of d are γ and its additive inverse. For some fields these are distinct, while for other fields they coincide, but for all fields they are the only square roots.

To study quadratic equations of the form $x^2 + c = 0$, we see that we must analyze the nature of square roots in K. Let us consider the general quadratic equation $x^2 + bx + c = 0$, for coefficients b and c lying in the field K. We may be tempted to use the quadratic formula to solve this equation, but before doing so, we need to figure out what the formula should mean in general and whether it is valid in general. For instance, the formula requires us to divide by 2 and multiply by 4. These numbers do not necessarily make sense for a general field K. We need to give interpretations to them.

We have already agreed that for a general field K, we will let 2 denote the element $1 + 1$, the sum of the multiplicative identity 1 of K with itself. Let us take 4 to be the product 2×2. We must deal with the possibility that $1 + 1$, or 2, may be 0 in K, in which case we cannot divide by 2. For such a field, the quadratic formula will not make any sense. We need to restrict our attention to fields in which $2 \neq 0$.

Exercise 13.2. Let K be a field with additive identity 0 and multiplicative identity 1. Write 2 for the sum $1 + 1$ and 4 for 2×2. Assume that $2 \neq 0$ in K, so that also $4 \neq 0$. In this exercise, we will mimic what was already done in Exercise 10.1.

1. Verify that for elements a and b of K,

$$(x + a)^2 = x^2 + 2ax + a^2,$$

and

$$x^2 + bx + \frac{b^2}{4}$$

is the square of a first-degree polynomial.

2. Show that solving the equation $x^2 + bx + c = 0$, where b and c are in K, is equivalent to solving an equation of the form $\left(x + \frac{b}{2}\right)^2 = \frac{d}{4}$ for a suitable element d of K. Write out the element d explicitly in terms of the coefficients b and c.

3. Deduce that if $d = 0$, then $x^2 + bx + c$ factors as $\left(x + \frac{b}{2}\right)^2$, and the one and only solution to $x^2 + bx + c = 0$ is $x = -\frac{b}{2}$.

4. Deduce that if d has no square root in K, then there is no solution to the equation $x^2 + bx + c = 0$, and therefore $x^2 + bx + c$ is irreducible in $K[x]$.

5. If d is nonzero and does have a square root, then there are two solutions to $x^2 + bx + c = 0$ in K. Write out these solutions explicitly in terms of b and c.

6. Conclude that the quadratic formula works for quadratic equations with coefficients in any field K in which $2 \neq 0$.

For a polynomial $x^2 + bx + c$ as in Exercise 13.2, the quantity $b^2 - 4c$ is called the polynomial's *discriminant*. As in the real case, the discriminant plays a crucial role in studying solutions of $x^2 + bx + c = 0$.

To use the quadratic formula for a field K, we need to know about square roots in K. Let us consider two special cases.

Exercise 13.3. Prove that every complex number $a + bi$ has a complex square root, that is, a complex number $r + si$ whose square is $a + bi$. (Hint: First notice that every real number a has a square root in $\mathbb{C}$. Thus, it remains to prove that every complex number $a + bi$ with $b \neq 0$ has a complex square root. In order to find a square root of $a + bi$, we must find real numbers r and s for which $(r + si)^2 = a + bi$. When you multiply out the left-hand side of this equation and start solving for r and s in terms of a and b, you will find it necessary to divide by r or s. Show that if either of r and s is zero, then b is zero, contrary to our assumption. Therefore, you can assume that r and s are both nonzero. Square $r + si$ to obtain an equation for a in terms of r and s and an equation for b in terms of r and s. Recall the standard approach to solving two equations in two unknowns: Use one equation to express one unknown in terms of the other, and then substitute in the other equation.

Follow this approach to obtain an equation for r. Notice that it is actually a quadratic equation for r^2. Use the quadratic formula to solve for r^2. Verify that you obtain a solution that is a *positive* real number. Therefore, you can take its square root to find a value for r that is a real number. You have already expressed s in terms of r. Thus you have found real values of r and s for which $(r + si)^2 = a + bi$. This proves that $a + bi$ has a square root.) Deduce that every quadratic polynomial $f(x)$ in $\mathbb{C}[x]$ has a root in $\mathbb{C}$ and that $f(x)$ factors as the product of two degree-one polynomials in $\mathbb{C}[x]$.

Exercise 13.4. The purpose of this exercise is twofold: to introduce another way of looking at complex numbers and to show that by changing one's point of view it is possible to make a complicated calculation easy. In the last exercise you proved that every complex number has a square root in $\mathbb{C}$. We will now prove the same fact, but the computations will be simplified by our looking at complex numbers in a different way, one that is equivalent to the shift in point of view from rectangular to polar coordinates in the Cartesian plane.

1. A complex number c can be written as $a + bi$ for real numbers a and b, allowing us to identify it with the point in the Cartesian plane with x-coordinate a and y-coordinate b. Define the *absolute value*, or *norm*, $|c|$ of the complex number $c = a + bi$ to be distance of the point (a, b) in the plane from the origin. Thus $|c| = \sqrt{a^2 + b^2}$.
2. Show that every complex number c can be written as the product of a positive real number r and a complex number whose absolute value is 1. (Hint: Divide and multiply by $|c|$.)
3. Argue that if $a + bi$ is a complex number with absolute value 1, then there is a real number θ such that $a = \cos\theta$ and $b = \sin\theta$.
4. From the previous two items, conclude that every complex number c can be written as $c = r(\cos\theta + i\sin\theta)$ for some nonnegative real number r and real number θ.
5. Show that if c is a complex number written as $c = r(\cos\theta + i\sin\theta)$ as above, then c has square roots $\pm\sqrt{r}\left(\cos\frac{\theta}{2} + i\sin\frac{\theta}{2}\right)$. (Hint: Use the trigonometric double-angle formulas.)

Exercise 13.5. We have proved that $\sqrt{2}$ is not rational. More generally, one can use the same argument to show that every positive integer n that is not the square of an integer has a square root $\sqrt{n}$ that is irrational. Using this, state a criterion describing which polynomials $x^2 + bx + c$ in $\mathbb{Z}[x]$ have roots in $\mathbb{Q}$ and which do not.

Exercise 13.6. We began this section with a look at two quadratic equations with coefficients in $\mathbb{F}_5$. Let us return to these examples. We will write elements of $\mathbb{F}_5$ without brackets.

1. Show that 1 and 4 have square roots in $\mathbb{F}_5$, but 2 and 3 do not.
2. Find the solutions to $x^2 + 2x + 2 = 0$ in $\mathbb{F}_5$ using the quadratic formula.
3. Show that $x^2 + 2x + 3 = 0$ has no solutions in $\mathbb{F}_5$.

4. Find all solutions to $x^2 + 3x + 1 = 0$.
5. Find all solutions to $x^2 + 3x + 3 = 0$.

13.3 Square Roots in Finite Fields

For any odd prime number p, the field $\mathbb{F}_p$ satisfies $2 \neq 0$, so the quadratic formula can be applied to quadratic polynomials in $\mathbb{F}_p[x]$. Thus the question of whether the polynomial $x^2 + bx + c$ in $\mathbb{F}_p[x]$ is irreducible or has a root in $\mathbb{F}_p$ reduces to the question of whether a certain element in $\mathbb{F}_p$, the discriminant $b^2 - 4c$ of $x^2 + bx + c$, has a square root in $\mathbb{F}_p$. It turns out to be surprisingly difficult to describe the elements of $\mathbb{F}_p$ that have square roots and the elements that do not. We will begin our examination of square roots in $\mathbb{F}_p$ by showing, using an elementary counting argument, that half of the $p-1$ nonzero elements of $\mathbb{F}_p$ have square roots and half do not. We can think of this as parallel to the situation in $\mathbb{R}$: Half the real numbers have square roots, the positive reals, and half do not have square roots, the negative reals.

Theorem 13.5. *For an odd prime number* p, *the field* $\mathbb{F}_p$ *contains* $\frac{p-1}{2}$ *nonzero elements that are squares of elements of* $\mathbb{F}_p$ *and* $\frac{p-1}{2}$ *nonzero elements that are not.*

Exercise 13.7. Prove Theorem 13.5. You can proceed as follows:

1. Observe that for every integer i between 1 and $p-1$ we have

$$-i \equiv p - i \pmod{p},$$

2. Deduce that the p distinct congruence classes of integers modulo p are represented by the p consecutive integers

$$-\frac{p-1}{2}, \quad -\frac{p-1}{2}+1, \quad \ldots, -1, \ 0, \ 1, \ 2, \ \ldots, \quad \frac{p-1}{2} - 1, \quad \frac{p-1}{2}.$$

In particular, the $p-1$ nonzero elements of $\mathbb{F}_p$ can be listed as

$$\left[-\frac{p-1}{2}\right], \quad \left[-\frac{p-1}{2}+1\right], \quad \ldots, [-1], [1], [2], \ldots, \quad \left[\frac{p-1}{2} - 1\right], \quad \left[\frac{p-1}{2}\right].$$

3. Recall that for every integer i, the additive inverse $-[i]$ of $[i]$ in $\mathbb{F}_p$ is $[-i]$. Thus $[-i]$ can be replaced by $-[i]$ in our list of the elements of $\mathbb{F}_p$, and the list of $p-1$ nonzero elements of $\mathbb{F}_p$ can be written as

$$\pm[1], \ \pm[2], \ \ldots, \ \pm\left[\frac{p-1}{2} - 1\right], \ \pm\left[\frac{p-1}{2}\right].$$

4. Suppose that b is a nonzero element of $\mathbb{F}_p$ and $c = b^2$. Review the results of Section 13.1 and deduce that $b \neq -b$, that $(-b)^2 = c$, and that b and $-b$ are the only elements of $\mathbb{F}_p$ whose squares equal c.

5. Conclude that the list

$$[1]^2, \quad [2]^2, \quad \ldots, \quad \left[\frac{p-1}{2} - 1\right]^2, \quad \left[\frac{p-1}{2}\right]^2$$

consists of $\frac{p-1}{2}$ distinct elements of $\mathbb{F}_p$ and that every nonzero square in $\mathbb{F}_p$ is on this list.

We have shown that half the nonzero elements of $\mathbb{F}_p$ for p an odd prime are squares and half are not. How can we decide which are which? This is a difficult problem. Let us begin by collecting some data.

Exercise 13.8. Determine which elements in the set $\{[1], [2], \ldots, [p-1]\}$ of nonzero elements of $\mathbb{F}_p$ are squares for each of the following values of p: 3, 5, 7, 11, 13, 19.

Let us treat what turns out to be the simplest nontrivial case, the question of whether the element $[-1]$ in $\mathbb{F}_p$ is a square. This element is the additive inverse of the multiplicative identity $[1]$. We want to know whether there is a positive integer m such that $[m]^2 = [-1]$ in $\mathbb{F}_p$. We can rephrase this in terms of integers and congruences. Is there a positive integer m such that

$$m^2 \equiv -1 \pmod{p}?$$

If this congruence is solvable, we will say that -1 is a square modulo p. Before stating a result that settles this question, let us look at some data in order to try making a guess.

We will consider all the odd primes up to some number, say 101, and decide for each such prime p whether -1 is a square or not a square modulo p. It is easily checked that the answer is no for $p = 3, 7, 11$ and yes for $p = 5, 13, 17$. For instance, $2^2 \equiv -1 \pmod 5$ and $5^2 \equiv -1 \pmod{13}$. We can keep going. Stop reading and do so now for some more primes.

What did you find? If you worked your way up to $p = 101$, you should have found that -1 is a square modulo the primes

$$5, \ 13, \ 17, \ 29, \ 37, \ 41, \ 53, \ 61, \ 73, \ 89, \ 97, \ 101,$$

but -1 is not a square modulo the primes

$$3, \ 7, \ 11, \ 19, \ 23, \ 31, \ 43, \ 47, \ 59, \ 67, \ 71, \ 79, \ 83.$$

Is there a pattern here? You may notice that the primes on the first list are all congruent to 1 modulo 4, and the primes on the second list are all congruent to 3 modulo 4. This suggests the following result:

Theorem 13.6. *Let p be an odd prime number. If $p \equiv 1 \pmod 4$, then $[-1]$ is a square in $\mathbb{F}_p$, while if $p \equiv 3 \pmod 4$, then $[-1]$ is not a square in $\mathbb{F}_p$.*

An odd prime number must be congruent to either 1 or 3 modulo 4. Thus, the contrapositive of the second half of Theorem 13.6 is the statement that if $[-1]$ is a square in $\mathbb{F}_p$, then $p \equiv 1 \pmod 4$. This can be proved as an application of Fermat's theorem, Theorem 7.7:

Exercise 13.9. Assume that p is an odd prime number such that $[-1]$ is a square in $\mathbb{F}_p$. We wish to prove that $p \equiv 1 \pmod 4$.

1. Review Fermat's theorem. Recall that it yields the congruence

$$a^{p-1} \equiv 1 \pmod p$$

   for every integer a not divisible by p.
2. Observe that $[-1]$ being a square in $\mathbb{F}_p$ means that there is an integer b such that $b^2 \equiv -1 \pmod p$. Thus we can assume that such a b exists.
3. Check that Fermat's theorem applies to b and conclude that

$$b^{p-1} \equiv 1 \pmod p.$$

4. Observe that $p - 1$ is even, so that $\frac{p-1}{2}$ is an integer. Then verify that

$$(-1)^{(p-1)/2} \equiv \left(b^2\right)^{(p-1)/2} = b^{p-1} \equiv 1 \pmod p.$$

   In particular, omitting the middle terms, we have shown that

$$(-1)^{(p-1)/2} \equiv 1 \pmod p.$$

5. Observe that -1 raised to an integer power equals either -1 or 1, so that in particular, $(-1)^{(p-1)/2}$ equals either 1 or -1. Combining this with the congruence $(-1)^{(p-1)/2} \equiv 1 \pmod p$, deduce that $(-1)^{(p-1)/2} = 1$.
6. Conclude that $\frac{p-1}{2}$ must be even, so that $p \equiv 1 \pmod 4$.

We have proved the second half of Theorem 13.6. Let us isolate the first half as a separate theorem:

Theorem 13.7. *Suppose p is a prime number satisfying $p \equiv 1 \pmod 4$. Then $[-1]$ is a square in $\mathbb{F}_p$.*

We will prove Theorem 13.7 in two steps. The first step is to prove Wilson's theorem, a famous, centuries-old congruence.

Exercise 13.10. Suppose p is an odd prime number.

1. Show that $x^2 - [1]$ has two distinct roots in $\mathbb{F}_p$, namely $[1]$ and $-[1]$.
2. Deduce that $[1]$ and $-[1]$ are the only elements of $\mathbb{F}_p$ that equal their own multiplicative inverses. In other words, if u in $\mathbb{F}_p$ is distinct from $[0]$, $[1]$, and $-[1]$, then $u \neq u^{-1}$.

3. Deduce that $[2] \times [3] \times \cdots \times [p-2] = [1]$ in $\mathbb{F}_p$. (Hint: Observe that each factor in this product pairs up with another factor to give the product $[1]$. You might first work this out explicitly in the cases where p equals 5, 7, and 11 to get a feeling for what is happening.)
4. Deduce that $[1] \times [2] \times [3] \times \cdots \times [p-1] = -[1]$ in $\mathbb{F}_p$.
5. Reinterpret the last result as the following congruence statement in $\mathbb{Z}$:

$$1 \times 2 \times 3 \times \cdots \times (p-1) \equiv -1 \pmod{p}.$$

This congruence is known as Wilson's theorem.

Exercise 13.11. Prove Theorem 13.7 as follows.

1. Since p is odd, $p-1$ is even, so $\frac{p-1}{2}$ is an integer. Show that it satisfies the relation

$$p - \frac{p-1}{2} = \frac{p-1}{2} + 1.$$

Then observe that therefore, we can rewrite

$$1 \times 2 \times 3 \times \cdots \times (p-1)$$

as the product of

$$1 \times 2 \times 3 \times \cdots \times \frac{p-1}{2}$$

and

$$(p-1) \times (p-2) \times (p-3) \times \cdots \times \left(p - \frac{p-1}{2}\right).$$

2. Notice that for each integer i, we have that $p - i \equiv -i \pmod{p}$. Deduce that

$$1 \times 2 \times 3 \times \cdots \times (p-1)$$
$$\equiv \left(1 \times 2 \times 3 \times \cdots \times \frac{p-1}{2}\right)\left((-1) \times (-2) \times (-3) \times \cdots \times \left(-\frac{p-1}{2}\right)\right)$$

modulo p.
3. Combining this last congruence with the congruence of Wilson's theorem, deduce that

$$(-1)^{(p-1)/2}\left(1 \times 2 \times 3 \times \cdots \times \frac{p-1}{2}\right)^2 \equiv -1 \pmod{p}.$$

4. Conclude that if $\frac{p-1}{2}$ is even, then -1 is congruent to the square of an integer modulo p.
5. Notice that $\frac{p-1}{2}$ is even precisely when 4 divides $p-1$, which means $p \equiv 1 \pmod{4}$. Therefore, you have proved that if $p \equiv 1 \pmod{4}$, then -1 is congruent to the square of an integer modulo p.
6. Pass to $\mathbb{F}_p$ and conclude that if $p \equiv 1 \pmod{4}$, then $[-1]$ is a square in $\mathbb{F}_p$.

Theorem 13.6 provides a procedure for deciding whether or not -1 is a square in the field $\mathbb{F}_p$. The other elements of $\mathbb{F}_p$ can be handled using the *law of quadratic reciprocity*, one of the most famous results in number theory. Gauss proved it when he was 19, and he published a proof a few years later in his book *Disquisitiones Arithmeticae*, published in 1801. This law is described in Section 17.2.

Theorem 13.6 can also be proved by taking a different, more comprehensive, approach. This approach sheds more light on the general question of squares in $\mathbb{F}_p$ and provides alternative proofs of the results we have obtained so far. It depends on the existence of an element in $\mathbb{F}_p$ known as a *primitive root*. Recall from Section 7.2 that an element r of a ring R is a root of unity if $r^n = 1$ for some positive integer n, and that the smallest such n is called the *order* of r. Fermat's theorem states that every nonzero element a in $\mathbb{F}_p$ satisfies $a^{p-1} = 1$. In particular, every nonzero a has an order, and the order is at most $p-1$. It is not hard to show that this order divides $p-1$. In studying examples, we saw that some elements may have order less than $p-1$ and some may have order $p-1$ itself. Let us review this.

Exercise 13.12. For each of the prime numbers $p = 3, 5, 7,$ 11, and 13, determine the orders of all the elements of $\mathbb{F}_p$. In each case, list all the elements of order $p-1$ and state how many of these there are.

Suppose that $\mathbb{F}_p$ does in fact have an element a of order $p-1$, as is the case in the examples above. Then $a^{p-1} = 1$, but $a^m \neq 1$ for every positive exponent m less than $p-1$. It follows that the elements $a, a^2, \ldots, a^{p-1}$ are all distinct (why?). Since $\mathbb{F}_p$ has only $p-1$ nonzero elements, we see that every nonzero element has the form a^i for some i. In other words, the powers of a provide a complete list of all the nonzero elements of $\mathbb{F}_p$. This leads to the following question:

Question 13.8 *Let p be an odd prime number.*

1. *Is there an element a in $\mathbb{F}_p$ of order $p-1$? In other words, is there an element a such that $a, a^2, a^3, \ldots, a^{p-1}$ is a complete list of the $p-1$ nonzero elements of $\mathbb{F}_p$?*
2. *Equivalently, is there a positive integer b relatively prime to p with the property that the powers $b, b^2, b^3, \ldots, b^{p-1}$ represent $p-1$ distinct congruence classes modulo p?*

Such an element a of $\mathbb{F}_p$ is called a *primitive root* of $\mathbb{F}_p$, and such an integer b is called a *primitive root* modulo p. The answer to the question is yes.

Theorem 13.9. *Let p be a prime number. Then $\mathbb{F}_p$ contains a primitive root.*

Even though Theorem 13.9 ensures that a primitive root exists in $\mathbb{F}_p$, neither its statement nor any known proof provides a formula that describes a particular primitive root in terms of p. Certainly, a primitive root can be found, given the knowledge that one exists. We can simply work our way through the finitely many elements of $\mathbb{F}_p$, computing their orders one by one until a primitive root appears.

The proof Theorem 13.9, although not difficult, involves ideas that take us in a different direction. Rather than pursuing this different direction now, we will postpone the proof. It can be found in Section 17.1. Assume in the remainder of this section that Theorem 13.9 has been proved. Let us use it to prove once again the results we have already obtained.

Exercise 13.13. Let p be an odd prime number and let a be a primitive root in $\mathbb{F}_p$. Thus $a^{p-1} = 1$, and $a, a^2, \ldots, a^{p-1}$ is a complete list of the nonzero elements of $\mathbb{F}_p$.

1. Show that if m is an even integer between 2 and $p - 1$, then a^m is the square of an element in $\mathbb{F}_p$. (Hint: You can write down a square root of a^m directly as a particular power of a.)
2. In contrast, prove that if m is an odd integer between 1 and $p - 1$, then a^m is not the square of an element in $\mathbb{F}_p$. (Hint: If a^m is a square, show that $a^m = a^n$ for some even integer n. Use this and cancellation to obtain a contradiction.)
3. Deduce anew that half of the $p - 1$ nonzero elements in $\mathbb{F}_p$ are squares and half are not.
4. Deduce that if b and c are two elements of $\mathbb{F}_p$ that are not squares, then their product bc is a square, and that if one of b and c is a square and the other is not, then the product bc is not a square.
5. Prove Theorem 13.6. (Hint: Recall that the only solutions in $\mathbb{F}_p$ to the equation $x^2 = [1]$ are $x = [1]$ and $x = -[1]$. Suppose a is a primitive root of $\mathbb{F}_p$. Describe $[1]$ and $-[1]$ as powers of a. Use the description of $-[1]$ as a power of a to prove the theorem.)

13.4 Quadratic Field Constructions

We can decide whether a quadratic polynomial $x^2 + bx + c$ with coefficients in one of the fields $\mathbb{C}$, $\mathbb{R}$, $\mathbb{Q}$, $\mathbb{F}_p$ (for p odd), has a root in that field, or whether the polynomial is irreducible. We simply calculate the discriminant $b^2 - 4c$ and decide whether this is a square in the field. If it is a square, we can use the quadratic formula to write down the two roots and to factor $x^2 + bx + c$. If it is not a square, there are no roots in the field, and we cannot factor $x^2 + bx + c$.

Suppose we are considering a particular quadratic polynomial $x^2 + bx + c$ with coefficients in a field K, and it turns out that $b^2 - 4c$ is not a square, so that $x^2 + bx + c$ does not have roots in K. What can we do?

The most familiar example is that of the polynomial $x^2 + 1$ over the field $\mathbb{R}$. Its discriminant is -4, which is not a square in $\mathbb{R}$, so we cannot find roots of $x^2 + 1$ in $\mathbb{R}$. We have learned that in this case we need not give up. Instead, we can simply create a root for the polynomial $x^2 + 1$. We then build a new, larger, field by expanding the field to include the new root and all other new numbers involving that root that are necessary for the new field to have the requisite properties of closure. This new root does not exist in the original field $\mathbb{R}$, but it does exist in the larger field.

In this example we built the larger field by choosing a symbol, i, for our square root of -1. Since a field must be closed under addition and multiplication, we were compelled to introduce all numbers of the form $a + bi$, where a and b are real. We then showed that we did not have to include any more numbers in our larger field because the resulting set of all such numbers $a + bi$, which we have denoted by $\mathbb{C}$ and called the complex numbers, is closed under addition, multiplication, and additive and multiplicative inverses. We proved this by first showing that addition in $\mathbb{C}$ is defined by the rule

$$(a + bi) + (c + di) = (a + c) + (b + d)i.$$

For multiplication, since the distributive law must hold, we obtained

$$(a + bi)(c + di) = ac + adi + bci + bdi^2,$$

and since we chose i to have the property that $i^2 = -1$, we could rewrite bdi^2 as $-bd$ and then rewrite the above equality as

$$(a + bi)(c + di) = (ac - bd) + (ad + bc)i.$$

We thereby were able to define multiplication in $\mathbb{C}$. Using these rules we easily checked that $\mathbb{C}$ is a ring. We also observed that the product $(a + bi)(a - bi)$ is a real number $a^2 + b^2$. In fact, it is a positive real number, unless $a = b = 0$. Using this fact, we found that every nonzero complex number $a + bi$ has a multiplicative inverse in $\mathbb{C}$, the complex number

$$\frac{a}{a^2 + b^2} - \frac{b}{a^2 + b^2} i.$$

Thus $\mathbb{C}$ is a field.

By passing from $\mathbb{R}$ to the larger field $\mathbb{C}$, we are able to find roots for $x^2 + 1$ and for many other polynomials as well. At the same time, we find that polynomials irreducible in $\mathbb{R}[x]$ have nontrivial factorizations in $\mathbb{C}[x]$. For instance, $x^2 + 1$ is irreducible in $\mathbb{R}[x]$, but factors as $(x + i)(x - i)$ in $\mathbb{C}[x]$.

This example suggests that when we are confronted with a quadratic polynomial $x^2 + bx + c$ with coefficients in a field K but with no roots in K, we need not think that there is nothing more we can do with the polynomial. Instead, we might be able to pass to a field extension L of K in which there are roots and over which the polynomial factors as the product of degree-one polynomials.

Let us illustrate this idea with the polynomial $x^2 - 2$ in $\mathbb{Q}[x]$. It has no roots in $\mathbb{Q}$. Suppose we wish to create a square root of 2. Of course, we know that a square root of 2 exists in $\mathbb{R}$, but suppose in the exercise below that we do not. Or to put it another way, the field $\mathbb{R}$ is much bigger than we need, and we are more interested in the abstract properties of a square root of 2 than we are in its role as a real number.

Exercise 13.14. Start with the field $\mathbb{Q}$ of rational numbers. The number 2 does not have a square root in $\mathbb{Q}$. Therefore, we invent a square root of 2, that is, a symbol γ with the property that $\gamma^2 = 2$. (We can think of γ as the real number $\sqrt{2}$, but let us work instead with γ as a new, abstract, entity, just as we have used i before when we wanted to work with a square root of -1.) Now we need to create a field that contains all of $\mathbb{Q}$, and γ as well. Since we need closure under addition, multiplication, and additive and multiplicative inverses, we will need at least the set K consisting of all expressions $a + b\gamma$, where a and b are rational numbers. Let us see whether the set K is sufficiently large. We define addition in K by the rule

$$(a + b\gamma) + (c + d\gamma) = (a + c) + (b + d)\gamma$$

and multiplication by

$$(a + b\gamma)(c + d\gamma) = ac + ad\gamma + bc\gamma + bd\gamma^2 = (ac + 2bd) + (ad + bc)\gamma.$$

Notice that we have used the fact that $\gamma^2 = 2$ to rewrite $bd\gamma^2$ as $2bd$, a rational number. It should be easy to see that K is a ring, that is, that it is closed under addition, multiplication, and additive inverses:

1. Check that K is a ring. Do not write out a proof of this.

But we want a field, and the question now is whether K is itself a field, or whether we have to include additonal elements to guarantee that every nonzero element in K has a multiplicative inverse:

2. Compute $(a + b\gamma)(a - b\gamma)$. Show that you get $a^2 - 2b^2$, a rational number.
3. Show that $a^2 - 2b^2$ cannot be 0 unless $a = b = 0$. (Hint: Suppose $a^2 - 2b^2 = 0$ but $b \neq 0$. Solve $a^2 - 2b^2 = 0$ for $\frac{a}{b}$.)
4. Assume that a and b are not both 0. Since $a^2 - 2b^2 \neq 0$, you can divide the product $(a+b\gamma)(a-b\gamma)$ by a^2-2b^2. Deduce that $a+b\gamma$ has a multiplicative inverse in K (what is it?) and that K is a field.

So by constructing a ring K containing a square root γ of 2, we get multiplicative inverses "for free."

5. Conclude that $x^2 - 2$ has roots γ and $-\gamma$ in K and that $x^2 - 2$ factors in $K[x]$ as $(x - \gamma)(x + \gamma)$.

Exercise 13.15. Suppose that m is an integer that is not the square of another integer. Review the steps in Exercise 13.14 and observe that they can all be carried out for m in place of 2. In other words, we form a set K of numbers of the form $a + b\gamma$, with $\gamma^2 = m$. Then we define addition and multiplication and see easily that we get a ring, and then we check that in fact, K is a field.

Let us try the same idea for the construction of new fields in a different setting. Consider the field $\mathbb{F}_3$, writing its three elements as 0, 1, and 2. We know that 1 is a square in $\mathbb{F}_3$, but 2 is not. Suppose we wish to construct a field extension of $\mathbb{F}_3$ that contains a square root of 2. Since $2 = -1$ in $\mathbb{F}_3$, we can also view the problem as that of trying to construct a square root of -1, just as we did starting from $\mathbb{R}$. Let us imitate what we have done in adding a square root of -1 to $\mathbb{R}$ and a square root of 2 to $\mathbb{Q}$.

Exercise 13.16. Start with the field $\mathbb{F}_3$. Form the set K consisting of all expressions $a + b\gamma$, where a and b are chosen from $\mathbb{F}_3$, and γ is some new formal symbol introduced to serve as a square root of 2 (or of -1); that is, $\gamma^2 = 2$. Define addition and multiplication in K by the following rules:

$$(a + b\gamma) + (c + d\gamma) = (a + c) + (b + d)\gamma;$$
$$(a + b\gamma)(c + d\gamma) = ac + ad\gamma + bc\gamma + bd\gamma^2 = (ac + 2bd) + (ad + bc)\gamma.$$

Notice, as before, that we have used the fact that $\gamma^2 = 2$ to rewrite $bd\gamma^2$ as $2bd$, an element of $\mathbb{F}_3$.

1. Check that K is a ring.
2. Observe that there are nine elements in K. Write a multiplication table for the eight nonzero elements of K.
3. Examine the table and observe from it that every nonzero element has an inverse.
4. Conclude that K is a field.
5. Alternatively, observe that we could have proceeded as before. First compute $(a + b\gamma)(a - b\gamma)$ and show that you get $a^2 - 2b^2$, which is the same as $a^2 + b^2$, an element of $\mathbb{F}_3$.
6. Show that $a^2 + b^2$ cannot be 0 unless $a = b = 0$.
7. Assume that a and b are not both 0. Since $a^2 + b^2 \neq 0$, you can divide the product $(a + b\gamma)(a - b\gamma)$ by $a^2 + b^2$. Deduce that $a + b\gamma$ has an inverse and conclude that K is a field.
8. Conclude that $x^2 - 2$ has roots γ and $-\gamma$ in K and that $x^2 - 2$ factors in $K[x]$ as $(x - \gamma)(x + \gamma)$.

You have constructed a field extension of $\mathbb{F}_3$ with nine elements that contains a square root for every element of $\mathbb{F}_3$. Let us call our new field $\mathbb{F}_9$

9. The field $\mathbb{F}_9$ expands our collection of finite fields beyond those with a prime number of elements. We have stated in Theorem 13.9 that each of the fields $\mathbb{F}_q$, with q prime, has a primitive root. Show that the field $\mathbb{F}_9$ also has a primitive root by writing down the powers $(1+\gamma)^i$ for $i = 1, 2, \ldots, 8$.

Exercise 13.17. Start with the field $\mathbb{F}_5$. Form the set K consisting of all expressions $a + b\gamma$, where a and b are chosen from $\mathbb{F}_5$, and γ is some new formal symbol introduced to serve as a square root of 2; that is, $\gamma^2 = 2$. Define addition and multiplication in K by the following rules:

$$(a + b\gamma) + (c + d\gamma) = (a + c) + (b + d)\gamma;$$
$$(a + b\gamma)(c + d\gamma) = ac + ad\gamma + bc\gamma + bd\gamma^2 = (ac + 2bd) + (ad + bc)\gamma.$$

1. Check that K is a ring.
2. Observe that there are twenty-five elements in K.
3. Compute $(a + b\gamma)(a - b\gamma)$ and show that you get $a^2 - 2b^2$, which is the same as $a^2 + 3b^2$, an element of $\mathbb{F}_5$.
4. Show that $a^2 + 3b^2$ cannot be 0 unless $a = b = 0$.
5. Assume that a and b are not both 0. Since $a^2 + 3b^2 \neq 0$, you can divide the product $(a+b\gamma)(a-b\gamma)$ by a^2+3b^2. Deduce that $a+b\gamma$ has a multiplicative inverse and conclude that K is a field.
6. Calculuate $(2\gamma)^2$ and observe that in building a field extension of $\mathbb{F}_5$ that contains a square root of 2, you have also constructed an extension that contains a square root of 3. You have constructed a field extension of $\mathbb{F}_5$ with twenty-five elements that contains a square root for every element of $\mathbb{F}_5$. Call this new field $\mathbb{F}_{25}$
7. Recall that we found earlier that the quadratic equation $x^2+2x+3 = 0$ has no solutions in $\mathbb{F}_5$. Show that it has two solutions in K. Use these solutions to factor $x^2 + 2x + 3$ in $K[x]$ as a product of degree-one polynomials.
8. Show that the field $\mathbb{F}_{25}$ has a primitive root by writing down the powers $(1 + 2\gamma)^i$ for $i = 1, 2, \ldots, 24$.

There is a general construction lurking. Let us see what it is.

Exercise 13.18. Suppose F is a field and suppose m is an element of F that is not a square; that is, there is no element of F whose square equals m. To put it still another way, there is no solution in F to the equation $x^2 - m = 0$. We wish to build a field extension K of F that contains a square root of m. Start by forming the set K consisting of expressions of the form $a + b\gamma$, where a and b are taken from F and γ is a new element introduced to be a square root of m; that is, we take γ to satisfy $\gamma^2 = m$. (Think of the four examples we have already worked with: F is $\mathbb{R}$, the number m is -1, and K is $\mathbb{C}$; F is $\mathbb{Q}$, the number m is 2, and K is the field built in Exercise 13.14; F is $\mathbb{F}_3$, the number m is 2, and K is the nine-element field built in Exercise 13.16; or F is $\mathbb{F}_5$, the element m is 2, and K is the field built in Exercise 13.17.) Define addition and multiplication in K by the rules

$$(a + b\gamma) + (c + d\gamma) = (a + c) + (b + d)\gamma$$

and

$$(a + b\gamma)(c + d\gamma) = ac + ad\gamma + bc\gamma + bd\gamma^2 = (ac + bdm) + (ad + bc)\gamma.$$

Recall that $\gamma^2 = m$.

1. Check that K is a ring. Do not write out a proof of this.
2. Compute $(a + b\gamma)(a - b\gamma)$. Show that you get $a^2 - mb^2$, an element of F.

3. Show that $a^2 - mb^2$ cannot be 0 unless $a = b = 0$.

4. Assume that a and b are not both 0. Since $a^2 - mb^2 \neq 0$, you can divide the product $(a + b\gamma)(a - b\gamma)$ by $a^2 - mb^2$. Deduce that $a + b\gamma$ has an inverse (what is it?) and conclude that K is a field.

5. Conclude that $x^2 - m$ has roots γ and $-\gamma$ in K, and that $x^2 - m$ factors in $K[x]$ as $(x - \gamma)(x + \gamma)$.

6. Assume that $2 \neq 0$ in F. Suppose that $x^2 + px + q$ is a quadratic polynomial in $F[x]$ whose discriminant $p^2 - 4q$ equals m. Observe that $x^2 + px + q$ has no roots in F, but describe two distinct roots of $x^2 + px + q$ in K.

The field K constructed in Exercise 13.18 is sometimes denoted by $F\left[\sqrt{m}\right]$. We have shown that for a field F and an element m of F that has no square root in F, the field $F\left[\sqrt{m}\right]$ is a field extension of F in which m has a square root. The last part of Exercise 13.18 shows further that for a field F in which $2 \neq 0$, we can use field extensions of the form $F\left[\sqrt{m}\right]$ to find solutions of general quadratic equations with coefficients in F. Thus, the quadratic formula can be used to find the roots of a quadratic polynomial even when its discriminant is not a square in F, provided that we are willing to pass to fields of the form $F\left[\sqrt{m}\right]$. Let us consider the special case in which F is one of the fields $\mathbb{F}_p$.

Exercise 13.19. Let p be a prime number congruent to 3 modulo 4 and write -1 for the additive inverse of 1 in $\mathbb{F}_p$. Recall from Theorem 13.6 that -1 is not a square in $\mathbb{F}_p$. Perform the construction of Exercise 13.18 on $\mathbb{F}_p$ and -1 to obtain a new field $\mathbb{F}_p\left[\sqrt{-1}\right]$ containing $\mathbb{F}_p$ in which -1 has a square root. Show that $\mathbb{F}_p\left[\sqrt{-1}\right]$ has p^2 elements.

Exercise 13.20. Let p be an odd prime number and let a be a primitive root in $\mathbb{F}_p$. Recall that this means that the elements $a, a^2, \ldots, a^{p-1}$ form a complete list of the nonzero elements of $\mathbb{F}_p$. Recall also that a^i is a square in $\mathbb{F}_p$ if i is even, and a^i is not a square if i is odd.

1. Perform the construction of Exercise 13.18 on $\mathbb{F}_p$ and a to obtain a new field $\mathbb{F}_p\left[\sqrt{a}\right]$ containing $\mathbb{F}_p$ in which a has a square root γ. Show that $\mathbb{F}_p\left[\sqrt{a}\right]$ has p^2 elements.

2. Show that every element of $\mathbb{F}_p$ has a square root in $\mathbb{F}_p\left[\sqrt{a}\right]$. Thus in building a field that contains a square root of a, we have succeeded in building a field with lots of square roots. Deduce that every polynomial $x^2 + bx + c$ in $\mathbb{F}_p[x]$ has a root in $\mathbb{F}_p\left[\sqrt{a}\right]$.

Exercise 13.21. The construction of Exercise 13.20 yields a finite field of size p^2 for each odd prime number p. What about the prime 2? Is there a field K of size 4 containing $\mathbb{F}_2$? We cannot use the idea of adjoining square roots to build such a field from $\mathbb{F}_2$, because there are no nonsquares in $\mathbb{F}_2$ in need of having square roots created. It turns out, though, that we have already built the desired field, namely, the last fruit ring we studied Section 6.3, the one with elements 0, 1, a, and b. Write K for this fruit ring. Notice that the elements 0 and 1 of K form the field $\mathbb{F}_2$, and so we may think of K as a field extension of $\mathbb{F}_2$.

1. Review the addition and multiplication tables for K and verify that it is indeed a field. Recall in particular that $a + 1 = b$ and $b + 1 = a$, so that $a + b + 1 = 0$. Recall also that $a^2 = b$ and $b^2 = a$.
2. Recall that the only degree-two polynomial in $\mathbb{F}_2[x]$ that is irreducible is $x^2 + x + 1$. Substitute the element a of the larger field K into $x^2 + x + 1$ and observe that a is a root. Similarly, check that b is a root. Thus, even though $x^2 + x + 1$ has no roots in $\mathbb{F}_2$, it has two distinct roots in the larger field K. Deduce that in $K[x]$, the polynomial $x^2 + x + 1$ factors as $(x + a)(x + b)$.

Fields of the type constructed in the last three exercises are not mere curiosities. The existence of finite fields other than the fields $\mathbb{F}_p$ has been of tremendous importance in recent decades in applications of mathematics to a wide variety of areas, including cryptography and computer science. In Section 17.3 we will take a brief look at the classification of finite fields; in particular, we will learn for which positive integers n there exists a finite field with n elements.

Polynomial Congruence Rings

14.1 A Construction of New Rings

We saw in Section 13.4 that we can construct a field extension $F[\sqrt{m}]$ of a field F in which m, an element of F that is not a square, has a square root. An equivalent way of looking at this construction is that we started with a field F and a quadratic (second-degree) polynomial $x^2 - m$ that does not have a root in F, and we found a field extension in which the polynomial does have a root.

With this construction as motivation, we raise the following more general question: For a field F and a positive-degree polynomial $m(x)$ in $F[x]$, if $m(x)$ has no root in F, is there a field extension K of F in which $m(x)$ does have a root?

If our field F happens to be a subfield of the complex numbers, such as $\mathbb{Q}$, then we know by the fundamental theorem of algebra (Theorem 10.7) that every polynomial with coefficients in F factors completely in $\mathbb{C}[x]$, and therefore has roots in the field extension $\mathbb{C}$.

Instead of considering subfields of the complex numbers, we would like to try to answer our question in greater generality, for an arbitrary field F, and in order to do so, we will proceed in two steps. First, we will introduce a ring construction that starts from a polynomial $m(x)$ of positive degree in $F[x]$ and produces a new ring $F[x]_{m(x)}$ analogous to the ring $\mathbb{Z}_m$, which was built from $\mathbb{Z}$ and a choice of integer $m > 1$ using the notion of congruence classes. Then we will show, in the case of $m(x)$ irreducible, that the ring $F[x]_{m(x)}$ is a field (that is, once again we get closure under multiplicative inverses for free). This process will take the entire chapter to complete.

Let us review the construction of Section 13.4. Starting with a field F and an element m of F that is not a square, we introduced a new element γ designed to serve as the square root of m. We then declared K to be the set of elements of the form $a + b\gamma$, where a and b are elements of F, and γ is the new element introduced and given the property $\gamma^2 = m$. We defined addition and multiplication in K by the rules

$$(a + b\gamma) + (c + d\gamma) = (a + c) + (b + d)\gamma$$

and

$$(a + b\gamma)(c + d\gamma) = ac + ad\gamma + bc\gamma + bd\gamma^2 = (ac + bdm) + (ad + bc)\gamma.$$

In this construction we can think of the elements of K as polynomials of degree at most 1 in the "variable" γ, with coefficients in F. The elements of K are not really polynomials, because elements of K do not multiply like polynomials, and γ is not really a variable, because it has the property that $\gamma^2 = m$. The product of two genuine polynomials of degrees s and t has degree $s + t$. In K, however, we have declared γ^2 to be the constant m, so that every element has degree either 0 or 1 in γ, and products do not obey the degree formula. It was a straightforward matter to verify that K is a ring. In order to show that K is a field, we explicitly showed that each nonzero $a + b\gamma$ in K has a multiplicative inverse, the element

$$\frac{a}{a^2 - mb^2} - \frac{b}{a^2 - mb^2}\gamma.$$

As a first step in extending this construction idea to other settings, suppose we want to build a field, or a ring at least, containing $\mathbb{Q}$ and a cube root of 2. Using as a guide our construction of a ring containing $\mathbb{Q}$ and a square root of 2, we begin by creating a new number γ designed to be a cube root of 2. We construct a new ring K with elements of the form $a + b\gamma + c\gamma^2$, where the coefficients a, b, and c are in $\mathbb{Q}$. Since we intend for γ to be a cube root of 2, we assign γ the property that it satisfies the equation $\gamma^3 = 2$ in K.

We now need to define addition and multiplication: For two elements $(a + b\gamma + c\gamma^2)$ and $(d + e\gamma + f\gamma^2)$ of K we define addition by the rule

$$(a + b\gamma + c\gamma^2) + (d + e\gamma + f\gamma^2) = (a + d) + (b + e)\gamma + (c + f)\gamma^2.$$

Multiplication is defined using the rule $\gamma^3 = 2$. If γ were a genuine polynomial variable, we would have

$$(a + b\gamma + c\gamma^2)(d + e\gamma + f\gamma^2)$$
$$= ad + (ae + bd)\gamma + (af + be + cd)\gamma^2 + (bf + ce)\gamma^3 + cf\gamma^4.$$

But γ is not a genuine variable: It satisfies $\gamma^3 = 2$, and therefore also $\gamma^4 = 2\gamma$. Replacing γ^3 and γ^4 in the above formula by 2 and 2γ, we obtain

$$(a + b\gamma + c\gamma^2)(d + e\gamma + f\gamma^2)$$
$$= (ad + 2bf + 2ce) + (ae + bd + 2cf)\gamma + (af + be + cd)\gamma^2.$$

We adopt this last equation as the multiplication rule for elements of K. Elements of K are polynomial-like expressions of degree at most two in γ, and the multiplication rule ensures that when we multiply two of these expressions we get another polynomial-like expression in γ of degree at most two.

Exercise 14.1. Let K be the collection of polynomial-like expressions in γ just introduced, with $\gamma^3 = 2$.

1. Check that K is a ring.
2. Observe that K contains $\mathbb{Q}$ and that γ is a cube root of 2 in K.
3. Is K a field? A nonzero element of K has the form $a + b\gamma + c\gamma^2$, with at least one of a, b, and c nonzero. We want to know whether such an element always has a multiplicative inverse. An inverse would have the form $d + e\gamma + f\gamma^2$ for rational numbers d, e, and f chosen in such a way that

$$\left(a + b\gamma + c\gamma^2\right)\left(d + e\gamma + f\gamma^2\right) = 1.$$

   Expand the left side and get three equations in the three unknowns d, e, and f, with a, b, and c regarded as known constants in $\mathbb{Q}$. Deduce that if you can solve these three equations, you can find an inverse to $a+b\gamma+c\gamma^2$. Do not try to solve these equations. Simply draw the conclusion that if you could show that these three equations are solvable for any choice of coefficients (a, b, c) besides $(0, 0, 0)$, then that would prove that K is a field.
4. The algebraic manipulations needed to compute the inverse in general are messy, but to see how it is done with a numerical example, find the multiplicative inverse of $2 + \gamma - \gamma^2$.

   Although we did not carry out the detailed calculations, in fact, the ring K of Exercise 14.1 is a field. Now that we have constructed fields containing square roots and cube roots of 2, let us try the more general construction of a ring containing $\mathbb{Q}$ and an nth root of 2 for *any* integer n greater than 1. Again, we begin with the creation of a new element γ that by definition satisfies $\gamma^n = 2$. With γ at our disposal we can now construct a new ring K whose elements have the form

$$a_0 + a_1\gamma + a_2\gamma^2 + \cdots + a_{n-1}\gamma^{n-1},$$

with all the coefficients a_i lying in $\mathbb{Q}$. In other words, the elements look like polynomials in γ of degree less than n with coefficients in $\mathbb{Q}$. We define addition in the obvious way. To define multiplication, we declare that $\gamma^n = 2$. It follows that $\gamma^{n+1} = 2\gamma$, that $\gamma^{n+2} = 2\gamma^2$, and so on, with $\gamma^{2n} = 4$. Then $\gamma^{2n+1} = 4\gamma$, and so on, with $\gamma^{3n} = 8$. What we see is that we can always write higher powers of γ in terms of the powers $1, \gamma, \ldots, \gamma^{n-1}$. Let us be more precise about this.

Exercise 14.2. Let K be the collection of polynomial-like expressions in γ just introduced, with $\gamma^n = 2$.

1. Show that for an arbitrary positive integer m one can write $\gamma^m = 2^q\gamma^r$ for unique nonnegative integers q and r with $r < n$. (Hint: Use the division theorem for integers to write $m = nq + r$.)

2. Suppose $a_0 + a_1\gamma + a_2\gamma^2 + \cdots + a_{n-1}\gamma^{n-1}$ and $b_0 + b_1\gamma + b_2\gamma^2 + \cdots + b_{n-1}\gamma^{n-1}$ are two elements of K. Using the result of part 1, show that you can define a multiplication rule for these two elements by treating them first as ordinary polynomials in γ and multiplying, then replacing the higher powers of γ by terms involving exponents less than n, so that the result is another element of K, a polynomial expression in γ of degree less than n.

3. Is K a field? Do not try to give a complete answer. Instead, think about the issue along the lines discussed in the previous exercise and show that the question can be reduced to the problem of solving a family of n linear equations in n unknowns.

We can think of the equation $\gamma^n = 2$ in the ring above as a "rewrite rule," telling us that every time we see an expression γ^n we can replace it with a 2. This allows us to convert any polynomial in γ with rational coefficients and degree n or greater to a polynomial in γ of degree less than n. Similarly, in constructing the complex numbers from the real numbers, we can view $i^2 = -1$ as a rewrite rule, telling us how to convert any polynomial in i with real coefficients into a polynomial of degree at most 1 in i. For instance, $i^7 - 4i^5 + 3i^4 + 2i^2$ becomes $-i - 4i + 3 - 2$, which is $-5i + 1$. Also, we can regard the construction of the ring $\mathbb{Z}_m$ as the process of applying the rewrite rule $m = 0$ to the ring $\mathbb{Z}$, allowing us to convert any integer to an integer between 0 and $m - 1$.

Let us consider as our next example the polynomial $x^3 - x - 1$ in $\mathbb{Q}[x]$. We can show by any of several methods that $x^3 - x - 1$ has no root in $\mathbb{Q}$, or equivalently in this case, that $x^3 - x - 1$ is irreducible in $\mathbb{Q}[x]$. We want to build a larger field K that contains $\mathbb{Q}$ and has a root of $x^3 - x - 1$ in it. Therefore, we introduce a new element γ that we define to be a root of $x^3 - x - 1$. Thus, this time, γ satisfies $\gamma^3 - \gamma - 1 = 0$, or equivalently,

$$\gamma^3 = \gamma + 1.$$

Let K be the set of polynomial-like expressions $a + b\gamma + c\gamma^2$, where the coefficients a, b, and c are rational numbers. Regard the equation $\gamma^3 = \gamma + 1$ as a rewrite rule that allows us to convert any polynomial of degree 3 or more in γ with rational coefficients to one of degree at most 2. Define addition in K by the rule

$$(a + b\gamma + c\gamma^2) + (d + e\gamma + f\gamma^2) = (a + d) + (b + e)\gamma + (c + f)\gamma^2.$$

To define multiplication, use the rewrite rule $\gamma^3 = \gamma + 1$.

Exercise 14.3. Let K be the set of elements $a + b\gamma + c\gamma^2$, where a, b, and c are rational numbers, as above.

1. Observe that if the rule $\gamma^3 = \gamma + 1$ is to hold in K, then also $\gamma^4 = \gamma^2 + \gamma$ must hold.

2. Using these two rules, multiply $\left(a + b\gamma + c\gamma^2\right)\left(d + e\gamma + f\gamma^2\right)$ to get a polynomial in γ of degree 4; then rewrite the result in terms of 1, γ, and γ^2. Take this as the definition of multiplication in K and check that K is a ring containing $\mathbb{Q}$ and a root of $x^3 - x - 1$.

3. Is K a field? Once again, show that this question can be reduced to the problem of solving three linear equations in three unknowns, but do not try to solve these equations.

We would like to introduce a general construction that includes as special cases the constructions of Sections 13.4 and this section. Start with a field F and a polynomial $m(x)$ in $F[x]$ of positive degree. We want to build a field extension K of F containing a root of $m(x)$. Let us build K as a ring and postpone the question of whether K is a field until Section 14.5.

We may as well assume that the polynomial $m(x)$ is monic, since $m(x)$ and its constant multiples have the same roots in any field extension of F. Since $m(x)$ is monic, it can be written in the form

$$x^n - a_{n-1}x^{n-1} - \cdots - a_1 x - a_0$$

for some elements a_i of F. Form the set K consisting of polynomial-like expressions in γ of degree less than n with coefficients in F. A typical element of K has the form

$$b_0 + b_1\gamma + b_2\gamma^2 + \cdots + b_{n-1}\gamma^{n-1}.$$

We can think of K as the set of polynomials of degree less than n in the "variable" γ. Addition is defined in the usual way, as if elements of K were ordinary polynomials.

If $f(x)$ and $g(x)$ are two polynomials in the ring $F[x]$, with degrees i and j that are both less than n, we have corresponding elements $f(\gamma)$ and $g(\gamma)$ in K, and we want a rule for their product $f(\gamma)g(\gamma)$. If $i + j < n$, we can take $f(\gamma)g(\gamma)$ to be the product as ordinary polynomials in γ. However, if $i + j \geq n$, then the ordinary polynomial product $f(x)g(x)$ has degree greater than or equal to n; replacing x by γ produces an expression in γ of degree too large to be an element of the set K. We need a rewrite rule that allows us to replace higher powers of γ, such as γ^n and γ^{n+1}, with expressions involving only powers of γ with exponents less than n.

To find a suitable rewrite rule, we recall that we want γ to be a root of the polynomial $m(x)$. In other words, we want γ to satisfy $m(\gamma) = 0$, which would mean that

$$\gamma^n - a_{n-1}\gamma^{n-1} - \cdots - a_1\gamma - a_0 = 0,$$

or equivalently,

$$\gamma^n = a_{n-1}\gamma^{n-1} + \cdots + a_1\gamma + a_0.$$

Let us take this as our rewrite rule. It allows us to rewrite γ^n as a polynomial-like expression in γ of degree less than n.

Once we adopt the rewrite rule for γ^n, we can rewrite higher powers of γ as expressions in K. For example, consider γ^{n+1}. Using the rewrite rule once, we get

$$\gamma^{n+1} = \gamma^n \gamma = \left(a_{n-1}\gamma^{n-1} + \cdots + a_1\gamma + a_0 \right)\gamma$$
$$= a_{n-1}\gamma^n + a_{n-2}\gamma^{n-1} + \cdots + a_1\gamma^2 + a_0\gamma.$$

This involves γ^n, to which we can apply the rewrite rule a second time to obtain

$$\gamma^{n+1} = a_{n-1}\left(a_{n-1}\gamma^{n-1} + \cdots + a_1\gamma + a_0 \right) + a_{n-2}\gamma^{n-1} + \cdots + a_1\gamma^2 + a_0\gamma$$
$$= \left(a_{n-1}^2 + a_{n-2} \right)\gamma^{n-1} + (a_{n-1}a_{n-2} + a_{n-3})\gamma^{n-2} + \cdots$$
$$+ (a_{n-1}a_1 + a_0)\gamma + a_{n-1}a_0.$$

In this way, we obtain an expression for γ^{n+1} as a polynomial in γ of degree at most $n-1$. We can continue to rewrite higher powers of γ as polynomials in γ of degree at most $n-1$ and use this to obtain a multiplication rule in K.

Exercise 14.4. Suppose F is a field, $m(x)$ is a monic polynomial in $F[x]$ of positive degree n, and K is the set of polynomial-like expressions in γ of degree less than n with coefficients in F.

1. Let $m(x) = x^n - a_{n-1}x^{n-1} - \cdots - a_1x - a_0$. Using the rewrite rule $\gamma^n = a_{n-1}\gamma^{n-1} + \cdots + a_1\gamma + a_0$, prove by induction that for every integer $t \geq n$ one can express γ^t as a polynomial-like expression in γ of degree less than n.
2. Conclude that the product of two elements of K is another element of K, so that K is a ring.

Thus starting with a field F and a monic polynomial of positive degree, we have succeeded in building a new ring K consisting of polynomials in the element γ with coefficients from F, subject to the rewrite rule $m(\gamma) = 0$. The ring K should be regarded as the polynomial analogue of the ring $\mathbb{Z}_m$. The analogy can be understood better if we introduce the notion of congruence for polynomials. We will do so in the next section.

14.2 Polynomial Congruences

We will introduce and study the basic properties of congruence for polynomials, proceeding in parallel with the discussion of congruences for integers in Section 4.1. Let F be a field, and let $m(x)$ be a polynomial in $F[x]$ of positive degree. Two polynomials $a(x)$ and $b(x)$ of $F[x]$ are *congruent modulo $m(x)$* if $a(x)$ and $b(x)$ have the same remainder upon division by $m(x)$; an equivalent condition for $a(x)$ and $b(x)$ to be congruent modulo $m(x)$ is that $m(x)$ divide

$b(x) - a(x)$. We denote the relation that "$a(x)$ is congruent to $b(x)$ modulo $m(x)$" by writing

$$a(x) \equiv b(x) \ (\mathrm{mod} \ m(x)).$$

For example,

$$x^7 + 2x^2 + 5 \equiv x^{11} - 8x^3 + 5 \ (\mathrm{mod} \ x)$$

in $\mathbb{Q}[x]$, since x divides the difference of the two polynomials. Another example is

$$x^5 + 3x^2 - 7 \equiv x^3 - 18x^2 + x + 13 \ (\mathrm{mod} \ x - 1)$$

in $\mathbb{Q}[x]$, since $x - 1$ divides the difference of the two polynomials.

Exercise 14.5. To gain some practice in working with polynomial congruences, carry out the following calculations.

1. Verify that

$$x^5 + 7x^3 - 5x^2 + 2x + 1$$
$$\equiv x^5 - x^4 + 4x^3 - 10x^2 - 7x - 5 \ (\mathrm{mod} \ x^3 + 3x + 2)$$

   in $\mathbb{Q}[x]$.
2. Find the value of the coefficient c in the field $\mathbb{F}_5$ that makes

$$x^4 + 3x^3 + 2x + 1 \equiv 3x^4 + x^2 + cx \ (\mathrm{mod} \ x^2 + x + 2)$$

   a valid congruence in $\mathbb{F}_5[x]$.
3. Is there a value of the coefficient c in the field $\mathbb{R}$ that makes

$$x^4 + 3x^3 + 2x + 1 \equiv 3x^4 + x^2 + cx \ (\mathrm{mod} \ x^2 + x + 2)$$

   a valid congruence in $\mathbb{R}[x]$? If so, what is it; if not, why?

Recall that for an integer $m > 1$, an integer a is congruent modulo m to a unique integer r between 0 and $m - 1$. If a is positive, r is obtained as the remainder when a is divided by m. The division theorem for polynomials, Theorem 9.5, has a similar statement as a consequence.

Theorem 14.1. *Let F be a field and let $m(x)$ be a polynomial in $F[x]$ of positive degree n. Every polynomial $a(x)$ in $F[x]$ is congruent modulo $m(x)$ to exactly one polynomial of degree less than n.*

Exercise 14.6. Prove Theorem 14.1.

We will call the unique polynomial $r(x)$ of degree less than d that is congruent to $a(x)$ modulo $m(x)$ the *lowest-degree residue* of $a(x)$ modulo $m(x)$, in analogy with the notion of an integer's least nonnegative residue. As an example, in $\mathbb{R}[x]$, the lowest-degree residue of $x^3 + 1$ modulo $x^2 + x + 2$ is $-x + 3$. After all, if we divide $x^3 + 1$ by $x^2 + x + 2$ using long division, we obtain a quotient of $x - 1$ and a remainder of $-x + 3$. This means that

$$x^3 + 1 = (x^2 + x + 2)(x - 1) + (-x + 3),$$

so that $x^3 + 1$ is congruent to $-x + 3$ modulo $x^2 + x + 2$.

Exercise 14.7. Determine the lowest-degree residues of

1. $x^2 - 4x + 77$ modulo $x^5 + x^4 + x + 1$ in $\mathbb{C}[x]$;
2. $x^{17} + 5x^2 - 3x + 498$ modulo x in $\mathbb{R}[x]$;
3. $x^4 - 1$ modulo $x^3 - 2x^2 + x - 2$ in $\mathbb{Q}[x]$;
4. $x^7 + x$ modulo $x^6 + 1$ in $\mathbb{F}_2[x]$.

Propositions 4.2, 4.3, and 4.4 provide evidence that integer congruences behave "sort of" like integer equalities. The same is true for polynomial congruences:

Proposition 14.2 *Let F be a field and let $m(x)$ be a positive-degree polynomial in $F[x]$.*

1. *Suppose $a(x)$, $b(x)$, and $c(x)$ are polynomials in $F[x]$ satisfying*

$$a(x) \equiv b(x) \;(\mathrm{mod}\; m(x)) \quad \text{and} \quad b(x) \equiv c(x) \;(\mathrm{mod}\; m(x)).$$

Then
$$a(x) \equiv c(x) \;(\mathrm{mod}\; m(x)).$$

2. *Suppose $a(x)$, $b(x)$, $e(x)$, and $f(x)$ are polynomials in $F[x]$ satisfying*

$$a(x) \equiv e(x) \;(\mathrm{mod}\; m(x)) \text{ and } b(x) \equiv f(x) \;(\mathrm{mod}\; m(x)).$$

Then

$$a(x)+b(x) \equiv e(x)+f(x) \;(\mathrm{mod}\; m(x)) \quad \text{and} \quad a(x)b(x) \equiv e(x)f(x) \;(\mathrm{mod}\; m(x))$$

3. *Suppose $a(x)$ and $b(x)$ are polynomials in $F[x]$ satisfying*

$$a(x) \equiv b(x) \;(\mathrm{mod}\; m(x)).$$

Then for any other polynomial $r(x)$, the congruence

$$r(x)a(x) \equiv r(x)b(x) \;(\mathrm{mod}\; m(x))$$

holds.

Exercise 14.8. Prove Proposition 14.2 by mimicking the proofs of Propositions 4.2, 4.3, and 4.4.

There is also a polynomial analogue of Theorem 4.5, which states that a factor can be canceled from a congruence under certain circumstances.

Theorem 14.3. *Let F be a field and let $r(x)$ and $m(x)$ be relatively prime polynomials in $F[x]$, with $m(x)$ of positive degree. If $a(x)$ and $b(x)$ are polynomials for which*

$$r(x)a(x) \equiv r(x)b(x) \;(\mathrm{mod}\; m(x)),$$

then
$$a(x) \equiv b(x) \;(\mathrm{mod}\; m(x)).$$

Exercise 14.9. Prove Theorem 14.3.

Recall that the collection of all integers congruent to a particular integer a modulo m is defined to be the congruence class of a modulo m. Theorem 4.1 can be interpreted as saying that every congruence class modulo m contains a unique nonnegative integer less than m. Suppose that F is a field and that $m(x)$ is a polynomial of positive degree in $F[x]$. The collection of all polynomials in $F[x]$ congruent to a particular polynomial $a(x)$ modulo $m(x)$ is called the *congruence class* of $a(x)$ modulo $m(x)$.

For example, in $\mathbb{R}[x]$, the congruence class of $x^2 - 1$ modulo $2x^3 - x + 1$ is the collection of all polynomials $b(x)$ satisfying

$$b(x) \equiv x^2 - 1 \pmod{2x^3 - x + 1}.$$

These polynomials can be described explicitly. For $b(x)$ to be in this congruence class, the difference $b(x) - (x^2 - 1)$ must be divisible by $2x^3 - x + 1$. This means that $b(x) - (x^2 - 1)$ is a polynomial multiple of $2x^3 - x + 1$, and thus $b(x)$ must have the form

$$p(x) \left(2x^3 - x + 1\right) + x^2 - 1,$$

where $p(x)$ is any polynomial in $F[x]$.

Exercise 14.10. Give a description of all the polynomials in each of the following congruence classes.

1. the congruence class of $x^5 + 3$ in $\mathbb{R}[x]$ modulo x;
2. the congruence class of $x^3 + x^2 + 1$ in $\mathbb{F}_2[x]$ modulo $x + 1$.

The following theorem gives a general description of any congruence class.

Theorem 14.4. *Let F be a field and let $m(x)$ be a polynomial in $F[x]$ of positive degree n. Let $q(x)$ be a polynomial in $F[x]$. Then the congruence class of $q(x)$ modulo $m(x)$ consists of all polynomials of the form $p(x)m(x) + q(x)$ for $p(x)$ an element of $F[x]$.*

Exercise 14.11. Prove Theorem 14.4.

Each individual polynomial in a congruence class of polynomials of $F[x]$ modulo $m(x)$ is called a *representative* of that congruence class. For integers we introduced the bracket notation to describe a congruence class in terms of one of its representatives. Thus, the congruence class represented by 2 modulo 5 is written as $[2]_5$, and the congruence class represented by 18 modulo 7 is $[18]_7$. Since the same congruence class has many representatives, the same congruence class can be written in many ways. For example, the congruence class that contains 2 modulo 5, which we wrote just above as $[2]_5$, can also be written as $[7]_5$, as $[-23]_5$, and as $[111\,883\,242]_5$.

Let us adopt a similar notation for polynomials, so that the congruence class of $a(x)$ in $F[x]$ modulo $m(x)$ will be written as $[a(x)]_{m(x)}$. For example, in $\mathbb{R}[x]$, the congruence class containing $x^3 + 2x + 1$ modulo $x^2 - x + 2$ will be written as $[x^3 + 2x + 1]_{x^2-x+2}$. (Check that this is also $[x - 1]_{x^2-x+2}$.)

Theorem 14.1 has the following interpretation in terms of congruence classes.

Theorem 14.5. *Let F be a field and let $m(x)$ be a polynomial in $F[x]$ of positive degree n. Every congruence class of polynomials in $F[x]$ has exactly one representative of degree less than n.*

Exercise 14.12. Prove Theorem 14.5.

14.3 Polynomial Congruence Rings

When we introduced the ring $\mathbb{Z}_m$, for m an integer greater than 1, we regarded it as the collection of special symbols $0, 1, \ldots, m-1$ added and multiplied subject to the "rewrite" rule $m = 0$. In Section 6.5 we came to understand $\mathbb{Z}_m$ as the collection of congruence classes of integers modulo m. To add two congruence classes $[a]_m$ and $[b]_m$, we simply pass to the congruence class $[a+b]_m$, and to multiply the same two congruence classes, we pass to $[a \cdot b]_m$. Proposition 4.3 ensures that these definitions of addition and multiplication make sense: Rephrased in terms of congruence classes, it states that if a and e represent a congruence class modulo m, and b and f represent a congruence class modulo m, then the congruence class of $a + b$ is the same as the congruence class of $e + f$, and the congruence class of ab is the same as the congruence class of ef. This is what makes addition and multiplication of congruence classes work, allowing us to make the collection of congruence classes modulo m into a ring, the ring $\mathbb{Z}_m$. (See Exercise 6.23.)

Proposition 14.2 yields analogous results for congruence classes of polynomials.

Exercise 14.13. Let F be a field and let $m(x)$ be a polynomial of positive degree in $F[x]$. Consider two polynomials $a(x)$ and $b(x)$ in $F[x]$.

1. Suppose $e(x)$ is a polynomial in the congruence class $[a(x)]_{m(x)}$ and $f(x)$ is a polynomial in the congruence class $[b(x)]_{m(x)}$. Show that

$$[e(x) + f(x)]_{m(x)} = [a(x) + b(x)]_{m(x)}$$

and

$$[e(x)f(x)]_{m(x)} = [a(x)b(x)]_{m(x)}.$$

2. Define addition and multiplication for the set of congruence classes of $F[x]$ modulo $m(x)$ by setting the sum of congruence classes

$$[a(x)]_{m(x)} + [b(x)]_{m(x)}$$

equal to the congruence class

$$[a(x) + b(x)]_{m(x)},$$

and the product of congruence classes

$$[a(x)]_{m(x)} \cdot [b(x)]_{m(x)}$$

equal to the congruence class

$$[a(x) \cdot b(x)]_{m(x)}.$$

3. Show that with respect to these rules of addition and multiplication, $[0]_{m(x)}$ is an additive identity and $[1]_{m(x)}$ is a multiplicative identity. Show further that the collection of congruence classes modulo $m(x)$ forms a ring.

We can write $F[x]_{m(x)}$ for the new ring we have constructed, the ring of congruence classes of polynomials in $F[x]$ modulo $m(x)$.

Exercise 14.14. Let us consider some simple examples of rings of congruence classes.

1. Show that there are four congruence classes in $\mathbb{F}_2[x]$ modulo x^2. Make addition and multiplication tables for the ring $\mathbb{F}_2[x]_{x^2}$. Using the tables, determine the units in the ring. Is $\mathbb{F}_2[x]_{x^2}$ a field?
2. Show that the ring $F[x]_x$ can be identified with the field F itself.
3. Show further that we can identify $F[x]_{m(x)}$ with F for *any* first-degree polynomial $m(x)$ in $F[x]$.

Exercise 14.15. Assume that $m(x)$ is a polynomial of positive degree in $F[x]$.

1. Show that in $F[x]_{m(x)}$, the collection of congruence classes of degree-zero polynomials (constants) is closed under addition and multiplication. Thus this collection forms a ring inside $F[x]_{m(x)}$.
2. Identify this ring with F.
3. Explain how this exercise generalizes part 3 of the previous exercise.

The preceding exercise shows that F sits naturally inside the ring $F[x]_{m(x)}$. Therefore, we may think of $F[x]_{m(x)}$ as an extension of F. For notational ease, it will be convenient to drop the brackets around congruence classes of elements of F when they are regarded as elements of $F[x]_{m(x)}$. In particular, we will write 0 for the congruence class $[0]_{m(x)}$, and 1 for the congruence class $[1]_{m(x)}$. Since these are the additive and multiplicative identities of $F[x]_{m(x)}$, this will cause no confusion.

Every element in the ring $F[x]_{m(x)}$ can be written as a polynomial expression in the congruence class of x, as we see in the next exercise.

Exercise 14.16. Continue to assume that $m(x)$ is a polynomial of positive degree in $F[x]$.

1. Using the definition of multiplication, show that $[x]^2_{m(x)} = [x^2]_{m(x)}$ and $[x]^3_{m(x)} = [x^3]_{m(x)}$. [Note: The notation $[x]^2_{m(x)}$ means $([x]_{m(x)})^2$.]

2. Use induction to show more generally that

$$[x]^t_{m(x)} = [x^t]_{m(x)}$$

for every positive integer t.

3. Suppose that $a_t x^t + a_{t-1} x^{t-1} + \cdots + a_1 x + a_0$ is a polynomial in $F[x]$. Show that

$$a_t [x]^t_{m(x)} + a_{t-1} [x]^{t-1}_{m(x)} + \cdots + a_1 [x]_{m(x)} + a_0$$
$$= \left[a_t x^t + a_{t-1} x^{t-1} + \cdots + a_1 x + a_0 \right]_{m(x)}$$

in $F[x]_{m(x)}$. (Hint: Use the definition of addition and induction.)

4. Suppose that $m(x)$ has degree n. For simplicity, write γ for the congruence class of x in $F[x]$ modulo $m(x)$; that is, $\gamma = [x]_{m(x)}$. Show that every element of the ring $F[x]_{m(x)}$ can be written uniquely in $F[x]_{m(x)}$ in the form

$$a_{n-1} \gamma^{n-1} + a_{n-2} \gamma^{n-2} + \cdots + a_1 \gamma + a_0.$$

(Hint: Recall that every polynomial in $F[x]$ is congruent modulo $m(x)$ to a unique polynomial of degree less than n. Equivalently, every congruence class in $F[x]$ contains a unique polynomial of degree less than n. Then use the preceding part of the exercise, substituting γ for $[x]_{m(x)}$.)

Suppose $f(T)$ is a polynomial with coefficients from F. We use the uppercase "T" here to suggest a variable, as opposed to an element of a specific ring. We are interested in the possibility of substituting particular elements of a ring in place of T. As in Exercise 14.16, let us use γ to denote the congruence class $[x]_{m(x)}$ of x. The above exercise ensures that we can substitute the congruence class γ for T, and that when we do, we obtain the identity

$$f(\gamma) = f([x]_{m(x)}) = [f(x)]_{m(x)}$$

in the ring $F[x]_{m(x)}$. The left-hand side of this identity is the polynomial expression in the congruence class γ. This is computed in the ring $F[x]_{m(x)}$, with the coefficients of $f(T)$ identified with congruence classes of degree-zero polynomials. The right-hand side is simply the congruence class of $f(x)$ in $F[x]_{m(x)}$. In particular, taking $f(T)$ to be the polynomial $m(T)$ itself, we find that

$$m(\gamma) = m([x]_{m(x)}) = [m(x)]_{m(x)} = [0]_{m(x)} = 0.$$

More generally, if $f(T)$ is any polynomial divisible by $m(T)$, we obtain

$$f(\gamma) = f([x]_{m(x)}) = [f(x)]_{m(x)} = [0]_{m(x)} = 0,$$

since $f(x)$ is congruent to 0 modulo $m(x)$. One consequence of this observation is that the congruence class γ satisfies the rewrite rule $m(\gamma) = 0$ in $F[x]_{m(x)}$.

We are now in a position to identify $F[x]_{m(x)}$ with the ring K constructed in Exercise 14.4.

Exercise 14.17. Suppose F is a field, $m(x)$ is a monic polynomial in $F[x]$ of positive degree n, and K is the ring of polynomial-like expressions in γ of degree less than n with coefficients in F, as constructed in Exercise 14.4. Identify each element $a(\gamma)$ of K with the congruence class $[a(x)]_{m(x)}$ in $F[x]_{m(x)}$. Under this identification, show that K and $F[x]_{m(x)}$ are the same ring.

We constructed the ring K of Exercise 14.4 in a concrete manner designed to ensure that it contained an element γ satisfying $m(\gamma) = 0$. However, the direct approach to the construction of K left us with no tools we could apply in order to answer the basic question of when K is a field. By identifying K with the ring $F[x]_{m(x)}$ of congruence classes, we make available the theory of congruences. We also introduce the possibility of drawing analogies from the family of rings of the form $\mathbb{Z}_m$ to our rings $F[x]_{m(x)}$.

In Theorem 6.2 we described the units of $\mathbb{Z}_m$ as the elements $[a]_m$ whose representatives a are relatively prime to m. This was proved in Exercise 6.21, where you were asked to draw the conclusion that $\mathbb{Z}_m$ is a field precisely when m is a prime number. You might expect that similar results hold for the ring $F[x]_{m(x)}$, and they do. To prove them, we will obtain some results on equations and congruences with polynomial unknowns.

14.4 Equations and Congruences with Polynomial Unknowns

In order to proceed further in our analysis of the new rings of the form $F[x]_{m(x)}$, we need some additional consequences of the polynomial Bézout's theorem. One important application of the integer Bézout's theorem was the following result, whose proof we will review.

Theorem 14.6. *Let a and b be integers with greatest common divisor d. If e is an integer, the equation*

$$aU + bV = e,$$

for unknowns U and V, has an integer solution if and only if d divides e. In particular, the equation

$$aU + bV = 1$$

has an integer solution if and only if a and b are relatively prime.

Proof. If there is an integer solution to the equation $aU + bV = e$, say $U = r$ and $V = s$, then the fact that d divides a and b (by hypothesis) implies that d divides $ar + bs$ (why?), which is e. Conversely, suppose d divides e, with $e = cd$ for some integer c. Bézout's theorem guarantees the existence of integers p and q satisfying $ap + bq = d$. We can multiply both sides of this equation by c to obtain a solution to $aU + bV = e$, namely, $U = pc$, $V = qc$.

Let us consider the polynomial analogue of Theorem 14.6. Suppose we wish to solve the equation

$$\left(x^2 + x + 1\right) U + \left(x^2 + 1\right) V = 1$$

in $\mathbb{Q}[x]$. Here U and V are unknowns, while $x^2 + x + 1$ and $x^2 + 1$ are constants, in the sense that they are known polynomials. We wish to find specific polynomials $r(x)$ and $s(x)$ in $\mathbb{Q}[x]$ so that we can substitute $r(x)$ for U and $s(x)$ for V in order to make the equation hold.

The polynomial Bézout's theorem (Theorem 12.10) tells us that we can do this if the two polynomials $x^2 + x + 1$ and $x^2 + 1$ are "relatively prime," that is, if they have the constant polynomial 1 as their greatest common divisor. Let us see why.

First we observe that $x^2 + x + 1$ and $x^2 + 1$ indeed have 1 as their greatest common divisor. By Bézout's theorem, there thus exist polynomials $r(x)$ and $s(x)$ in $\mathbb{Q}[x]$ such that

$$\left(x^2 + x + 1\right) r(x) + \left(x^2 + 1\right) s(x) = 1.$$

The Euclidean algorithm allows us to find such polynomials $r(x)$ and $s(x)$ explicitly. Carrying out the Euclidean algorithm for the pair $x^2 + x + 1$ and $x^2 + 1$, we find that

$$x^2 + x + 1 = \left(x^2 + 1\right) \cdot 1 + x$$

and

$$x^2 + 1 = x \cdot x + 1.$$

Working backwards, we obtain from this that

$$\left(x^2 + x + 1\right)(-x) + \left(x^2 + 1\right)(x + 1) = 1.$$

Thus we obtain the solution $U = -x$ and $V = x + 1$.

In general, if we are working in a field F and are considering polynomials $a(x)$ and $b(x)$ in $F[x]$ with greatest common divisor $d(x)$, Bézout's theorem states that there exist polynomials $r(x)$ and $s(x)$ in $F[x]$ satisfying

$$a(x)r(x) + b(x)s(x) = d(x).$$

We can rephrase this to say that there is a solution in $F[x]$ to the polynomial equation

$$a(x)U + b(x)V = d(x).$$

Here U and V are unknowns, and the polynomials $a(x)$, $b(x)$, and $d(x)$ are the constants of the equation, in the sense that they are known polynomials. Just as in the example above, Bézout's theorem guarantees that solutions exist, and the Euclidean algorithm gives us a way to find them.

Theorem 14.7. *Let F be a field, let $a(x)$ and $b(x)$ be polynomials in $F[x]$ with greatest common divisor $d(x)$, and let $e(x)$ be a polynomial in $F[x]$. Then the equation*

$$a(x)U + b(x)V = e(x)$$

has a polynomial solution if and only if $d(x)$ divides $e(x)$. In particular, the equation

$$a(x)U + b(x)V = 1$$

has a polynomial solution if and only if $a(x)$ and $b(x)$ are relatively prime.

Exercise 14.18. Prove Theorem 14.7 by imitating the proof of Theorem 14.6.

Exercise 14.19. Decide which of the equations below are solvable. If an equation is solvable, solve it, and if not, explain why not.

1. $\left(x^2 + 1\right)U + x^3V = 1$, in $\mathbb{Q}[x]$.
2. $x^2\,U + x^3\,V = x^4 + 5$, in $\mathbb{F}_7[x]$.
3. $\left(x^3 + 1\right)U + \left(x^3 + x^2 + x + 1\right)V = x^2 + 1$, in $\mathbb{F}_2[x]$.

One can study not just equations with polynomial coefficients and unknowns, but also congruences with polynomial coefficients and unknowns. For example, one can try to solve the congruence

$$\left(x^3 + 3x + 1\right)U \equiv 1 \ \left(\bmod\ x^4 + 1\right)$$

in $\mathbb{R}[x]$. A solution is a polynomial that we can substitute for U in order to obtain a genuine congruence.

Exercise 14.20. Decide whether the polynomial congruence

$$\left(x^3 + 3x + 1\right)U \equiv 1 \ \left(\bmod\ x^4 + 1\right)$$

is solvable in $\mathbb{R}[x]$. If it is, solve it. (Hint: Convert the congruence to an equality involving another unknown polynomial V. Then proceed as you did in the previous exercise.)

Polynomial congruences of the form

$$a(x)U \equiv 1 \ (\bmod\ m(x))$$

are of special importance, just as integer congruences of the form $aU \equiv 1 \ (\bmod\ m)$ are important. We found in Theorem 4.6 that such an integer congruence is solvable if and only if $(a, m) = 1$. The analogous result holds for polynomials:

Theorem 14.8. *Let F be a field. Let $a(x)$ and $m(x)$ be polynomials in $F[x]$ with $m(x)$ of positive degree. The congruence*

$$a(x)U \equiv 1 \ (\bmod\ m(x))$$

is solvable if and only if $(a(x), m(x)) = 1$.

Exercise 14.21. Prove Theorem 14.8. (Hint: Interpret Theorem 14.7 in terms of congruences.) Then prove the corollary below.

Corollary 14.9 *Let F be a field. Suppose $m(x)$ is an irreducible polynomial in $F[x]$ and $a(x)$ is a nonzero polynomial in $F[x]$ of degree less than the degree of $m(x)$. There exists a polynomial $r(x)$ in $F[x]$ such that*

$$a(x)r(x) \equiv 1 \pmod{m(x)}.$$

14.5 Polynomial Congruence Fields

Let F be a field, and let $m(x)$ be a polynomial of positive degree. We would like to know whether the ring $F[x]_{m(x)}$ is a field, or more precisely, we would like to know what conditions must be placed on the polynomial $m(x)$ to ensure that the ring $F[x]_{m(x)}$ is a field. The integer analogue would be to begin with an integer m greater than 1 and ask what condition on m ensures that the ring $\mathbb{Z}_m$ is a field. More generally, we might ask which elements of $F[x]_{m(x)}$ are units, as we asked which elements of $\mathbb{Z}_m$ are units. Recall that we saw in Theorem 6.2 that the units in $\mathbb{Z}_m$ are the congruence classes represented by integers a relatively prime to m. Here is the polynomial analogue, which we are now in a position to prove:

Theorem 14.10. *Let F be a field. Let $a(x)$ and $m(x)$ be polynomials in $F[x]$ with $m(x)$ of positive degree. The congruence class $[a(x)]_{m(x)}$ is a unit in $F[x]_{m(x)}$ if and only if $(a(x), m(x)) = 1$.*

Exercise 14.22. Prove Theorem 14.10. (Hint: Use Theorem 14.8 and the definition of multiplication in $F[x]_{m(x)}$.)

With Theorem 14.10 it becomes a simple matter to determine which polynomial congruence rings $F[x]_{m(x)}$ are fields. We first consider the case of a polynomial $m(x)$ that is not irreducible. In the integer case, if m is not prime, then the ring $\mathbb{Z}_m$ is not a field. Let us review one way to prove this. Since m is not prime, $m = ab$ for integers a and b satisfying $1 < a < m$ and $1 < b < m$. Therefore, neither $[a]_m$ nor $[b]_m$ is zero in $\mathbb{Z}_m$, but $[a]_m[b]_m = [m]_m = 0$. This means that $\mathbb{Z}_m$ has zero-divisors, and since a field cannot have zero-divisors, $\mathbb{Z}_m$ is not a field. The same argument shows that $F[x]_{m(x)}$ is not a field if the polynomial $m(x)$ is not irreducible in $F[x]$.

Exercise 14.23. Let F be a field and suppose $m(x)$ is a polynomial of positive degree that is not irreducible.

1. Since $m(x)$ is not irreducible, it has a nontrivial factorization $a(x)b(x)$ in $F[x]$. Explain why $[a(x)]_{m(x)}$ and $[b(x)]_{m(x)}$ are nonzero elements of $F[x]_{m(x)}$.

2. Show that

$$[a(x)]_{m(x)}[b(x)]_{m(x)} = 0$$

in $F[x]_{m(x)}$.

3. Conclude that $F[x]_{m(x)}$ has zero-divisors and is not a field.

We have just shown in Exercise 14.23 that if $m(x)$ is not irreducible, then $F[x]_{m(x)}$ is not a field. Is the converse true? That is, if $m(x)$ is an irreducible polynomial in $F[x]$, is $F[x]_{m(x)}$ a field? We have already handled one family of special cases: If $m(x)$ has the form $x^2 - a$, and a has no square roots in F, then $x^2 - a$ is irreducible in $F[x]$. The ring we constructed in Exercise 13.18 is $F[x]_{m(x)}$, and we showed that it is indeed a field. To handle the general case, we use Theorem 14.10.

Exercise 14.24. Let F be a field and suppose $m(x)$ is an irreducible polynomial in $F[x]$. Show that $F[x]_{m(x)}$ is a field.

By solving Exercises 14.23 and 14.24 we have proved the following theorem, the polynomial analogue of the familiar theorem that the ring $\mathbb{Z}_m$ is a field if and only if m is prime.

Theorem 14.11. *Suppose that F is a field and that $m(x)$ is a polynomial of positive degree in $F[x]$. The ring $F[x]_{m(x)}$ is a field if and only if $m(x)$ is irreducible.*

Recall the question with which we began this chapter: For a field F and a positive-degree polynomial $m(x)$ in $F[x]$, if $m(x)$ has no root in F, is there a field extension K of F in which $m(x)$ does have a root? Of course, we can assume that $m(x)$ is irreducible, for if an irreducible factor of $m(x)$ in $F[x]$ has a root in a field extension K, so does $m(x)$ itself. We have at last obtained an answer.

Theorem 14.12. *Suppose that F is a field and that $m(x)$ is an irreducible polynomial in $F[x]$. The ring $F[x]_{m(x)}$ is a field extension of F in which the element $[x]_{m(x)}$ is a root of $m(x)$.*

Exercise 14.25. Prove Theorem 14.12.

Part III

All Together Now

15

Euclidean Rings

15.1 Factoring Elements in Rings

The polynomial rings $K[x]$ over fields K share many features with the ring of integers $\mathbb{Z}$. Let us examine the parallels between them more systematically, with two purposes in mind: to gain deeper insight into the results we have obtained and to be in a better position to prove similar theorems for other rings.

The starting point is that each ring, $\mathbb{Z}$ and $K[x]$, has a measure of size. The size of an integer is its absolute value. The size of a polynomial is its degree. In order to talk about $\mathbb{Z}$ and $K[x]$ simultaneously, and other rings as well, let us use a uniform notation. We shall write R for the ring, whichever one it is, and use letters such as r for integers or polynomials in R. In particular, we shall no longer write polynomials with reference to the variable x. Thus, if $R = \mathbb{Z}$, then r is an integer, and if $R = K[x]$, then r is a polynomial. When we do not care which ring we are working in, we shall call r an element rather than an integer or a polynomial.

Let us introduce a common notation for absolute value and degree. Write $N(r)$ for the absolute value of r if r is an integer, and for the degree of r if r is a polynomial. Rather than speaking of absolute value or degree, we shall call $N(r)$ the *size* of r or the *norm* of r. Recall the convention we adopted that the size of the zero polynomial is $-\infty$. Using the common notion of size and the common notation for it, we can formulate the following statement simultaneously for $\mathbb{Z}$ and $K[x]$.

Theorem 15.1. *Let R be the ring of integers or the ring of polynomials over a field and let N be the measure of size for R.*

1. *The unique element of R of smallest size is 0.*
2. *The elements of R of the second-smallest size are precisely the units of R.*
3. *The elements of R of the third-smallest size are irreducible.*
4. *If a and b are nonzero elements of R, then*

$$N\left(a\right) \leq N\left(ab\right).$$

Moreover, equality holds if and only if b is a unit.

Exercise 15.1. Prove Theorem 15.1. You will find that the first three statements of Theorem 15.1 have been proved already in our separate treatments of $\mathbb{Z}$ and $K[x]$. The last statement may look unfamiliar, but it follows easily from stronger results that we already know for $\mathbb{Z}$ and $K[x]$: For integers, the absolute value obeys the rule $|ab| = |a||b|$, or $N\left(ab\right) = N\left(a\right)N\left(b\right)$; for nonzero polynomials, the degree obeys the rule $N\left(ab\right) = N\left(a\right) + N\left(b\right)$.

The inequality $N\left(a\right) \leq N\left(ab\right)$ of Theorem 15.1 is weaker than the above-mentioned equality, but it has the virtue that it holds for $\mathbb{Z}$ and $K[x]$ simultaneously. Moreover, this weaker inequality and the other parts of Theorem 15.1 are all that is needed to prove the following result. We have already proved it separately for $\mathbb{Z}$ and $K[x]$. Using Theorem 15.1, we can provide a simultaneous proof for both rings.

Theorem 15.2. *Let R be the ring of integers or the ring of polynomials over a field. Suppose r is an element of R that is not zero or a unit.*

1. *If $r = ab$ is a nontrivial factorization of r, then $N\left(a\right) < N\left(r\right)$ and $N\left(b\right) < N\left(r\right)$.*
2. *Either r is irreducible, or r is a product of irreducible elements.*

Exercise 15.2. Prove Theorem 15.2 using Theorem 15.1 but not any additional specific properties of $\mathbb{Z}$ or $K[x]$. In particular, your argument should not mention integers or polynomials explicitly. Rather, it should make reference only to elements of the ring R, as in the statements of Theorems 15.1 and 15.2. For the first statement, use the fourth part of Theorem 15.1. For the second statement, make an induction argument on the size of r.

Since you have used only the conclusions of Theorem 15.1 in your proof of Theorem 15.2, any ring satisfying the conclusions of Theorem 15.1 should also satisfy the conclusion of Theorem 15.2. Thus, if we wish to prove for another ring R that the conclusion of Theorem 15.2 holds, we can do so by proving that the conclusions of Theorem 15.1 hold for R.

As an example, let us consider a family of rings that includes and generalizes the Gaussian integers $\mathbb{Z}[i]$. Suppose m is a positive integer and assume that m is *square-free*; that is, m is not divisible by the square of any integer besides 1. Write $\mathbb{Z}\left[\sqrt{-m}\right]$ for the collection of complex numbers of the form $r + s\sqrt{-m}$, where r and s are integers. You can easily check that $\mathbb{Z}\left[\sqrt{-m}\right]$ is a ring, with multiplication given by

$$\left(r + s\sqrt{-m}\right)\left(t + u\sqrt{-m}\right) = rt + su\left(-m\right) + ru\sqrt{-m} + st\sqrt{-m}$$
$$= \left(rt - msu\right) + \left(ru + st\right)\sqrt{-m}.$$

In the case $m = 1$, the ring $\mathbb{Z}[\sqrt{-1}]$ is the usual ring $\mathbb{Z}[i]$ of Gaussian integers.

A natural notion of size is already available for $\mathbb{Z}[\sqrt{-m}]$, namely, the *absolute value*, or *norm*, for complex numbers introduced in Exercise 13.4. Suppose that a and b are real numbers. Then the *norm* of the complex number $a + bi$ is defined to be $(a + bi)(a - bi)$, which equals $a^2 + b^2$. Notice that if we identify the complex number $a + bi$ with the point (a, b) in the Cartesian plane, as we did in Exercise 13.4, then its norm is the square of the distance from the origin to (a, b). For an element $r + s\sqrt{-m}$ of $\mathbb{Z}[\sqrt{-m}]$, we define its *size* to be its norm:

$$N\left(r + s\sqrt{-m}\right) = \left(r + s\sqrt{-m}\right)\left(r - s\sqrt{-m}\right) = r^2 + ms^2.$$

Notice that this norm is 0 if $r + s\sqrt{-m} = 0$ and is a positive integer otherwise. In $\mathbb{Z}[\sqrt{-m}]$, as in any integral domain, a factorization bc of a nonzero, nonunit element a of $\mathbb{Z}[\sqrt{-m}]$ is called *trivial* if one of the factors b or c is a unit and *nontrivial* otherwise. The element a is *irreducible* if it is not zero, is not a unit, and has only trivial factorizations.

We wish to prove that a version of Theorem 15.1 holds for the ring $\mathbb{Z}[\sqrt{-m}]$. Here is the appropriate statement:

Theorem 15.3. *Let m be a square-free integer, let R be the ring $\mathbb{Z}[\sqrt{-m}]$, and let N be the norm function on R.*

1. *The unique element of R of smallest size is 0.*
2. *The elements of R of the second-smallest size are precisely the units of R.*
3. *The elements of R of the third-smallest size are irreducible.*
4. *For any two nonzero elements a and b of R,*

$$N(a) \leq N(ab).$$

Moreover, equality holds if and only if b is a unit.

To prove Theorem 15.3, we will use a basic fact about norms of complex numbers. Let us assemble some information on the norm:

Exercise 15.3. Suppose a and b are complex numbers.

1. Verify that $N(ab) = N(a)N(b)$.
2. Check that if a is in $\mathbb{Z}[\sqrt{-m}]$, then $N(a)$ is an integer.
3. Deduce that if a and b are both in $\mathbb{Z}[\sqrt{-m}]$, then $N(a)N(b)$ is a factorization of the integer $N(ab)$ as a product of two integers.

Rather than proving Theorem 15.3, we will prove the following refinement of it:

Theorem 15.4. *Let m be a square-free integer, let R be the ring $\mathbb{Z}[\sqrt{-m}]$, and let N be the norm function on R.*

1. *The unique element of R of smallest size is 0. Its size is 0.*

2. *The elements of R of second-smallest size are precisely the units of R.*
   This second-smallest size is 1. If $m = 1$, the units are 1, -1, i, and $-i$.
   If $m > 1$, the units are 1 and -1.
3. *The elements of R of the third-smallest size are irreducible. If $m = 1$, the*
   third smallest size is 2, and the elements of this size are the four elements
   $\pm 1 \pm i$. If $m = 2$ or $m = 3$, the third-smallest size is m itself, and the
   elements of this size are $\pm\sqrt{-m}$. If $m \geq 5$, the third-smallest size is 4,
   and the elements of this size are ± 2.
4. *For any two nonzero elements a and b of R,*

$$N\left(a\right) N\left(b\right) = N\left(ab\right).$$

In particular,
$$N\left(a\right) \leq N\left(ab\right),$$

and equality holds if and only if b is a unit.

Exercise 15.4. Prove Theorem 15.4 and conclude that Theorem 15.3 holds as well.

Exercise 15.5. We have observed that a ring satisfying the conclusions of Theorem 15.1 should satisfy the conclusion of Theorem 15.2. Verify this for the rings $\mathbb{Z}\left[\sqrt{-m}\right]$ by proving Theorem 15.5 below using Theorem 15.3. Your proofs of Theorems 15.2 and 15.5 should be essentially identical.

Theorem 15.5. *Let m be a square-free integer, let R be the ring $\mathbb{Z}\left[\sqrt{-m}\right]$, and suppose r is an element of R that is not zero or a unit.*

1. *If $r = ab$ is a nontrivial factorization of r, then $N\left(a\right) < N\left(r\right)$ and $N\left(b\right) < N\left(r\right)$.*
2. *Either r is irreducible, or r is a product of irreducible elements.*

For a ring in which every nonzero, nonunit element r factors as a product of irreducible elements, such as the rings of the form $\mathbb{Z}\left[\sqrt{-m}\right]$, it is natural to ask whether the factorization of r as a product of irreducible elements is unique. What we mean by unique needs to be clarified, but we provided such a clarification in Chapter 12. We are led to the following question:

Question 15.6 *Let m be a square-free integer, and let R be the ring $\mathbb{Z}\left[\sqrt{-m}\right]$. Suppose $p_1 p_2 \cdots p_m$ and $q_1 q_2 \cdots q_n$ are two irreducible factorizations of a nonzero, nonunit element a of R. Is it true that $m = n$ and that the order of the factors in the second factorization can be changed so that for each index i there is a unit u_i of R such that $q_i = u_i p_i$?*

The answer to Question 15.6 is not always yes. The discovery that unique factorization need not hold in rings of this type was made in the mid-nineteenth century. Let us examine one such example.

Exercise 15.6. We will show that unique factorization fails in the ring $\mathbb{Z}\left[\sqrt{-5}\right]$.

1. Show that $\mathbb{Z}\left[\sqrt{-5}\right]$ has no elements of norm 2, 3, 7, or 8.
2. Find all elements in $\mathbb{Z}\left[\sqrt{-5}\right]$ of norms 4, 5, 6, and 9.
3. Prove that if a in $\mathbb{Z}\left[\sqrt{-5}\right]$ has norm equal to 4, 6, or 9, then a must be irreducible. (Hint: Suppose a factors as bc. Show that either b or c has norm 1 and conclude that the factorization is trivial.)
4. Deduce that the elements 2, 3, $1 + \sqrt{-5}$, and $1 - \sqrt{-5}$ are irreducible in $\mathbb{Z}\left[\sqrt{-5}\right]$.
5. Conclude that 2×3 and $\left(1 + \sqrt{-5}\right)\left(1 - \sqrt{-5}\right)$ are factorizations of 6 in $\mathbb{Z}\left[\sqrt{-5}\right]$ as products of irreducible elements.
6. Conclude that the answer to Question 15.6 for $\mathbb{Z}\left[\sqrt{-5}\right]$ is no. Thus, unique factorization fails for $\mathbb{Z}\left[\sqrt{-5}\right]$.

Even though unique factorization need not hold for rings of the form $\mathbb{Z}\left[\sqrt{-m}\right]$, it does hold for certain values of m, including $m = -1$. We will prove in Section 15.3 that unique factorization holds for $\mathbb{Z}[i]$, after developing a general approach in Section 15.2 to proving unique factorization theorems.

15.2 Euclidean Rings

Our review in Section 12.1 of the proof of the fundamental theorem of arithmetic and our subsequent proof of unique factorization for polynomial rings $K[x]$ revealed that the starting point in both cases for proving unique factorization is a division theorem. The division theorem for $\mathbb{Z}$ and $K[x]$, with the part about uniqueness omitted, can be given the following common formulation.

Theorem 15.7. *Let R be the ring of integers or the ring of polynomials over a field and let N be the measure of size on elements of R. For any two nonzero elements a and b of R, there exist elements q and r such that*

$$b = aq + r$$

and $N(r) < N(a)$.

In the case that R is a ring of polynomials over a field, Theorem 15.7 was proved as part of Theorem 9.5. In the case that R is the ring of integers, Theorem 15.7 is an extension to all nonzero integers of the usual division theorem. The generalization is easily proved. Suppose, for example, that b is negative and a is positive. Then the division theorem yields nonnegative integers q and r such that $-b = aq + r$. Therefore, $b = a(-q) + (-r)$, with $N(-r) = r < a = N(a)$.

Let us restate simultaneously as one theorem several results that we have proved separately for $\mathbb{Z}$ and for polynomial rings $K[x]$ over fields.

Theorem 15.8. *Let R be the ring* $\mathbb{Z}$ *of integers or the ring* $K[x]$ *of polynomials over a field K and let N be the measure of size on elements of R.*

1. *The unique element of R of smallest size is* 0.
2. *For any two nonzero elements a and b of R,*

$$N\left(a\right) \leq N\left(ab\right).$$

3. *For any two nonzero elements a and b of R, there exist elements q and r such that*
$$b = aq + r$$

and $N\left(r\right) < N\left(a\right)$.

A ring R is called a *Euclidean ring* if it satisfies the following three properties:

A. There is a *norm function* N assigning to every nonzero element a of R a nonnegative integer $N\left(a\right)$ and assigning to 0 a value $N\left(0\right)$ less than the norm of every nonzero element of R.
B. For any two nonzero elements a and b of R,

$$N\left(a\right) \leq N\left(ab\right).$$

C. For any two nonzero elements a and b of R, there exist elements q and r such that
$$b = aq + r$$

and $N\left(r\right) < N\left(a\right)$.

Properties A and B are mild assumptions on the norm function N. Property C is the crucial one. It says, in effect, that R satisfies a division theorem. Theorem 15.8 states that the rings $\mathbb{Z}$ and $K[x]$ are Euclidean rings. More precisely, $\mathbb{Z}$ is a Euclidean ring with respect to the absolute value function, and $K[x]$ is a Euclidean ring with respect to the degree function.

The reason to introduce the notion of a Euclidean ring is that we can try to use the Properties A, B, and C that define a Euclidean ring to prove theorems that apply to all such rings at once. Any theorem that we prove for a Euclidean ring must hold simultaneously for $\mathbb{Z}$ and $K[x]$, as well as for any other ring that we might later study and show to be a Euclidean ring. Thus, rather than giving proofs of similar-looking theorems over and over again, once for each ring, we need only state and prove a single theorem that applies simultaneously to all Euclidean rings. This allows us to treat $\mathbb{Z}$ and $K[x]$ together and to obtain theorems for new Euclidean rings with no additional work.

An example of a Euclidean ring other than $\mathbb{Z}$ and $K[x]$ is the ring $\mathbb{Z}[i]$ of Gaussian integers. To prove this, we need a division theorem for $\mathbb{Z}[i]$.

Theorem 15.9. *Let R be the ring $\mathbb{Z}[i]$ of Gaussian integers and let N be the norm function on R. For every two nonzero elements a and b of R, there exist elements q and r such that*

$$b = aq + r$$

and $N(r) < N(a)$.

Exercise 15.7. Prove the division theorem for $\mathbb{Z}[i]$ by following the outline below. The idea is simple, but a large quantity of notation is required in order to name a large cast of numbers.

1. Think of the number q we need to find as the quotient of b divided by a and think of r as the remainder. Recall that the set of complex numbers $\mathbb{Q}[i]$ of the form $f + gi$, where f and g are both rational, is a field. It is built from $\mathbb{Q}$ by our standard method of constructing a field in which we create a square root, in this case starting from $\mathbb{Q}$ and forming a square root of -1. Therefore, a nonzero Gaussian integer a, though not necessarily a unit in $\mathbb{Z}[i]$, is a unit in $\mathbb{Q}[i]$. This means we can divide by a in $\mathbb{Q}[i]$, so that there is a genuine quotient b/a, or ba^{-1}, in $\mathbb{Q}[i]$. Suppose this quotient is $f + gi$. In other words, f and g are rational numbers such that

$$b = a(f + gi).$$

   Since f and g may not be integers, $f + gi$ may not be in $\mathbb{Z}[i]$.

2. We want to find a "quotient" b/a in $\mathbb{Z}[i]$. Unless f and g are integers, we will not be able to find an actual quotient. This is why we must allow for a remainder. If there is to be a Gaussian integer q that serves as a quotient, it should be close to the actual quotient $f + gi$ in $\mathbb{Q}[i]$. Compare the situation in the integers. For two nonzero integers a and b, the quotient b/a exists in $\mathbb{Q}$. When we write $b = aq + r$, the integer we take for q is the largest integer less than or equal to the rational number b/a. It is an *approximation* to the actual quotient. Let us do something similar here. We will choose q to be an approximation in $\mathbb{Z}[i]$ to the actual quotient $f + gi$. To do so, choose an integer m within $\frac{1}{2}$ of f and an integer n within $\frac{1}{2}$ of g. Notice that m, which may be smaller than f or larger than f, is uniquely determined unless f is exactly halfway between two integers, and similarly for n. Set $q = m + ni$. Then q is a Gaussian integer, as is $b - aq$. Set $r = b - aq$, so that $b = aq + r$. This r is our remainder. Show that

$$r = a((f - m) + (g - n)i).$$

   (Notice that although $(f - m) + (g - n)i$ may not be a Gaussian integer, its product with a is.)

3. To conclude the proof of the division theorem, you need only check that our putative remainder r satisfies $N(r) < N(a)$. Do so.

Exercise 15.8. Use the division theorem for $\mathbb{Z}[i]$ to prove Theorem 15.10 below.

Theorem 15.10. $\mathbb{Z}[i]$ *is a Euclidean ring.*

We now have three different kinds of Euclidean rings: the ring $\mathbb{Z}$ of integers, the ring $\mathbb{Z}[i]$ of Gaussian integers, and the polynomial rings $K[x]$ over a field K. Any theorem we prove for Euclidean rings will apply to all three at once. As an example of the use of Properties A, B, and C to prove something that can be applied to all Euclidean rings, let us prove the following result.

Theorem 15.11. *Suppose R is a Euclidean ring.*

1. *R is an integral domain; that is, R has no zero-divisors.*
2. *Cancellation holds in R; that is, for a a nonzero element of R, if $ab = ac$, then $b = c$.*
3. *For any two nonzero elements a and b of R, if $N(a) = N(ab)$, then b is a unit.*

Proof. For the first part, suppose a and b are nonzero. We recall that 0 is the unique element of R of smallest norm. Therefore, $N(0) < N(a) \leq N(ab)$. Since $N(0) \neq N(ab)$, the product ab cannot be 0. The second part is really a statement about integral domains in general. If $ab = ac$, then $a(b - c) = 0$. But since R has no zero-divisors and a is nonzero, $b - c$ must be zero. Thus $b = c$.

The proof of the third part is more complicated. Assume that $N(a) = N(ab)$. Let us try to divide a by ab. By Property C, there exist elements q and r in R such that $a = (ab)q + r$ and $N(r) < N(ab)$. Let us rewrite the equation as $a - abq = r$, or as $a(1 - bq) = r$. We know that a is nonzero. Is it possible that $1 - bq$ is also nonzero? If it were, then Property B would imply that

$$N(a) \leq N(a(1 - bq)) = N(r).$$

But this is impossible, since $N(r) < N(a)$. Therefore, $1 - bq$ must be zero, and $1 = bq$, proving that b is a unit.

Since a Euclidean ring R is an integral domain, the notion of irreducibility can be introduced. A factorization bc of a nonzero, nonunit element a of R is *trivial* if one of the factors b or c is a unit, and is *nontrivial* otherwise. The element a is *irreducible* if it is not zero, is not a unit, and has only trivial factorizations. Here is another example of a family of results that we have proved separately for different rings, but that can be proved simultaneously for all Euclidean rings.

Theorem 15.12. *Suppose R is a Euclidean ring.*

1. *The elements of R of the second-smallest size are precisely the units of R.*
2. *The elements of R of the third-smallest size are irreducible.*
3. *Every nonzero, nonunit element of R is irreducible or a product of irreducible elements.*

Exercise 15.9. Suppose R is a Euclidean ring. Prove Theorem 15.12. You can proceed as follows.

1. Show that all units have the same norm by showing that $N(u) = N(1)$ for every unit u. There are two arguments to make. First show that $N(1) \leq N(u)$, then show that $N(u) \leq N(1)$. In both cases, you can use Property B.
2. Prove that if a is nonzero and its norm is the smallest possible among the norms of nonzero elements of R, then a is a unit. To do this, use the division theorem (Property C) for R to divide 1 by a, obtain a quotient and remainder, and show that the remainder must be zero. Conclude that you have proved the first part of the theorem.
3. Prove the second part in the same way that you have already proved similar statements, such as the third part of Theorem 15.3.
4. Prove the third part in the same way that you have already proved Theorems 15.2 and 15.5.

15.3 Unique Factorization

We saw in Chapter 12 that we can obtain the unique factorization theorems for $\mathbb{Z}$ and $K[x]$ by proving a specific sequence of theorems starting from the division theorem. What we will next see is that we can rephrase the proofs we gave for $\mathbb{Z}$ and $K[x]$ so that the arguments use the language and the conclusions of Theorem 15.8 but nothing more. The conclusions of Theorem 15.8 are those that define a Euclidean ring. Thus what we will see is that every Euclidean ring satisfies a unique factorization theorem. This will apply in particular to the ring $\mathbb{Z}[i]$. The starting point is a Euclidean algorithm for Euclidean rings, and this depends on having a notion of greatest common divisor in a Euclidean ring.

Before proceeding in general, let us place into a common framework the notion of greatest common divisor for $\mathbb{Z}$ and $K[x]$. This notion is quite intuitive in the case of two integers a and b. In this case the greatest common divisor is just that: the largest integer that is a common divisor of a and b. For example, the greatest common divisor of 30 and 42 is 6. When we work with both positive and negative integers, we measure their size not by themselves but by their absolute values, so that negative integers have the same size as their positive counterparts. For instance, 6 and -6 have the same size. Therefore, perhaps we should regard not just 6 but also -6 as a greatest common divisor of 30 and 42. Both are common divisors, and both have the largest possible size among the common divisors ± 1, ± 2, ± 3, and ± 6 of 30 and 42. There is no particular reason to prefer 6 to -6. From this point of view, a pair of integers a and b has *two* greatest common divisors, the usual positive one d and its negative $-d$. Each is a common divisor of a and b of largest possible size.

For a field K, we defined a greatest common divisor of two polynomials $a(x)$ and $b(x)$ in $K[x]$ to be a common divisor of largest size. There can be many of these, but they all differ from each other by nonzero constant multiples, the units of $K[x]$. Similarly, the two greatest common divisors of integers a and b differ from each other by a factor of -1, the one unit in $\mathbb{Z}$ in addition to 1. We see that in both cases the common statement applies that the greatest common divisors of two elements all differ from each other by unit multiples.

Suppose more generally that R is a Euclidean ring. For two nonzero elements a and b of R, define a *greatest common divisor* of a and b to be an element d that is a common divisor and that has the largest possible size $N(d)$ among common divisors of a and b.

We can now review all the theorems we proved for $\mathbb{Z}$ and $\mathbb{K}[x]$, stating them in a form that makes sense for every Euclidean ring R. As you read the results below, you can think of the ring R as $\mathbb{Z}$, as the ring $K[x]$ of polynomials with coefficients in a field K, or more generally as any Euclidean ring, including $\mathbb{Z}[i]$. We have proved these theorems in the first two cases. What we shall see is that the proofs we have already given in these two cases apply to the general case of all Euclidean rings as well, with wording changed as necessary. This has two enormous benefits: We need to prove the theorems only once, and they will apply simultaneously to $\mathbb{Z}$ and $K[x]$; having proved the theorems not just for these specific rings but also for arbitrary Euclidean rings, we can apply them to other Euclidean rings, such as $\mathbb{Z}[i]$.

The first result that we proved for $\mathbb{Z}$ and $\mathbb{K}[x]$ after the division theorem was the following result about greatest common divisors, the basis for the Euclidean algorithm.

Theorem 15.13. *Suppose R is a Euclidean ring. Let a and b be nonzero elements in R, and let q and r be elements of R such that $b = aq + r$ and $N(r) < N(a)$. Then the greatest common divisors of a and b coincide with the greatest common divisors of r and a.*

Let us not worry about the proof of Theorem 15.13, or the proofs of the theorems to follow, until we have obtained a complete list of the theorems we wish to prove.

If a and b are two nonzero elements of a Euclidean ring R, how might we set about finding a greatest common divisor? If a divides b, then a is a greatest common divisor of a and b. If not, we use the division theorem (Property C) to write b as $aq_1 + r_1$ and replace the pair b and a with the pair a and r_1. By Theorem 15.13, both pairs have the same greatest common divisors. If r_1 divides a, it is a greatest common divisor. If not, we use the division theorem to write a as $r_1 q_2 + r_2$. Now we replace the pair a and r_1 with the pair r_1 and r_2. The pairs again have the same common divisors, by another appeal to Theorem 15.13. If r_2 divides r_1, it is a greatest common divisor. If not, we proceed once again, writing r_1 as $r_2 q_3 + r_3$, and so on. Eventually, we reach a remainder r_n that divides the preceding remainder r_{n-1} (why?), so that the next remainder r_{n+1} would be 0. This sequence of calculations, with the pair of elements a and b as the *input* and the final nonzero remainder r_n as the *output*, is the *Euclidean algorithm* for R.

As in the cases of $\mathbb{Z}$ and $K[x]$, one can prove by induction on the number n of steps required for the algorithm to terminate that r_n is a greatest common divisor of a and b. Then one can prove Bézout's theorem below, using a similar induction and working backwards through the Euclidean algorithm.

Theorem 15.14 (Bézout's theorem for Euclidean rings). *Suppose R is a Euclidean ring. Let d be the greatest common divisor of elements a and b of R produced by the Euclidean algorithm. There exist elements r and s in R such that*

$$d = ar + bs.$$

The proof of Bézout's theorem shows both that r and s exist and that they can be computed explicitly. Using Bézout's theorem, we can describe the relationship among the greatest common divisors of a and b.

Corollary 15.15 *Suppose R is a Euclidean ring. Let d be the greatest common divisor of elements a and b of R produced by the Euclidean algorithm.*

1. *Every common divisor of a and b divides d. In particular, every greatest common divisor of a and b divides d.*
2. *All the greatest common divisors of a and b differ from each other by unit multiples.*

As usual, we say that two elements a and b of R are *relatively prime* if 1 is one of their greatest common divisors. Equivalently, the only common divisors of relatively prime elements a and b are units. Bézout's theorem yields the following result by the usual proof.

Theorem 15.16. *Suppose R is a Euclidean ring. Suppose also that a and b are relatively prime elements of R and that c is an element of R such that a divides the product bc. Then a divides c.*

In turn, Theorem 15.16 yields Theorem 15.17 below.

Theorem 15.17. *Let R be a Euclidean ring. Let p be an irreducible element of R and suppose p divides the product bc of elements b and c of R. Then p divides b, or p divides c. More generally, suppose p divides $a_1 a_2 \cdots a_n$. Then p divides one of the factors a_i.*

From Theorem 15.17, we can deduce the following result in the usual way:

Theorem 15.18. *Suppose R is a Euclidean ring. Suppose also that $p_1 \cdots p_m$ and $q_1 \cdots q_n$ are two factorizations of a nonzero, nonunit element a in R as a product of irreducible elements:*

$$a = p_1 \cdots p_m = q_1 \cdots q_n.$$

Then for some index j and some unit u, we have $q_j = up_m$. Hence

$$p_1 \cdots p_{m-1} = u q_1 \cdots q_{j-1} q_{j+1} \cdots q_n.$$

Reordering the q_i's if necessary, we obtain

$$p_1 \cdots p_{m-1} = u q_1 \cdots q_{n-1}.$$

In the setting of Theorem 15.18, we can keep going, reordering again if necessary to deduce that q_{n-1} is a unit multiple of p_{m-1}, and so on. Eventually, we find that $m = n$ and that for a suitable ordering, each q_i is a unit multiple of p_i. In order to make this rigorous, we make an induction argument on the number of factors m in one of the factorizations of a, proving the unique factorization theorem below.

Theorem 15.19. *Suppose R is a Euclidean ring. Suppose also that*

$$p_1 p_2 \cdots p_m \quad \text{and} \quad q_1 q_2 \cdots q_n$$

are two irreducible factorizations of a nonzero, nonunit element a of R. Then $m = n$, and the order of the factors in the second factorization can be changed so that for each index j there is a unit u_j of R such that $q_j = u_j p_j$.

Exercise 15.10. Review the proofs for $\mathbb{Z}$ and $K[x]$ of all the theorems above, starting with Theorem 15.13 and ending with Theorem 15.19. In performing the review, observe that the proof of each theorem uses only the Properties A, B, and C that define a Euclidean ring and the results of Theorems 15.11 and 15.12.

1. Conclude from your review that all the theorems from Theorem 15.13 to Theorem 15.19 hold for every Euclidean ring R.
2. In particular, write out detailed proofs of Theorem 15.13, Corollary 15.15, and Theorem 15.17.

Exercise 15.11. Using Theorem 15.10, conclude that all the theorems from Theorem 15.13 to Theorem 15.19 hold for $\mathbb{Z}[i]$. In particular, conclude that the unique factorization theorem below holds for $\mathbb{Z}[i]$.

Theorem 15.20. *Suppose that*

$$p_1 p_2 \cdots p_m \quad \text{and} \quad q_1 q_2 \cdots q_n$$

are two irreducible factorizations of a nonzero, nonunit Gaussian integer a of R. Then $m = n$, and the order of the factors in the second factorization can be changed so that for each index j the elements p_j and q_j either equal each other or differ from each other by multiplication by -1, i, or $-i$.

We know that in contrast to $\mathbb{Z}[i]$, the ring $\mathbb{Z}\left[\sqrt{-5}\right]$ does not satisfy unique factorization. In $\mathbb{Z}\left[\sqrt{-5}\right]$ there are two factorizations of 6 as a product of irreducible elements that cannot be obtained from each other by change of order and unit multiplication. Therefore, in contrast to $\mathbb{Z}[i]$, the ring $\mathbb{Z}\left[\sqrt{-5}\right]$ cannot be a Euclidean ring. Properties A and B in the definition of a Euclidean ring are satisfied by $\mathbb{Z}\left[\sqrt{-5}\right]$. What fails is Property C, the division theorem. This highlights how important the division theorem is.

The approach we have taken in this section is an illustration of the *axiomatic method*, one of the hallmarks of mathematics. By closely examining certain objects, we find what common properties account for many of the theorems about them. In the case of the rings $\mathbb{Z}$ and $K[x]$, we used this process of *abstraction* to identify the Properties A, B, and C that we used to define a Euclidean ring. Then, working merely from these properties, or *axioms*, we are able to prove a family of theorems. The power of the method is that these theorems are applicable to every other object that satisfies the same axioms, that is, to every other Euclidean ring, such as $\mathbb{Z}[i]$. This approach to the development of new mathematical information is one of the most profound ideas of intellectual history.

The Ring of Gaussian Integers

16.1 The Irreducible Gaussian Integers

We have proved that the ring $\mathbb{Z}[i]$ is a Euclidean ring and therefore that $\mathbb{Z}[i]$ satisfies a unique factorization theorem. Since every nonzero, nonunit Gaussian integer is a product of irreducible Gaussian integers, it would be desirable to know which Gaussian integers are irreducible. Our first result provides a collection of irreducible Gaussian integers. Recall that we have defined the norm N of a complex number $a + bi$ as $N(a + bi) = a^2 + b^2$.

Theorem 16.1. *Let p be a prime number. Suppose r is a Gaussian integer satisfying $N(r) = p$. Then r is irreducible in $\mathbb{Z}[i]$. In particular, if a and b are integers such that $a^2 + b^2 = p$, then the Gaussian integers $\pm a \pm bi$ and $\pm b \pm ai$ are irreducible.*

Exercise 16.1. Prove Theorem 16.1. (Hint: For the first part, suppose st is a factorization of r. You must show that this factorization is trivial. Apply the norm to obtain $p = N(s)N(t)$. Deduce that either $N(s)$ or $N(t)$ equals 1 and conclude that either s or t is a unit.)

We have just shown that a Gaussian integer whose norm is a prime number is irreducible in $\mathbb{Z}[i]$. Next we will prove that the only other possibility for the norm of an irreducible Gaussian integer is the square of a prime number. This will provide a rough classification of the irreducible elements of $\mathbb{Z}[i]$ into two families, those with prime norm and those with prime-squared norm. But there is a significant difference between these two classes: Theorem 16.1 tells us that *every* Gaussian integer with prime norm is irreducible. However, a Gaussian integer whose norm is the square of a prime may or may not be irreducible! Our proof that the only irreducible Gaussian integers are those with prime or prime-squared norm uses the unique factorization theorem for $\mathbb{Z}[i]$ and the following result:

Proposition 16.2 *Let r be a Gaussian integer that is neither zero nor a unit. Then r is irreducible in $\mathbb{Z}[i]$ if and only if its complex conjugate $\bar{r}$ is irreducible in $\mathbb{Z}[i]$.*

Exercise 16.2. Prove Proposition 16.2. (Hint: Suppose r has a nontrivial factorization $r = xy$. Recall that $\bar{r} = \bar{x}\,\bar{y}$. Show that this is a nontrivial factorization of $\bar{r}$. Deduce that if $\bar{r}$ is irreducible, so is r. Then reverse the roles of r and $\bar{r}$ to complete the proof.)

Theorem 16.3. *Let r be an irreducible Gaussian integer. Then one of the following holds:*

1. *There is a prime number p such that $N(r) = p$. In this case, r equals $a + bi$ for a pair of nonzero integers a and b satisfying $a^2 + b^2 = p$.*
2. *There is a prime number p such that $N(r) = p^2$. In this case, r equals one of the four numbers p, $-p$, pi, or $-pi$.*

Exercise 16.3. Prove Theorem 16.3. You can follow the outline below.

1. First observe that $r\bar{r}$ is a factorization of $N(r)$ in $\mathbb{Z}[i]$ as a product of irreducible Gaussian integers. Use the unique factorization theorem to deduce that every factorization of $N(r)$ in $\mathbb{Z}[i]$ as a product of irreducible Gaussian integers has two factors.
2. Observe that since r is not 0 or a unit of $\mathbb{Z}[i]$, its norm $N(r)$ is an integer greater than 1. Introduce notation for a prime factorization of $N(r)$ in $\mathbb{Z}$, say $N(r) = p_1 \cdots p_t$. Be aware that the primes p_j may or may not be irreducible in $\mathbb{Z}[i]$; nothing is assumed about this. (Recall as an example that 2 is prime in $\mathbb{Z}$, but it is not irreducible in $\mathbb{Z}[i]$, since it factors as $2 = (1 + i)(1 - i)$.) In any case, each prime p_j is a Gaussian integer $(p_j = p_j + 0i)$ and therefore factors uniquely in $\mathbb{Z}[i]$ as a product of one or more Gaussian integers. Argue that there must exist a factorization of $N(r)$ in $\mathbb{Z}[i]$ as a product of at least t irreducible Gaussian integers, and that therefore, by the first part, t equals 1 or 2.
3. Suppose that $t = 2$. Then $N(r) = r\bar{r} = p_1 p_2$. Using the unique factorization theorem, deduce that r differs from either p_1 or p_2 by multiplication by a unit of $\mathbb{Z}[i]$. Conclude that there is a prime number p in $\mathbb{Z}$ such that r equals one of the four numbers p, $-p$, pi, or $-pi$. Notice that in all four of these cases, $N(r) = p^2$.
4. Suppose that $t = 1$. To simplify notation, write p_1 simply as p. Thus $N(r) = p$. Write r as $a + bi$ for integers a and b. Observe that if either a or b is 0, then $N(r)$ cannot be a prime number. Thus a and b are both nonzero. Observe next that

$$p = N(r) = r\bar{r} = (a + bi)(a - bi) = a^2 + b^2.$$

We have found that there are two kinds of irreducible Gaussian integers: ordinary prime numbers p that do not factor nontrivially in $\mathbb{Z}[i]$, along with their unit multiples $-p$, pi, and $-pi$, and Gaussian integers $a + bi$ whose norms $a^2 + b^2$ are prime numbers. We would like to be able to characterize both types of irreducible Gaussian integers, and we will be guided by the following questions: (1) What are the primes p that do not factor nontrivially in $\mathbb{Z}[i]$? (2) What primes p are the norms of Gaussian integers (or equivalently, what primes p can be expressed in the form $a^2 + b^2$ for ordinary integers a and b)? (3) For a particular prime p that factors in $\mathbb{Z}[i]$, what are the possible integers a and b such that $a^2 + b^2 = p$?

Let us answer question (3) first: We will start with prime numbers p in $\mathbb{Z}$ and try to factor them nontrivially in $\mathbb{Z}[i]$. Some prime numbers will remain irreducible in $\mathbb{Z}[i]$, but others will not. For instance, as we have already seen, the primes 2 and 5 factor nontrivially in $\mathbb{Z}[i]$:

$$2 = (1 + i)(1 - i) \text{ and } 5 = (2 + i)(2 - i).$$

Exercise 16.4. Let p be a prime number.

1. Deduce from Theorems 16.1 and 16.3 that either p is irreducible in $\mathbb{Z}[i]$ or p factors as the product $(a + bi)(a - bi)$ of two conjugate irreducible Gaussian integers.
2. Conclude that p is irreducible in $\mathbb{Z}[i]$ if and only if the equation $p = x^2 + y^2$ is not solvable in $\mathbb{Z}$. Conclude that equivalently, p factors nontrivially in $\mathbb{Z}[i]$ if and only if the equation $p = x^2 + y^2$ is solvable in $\mathbb{Z}$. Observe further that in this case, each pair of integers (a, b) satisfying the equation $p = a^2 + b^2$ corresponds to a nontrivial factorization of p as $(a+bi)(a-bi)$.
3. Deduce that if p factors nontrivially in $\mathbb{Z}[i]$, then the irreducible Gaussian integers $a + bi$ occurring as factors of p correspond to the solutions (a, b) to the equation $x^2 + y^2 = p$.
4. If $p = 2$, observe that the only solutions to $x^2 + y^2 = 2$ are $(\pm 1, \pm 1)$. Conclude that the four Gaussian integers $\pm 1 \pm i$ are irreducible and that they are the only irreducible Gaussian integers dividing 2.
5. If p is odd and (a, b) satisfies $a^2 + b^2 = p$, explain why $|a| \neq |b|$. Conclude that associated with (a, b) are eight distinct solutions to $x^2 + y^2 = p$, namely, $(\pm a, \pm b)$ and $(\pm b, \pm a)$. (Why are they distinct?)
6. Observe that these eight solutions can be grouped into four pairs corresponding to four factorizations of p as a product of two irreducible Gaussian integers. Explain why this does not violate the uniqueness of factorizations of p as products of irreducible elements of $\mathbb{Z}[i]$.
7. In this same situation, use the uniqueness of factorizations as products of irreducible Gaussian integers to prove that the eight solutions to $x^2 + y^2 = p$ must be all of the solutions; there can be no others.

Some of what we have discovered in the preceding exercise can be summarized by the following theorem.

Theorem 16.4. *Let p be a prime number in $\mathbb{Z}$.*

1. *There are four solutions $(\pm 1, \pm 1)$ to the equation $x^2 + y^2 = 2$, corresponding to which there are four irreducible Gaussian integers $\pm 1 \pm i$ dividing 2. These four Gaussian integers fall into two pairs whose product is 2:*

$$2 = (1 + i)(1 - i) = (-1 + i)(-1 - i).$$

2. *If p is an odd prime number for which the equation $x^2 + y^2 = p$ is not solvable in $\mathbb{Z}$, then p is irreducible in $\mathbb{Z}[i]$, as are $-p$, pi, and $-pi$.*

3. *If p is an odd prime number for which the equation $x^2 + y^2 = p$ is solvable in $\mathbb{Z}$, then this equation has exactly eight solutions. There are integers a and b satisfying $b \neq \pm a$ such that the eight solutions are $(\pm a, \pm b)$ and $(\pm b, \pm a)$. Corresponding to these solutions are eight irreducible Gaussian integers, $\pm a \pm bi$ and $\pm b \pm ai$, that divide p. These eight Gaussian integers fall into four pairs whose product is p, yielding factorizations of p as a product of irreducible Gaussian integers such as the factorization*

$$p = (a + bi)(a - bi).$$

Theorem 16.4 describes all the ways in which a prime number in $\mathbb{Z}$ can factor in $\mathbb{Z}[i]$. When combined with Theorem 16.3, it provides a classification of irreducible Gaussian integers, as described by the theorem below.

Theorem 16.5. *Every irreducible Gaussian integer r has one of the following three forms:*

1. *r equals $\pm p$ or $\pm pi$ for an odd prime number p in $\mathbb{Z}$ that remains irreducible in $\mathbb{Z}[i]$;*
2. *r is one of the eight irreducible factors $\pm a \pm bi$ or $\pm b \pm ai$ of an odd prime number p in $\mathbb{Z}$ that factors nontrivially in $\mathbb{Z}[i]$, with a and b integers satisfying $a^2 + b^2 = p$;*
3. *r is one of the four irreducible factors $\pm 1 \pm i$ of 2.*

Exercise 16.5. Prove Theorem 16.5.

What keeps Theorem 16.5 from being a full-fledged classification of the irreducible Gaussian integers is that we still need to determine which prime numbers p are irreducible in $\mathbb{Z}[i]$ and which factor nontrivially. That is, we have still to answer questions (1) and (2) above. Our work thus far shows that the problem of determining the irreducibility of a given prime number p in $\mathbb{Z}[i]$ is equivalent to the problem of determining the solvability of the equation $x^2 + y^2 = p$ in $\mathbb{Z}$. This is summarized in the following theorem.

Theorem 16.6. *The prime number p is irreducible in $\mathbb{Z}[i]$ if and only if the equation $x^2 + y^2 = p$ has no integer solutions.*

Theorem 16.6 unites a classical problem in number theory that could have been considered two thousand years ago, the solvability of the equation $x^2 + y^2 = p$, and an algebraic problem that is much more modern, the irreducibility in $\mathbb{Z}[i]$ of p. We will determine which primes are which, thereby answering our two remaining questions, in Section 16.3.

16.2 Gaussian Congruence Rings

In Section 6.4 we introduced the ring $\mathbb{Z}_m$, and we proved there that $\mathbb{Z}_m$ is a field if and only if m is prime. In Section 14.3 we introduced the ring $F[x]_{m(x)}$, and we proved in Section 14.5 that $F[x]_{m(x)}$ is a field if and only if $m(x)$ is irreducible. This suggests to us that we might learn something interesting by doing something similar in the ring $\mathbb{Z}[i]$.

Let us, then, take a nonzero, nonunit element m in $\mathbb{Z}[i]$ and construct a ring $(Z[i])_m$, consisting of equivalence classes of elements of $\mathbb{Z}[i]$ in which two elements of $\mathbb{Z}[i]$ belong to the same class if they differ by a multiple of m. We can then ask whether there is a meaningful relationship between $(\mathbb{Z}[i])_m$ being a field and m being irreducible in $\mathbb{Z}[i]$. For simplicity, we are going to consider only the special case in which m is a positive integer, which is the only case that we need.

Fix an integer $m > 1$. To create the new ring $(\mathbb{Z}[i])_m$, roughly speaking, we will apply the rewrite rule $m = 0$ to $\mathbb{Z}[i]$. The ring $\mathbb{Z}[i]$ is the set of complex numbers of the form $c + di$, where c and d are integers and $i^2 = -1$. The rewrite rule $m = 0$ allows us to replace c and d by integers lying between 0 and $m - 1$. We can therefore regard the coefficients as elements of $\mathbb{Z}_m$. This leads to the following definition: The ring $(\mathbb{Z}[i])_m$ is the set of elements of the form $c + di$, where c and d are in $\mathbb{Z}_m$ and $i^2 = -1$. This definition allows us to simplify the notation for the name of our new ring: From now on, we shall write $\mathbb{Z}_m[i]$ instead of $(\mathbb{Z}[i])_m$.

Addition and multiplication in $\mathbb{Z}_m[i]$ are given by the rules

$$(c + di) + (e + fi) = (c + e) + (d + f)i$$

and

$$(c + di)(e + fi) = (ce - df) + (cf + de)i.$$

Notice that this is exactly how we define addition and multiplication in $\mathbb{Z}[i]$. What is different is that for $\mathbb{Z}_m[i]$ the coefficients c, d, e, and f are elements of $\mathbb{Z}_m$, not integers. For example, if we take $m = 12$, then in the ring $\mathbb{Z}_{12}[i]$ we have $([7] + [11]i) + ([3] + [9]i) = [10] + [8]i$, and $([7] + [11]i)([3] + [9]i) = [6]$. (Check this!)

Exercise 16.6. Let us examine the two smallest rings of the form $\mathbb{Z}_m[i]$.

1. According to the definitions, the ring $\mathbb{Z}_2[i]$ consists of all elements of the form $a + bi$, with a and b in $\mathbb{Z}_2$ (also known as $\mathbb{F}_2$). Deduce that $\mathbb{Z}_2[i]$ consists of the four elements 0, 1, i, and $1 + i$.
2. Using these four elements, make addition and multiplication tables for $\mathbb{Z}_2[i]$, the way we did for fruit rings in Section 6.3.
3. Review the multiplication table and answer the following questions:
   (a) Are there zero-divisors in $\mathbb{Z}_2[i]$?
   (b) Does every nonzero element of $\mathbb{Z}_2[i]$ have a multiplicative inverse?
   (c) Is $\mathbb{Z}_2[i]$ a field?

4. Perform a similar analysis for the ring $\mathbb{Z}_3[i]$, starting with the observation that it contains nine distinct elements. List these elements. Do not bother with the addition table, but make a multiplication table for $\mathbb{Z}_3[i]$. Use the table to answer the following questions:
   (a) Are there zero-divisors in $\mathbb{Z}_3[i]$?
   (b) Does every nonzero element of $\mathbb{Z}_3[i]$ have a multiplicative inverse?
   (c) Is $\mathbb{Z}_3[i]$ a field?

We would like to know in general when $\mathbb{Z}_m[i]$ is a field. If m is not a prime number, then $\mathbb{Z}_m[i]$ cannot be a field. We can show this by a now-familiar argument. If m is not prime, then it factors nontrivially as a product ab of two positive integers. Then $[a][b] = 0$ in $\mathbb{Z}_m[i]$, but $[a]$ and $[b]$ are not themselves 0 in $\mathbb{Z}_m[i]$. This means that $\mathbb{Z}_m[i]$ has zero-divisors, and thus it is not a field.

Suppose that m is a prime number. Recall that m may or may not be irreducible in $\mathbb{Z}[i]$. If m is not irreducible, then $\mathbb{Z}_m[i]$ is again not a field. After all, as we have seen, m has a factorization in $\mathbb{Z}[i]$ of the form $(a + bi)(a - bi)$ for some choice of positive integers a and b. These positive integers must be less than m, since $m = a^2 + b^2$. Therefore, $[a]$ and $[b]$ are nonzero in $\mathbb{Z}_m[i]$, and $([a] + [b]i)([a] - [b]i)[m] = [0]$. This shows that $\mathbb{Z}_m[i]$ has zero-divisors.

The only possible case, then, in which $\mathbb{Z}_m[i]$ can be a field is that in which m is a prime number in $\mathbb{Z}$ that is irreducible in $\mathbb{Z}[i]$. To deal with this case, we need a consequence of Bézout's theorem, familiar to us for the Euclidean rings $\mathbb{Z}$ and $K[x]$. Rather than stating it in full generality, let us just state what we need.

Theorem 16.7. *Suppose a is nonzero in $\mathbb{Z}[i]$ and p is a prime integer. If a and p are relatively prime in $\mathbb{Z}[i]$, then there exist Gaussian integers r and s such that*

$$ar - ps = 1.$$

In particular, if p is irreducible in $\mathbb{Z}[i]$, and p does not divide a, then there exist Gaussian integers r and s such that

$$ar - ps = 1.$$

Exercise 16.7. Prove Theorem 16.7 as an application of Theorem 15.14, Bézout's theorem for Euclidean rings.

Just as one might have expected, it turns out that if p is a prime number that is irreducible in $\mathbb{Z}[i]$, then the ring $\mathbb{Z}_p[i]$ is a field, and we have the following theorem:

Theorem 16.8. *Given a prime number p, the ring $\mathbb{Z}_p[i]$ is a field if and only if p is irreducible in $\mathbb{Z}[i]$.*

Exercise 16.8. Prove Theorem 16.8 following the outline below: Let p be a prime number that is irreducible in $\mathbb{Z}[i]$. We wish to show that $\mathbb{Z}_p[i]$ is a field. Let $[c] + [d]i$ be a nonzero element of $\mathbb{Z}_p[i]$, with $[c]$ and $[d]$ in $= \mathbb{Z}_p$. (Thus we may take c and d to be integers representing their congruence classes.) We need to prove that $[c] + [d]i$ is a unit.

1. Notice that $[c] + [d]i$ is a unit if one of $[c]$ and $[d]$ is $[0]$ and the other is not.

2. Having taken care of the case in which one of $[c]$ and $[d]$ is the zero congruence class in $\mathbb{Z}_p$, suppose now that $[c]$ and $[d]$ are both nonzero elements of $\mathbb{Z}_p[i]$. Observe that in $\mathbb{Z}[i]$, the prime p cannot divide $c + di$ (why?), so that p and $c + di$ are relatively prime.

3. Deduce that in this case, by Theorem 16.7, there exist Gaussian integers r and s such that
$$(c + di)r = 1 + ps.$$

4. Suppose $r = e + fi$ for integers e and f. Deduce that
$$\big([c] + [d]i\big)\big([e] + [f]i\big) = 1$$
in $\mathbb{Z}_p[i]$.

5. Conclude that $\mathbb{Z}_p[i]$ is a field.

For the remainder of this chapter we will be interested in rings of the form $\mathbb{Z}_m[i]$ only for integers m that are prime. Let us change notation, then, since for a prime number p we are accustomed to writing the ring $\mathbb{Z}_p$ as $\mathbb{F}_p$, where the $\mathbb{F}$ reminds us that $\mathbb{F}_p$ is a Field. By analogy, let us also write $\mathbb{F}_p[i]$ for the ring $\mathbb{Z}_p[i]$.

We still do not know which prime numbers p are irreducible in $\mathbb{Z}[i]$. In order to decide this, we will make use of an alternative description of $\mathbb{F}_p[i]$. We have constructed $\mathbb{F}_p[i]$ as follows. We started with $\mathbb{Z}$, to which we adjoined a new number i whose square is -1 in order to obtain the ring $\mathbb{Z}[i]$. Then we passed to $\mathbb{F}_p[i]$ by setting $p = 0$. Let us describe this in more detail, with the addition of an extra stage. Start with $\mathbb{Z}$ and pass to the polynomial ring $\mathbb{Z}[x]$. Introduce the rewrite rule $x^2 = -1$, allowing us to work only with polynomials in x of degree at most 1. This makes x a square root of -1. For convenience, replace x with the symbol i. So far, we have constructed $\mathbb{Z}[i]$. Introduce the rewrite rule $p = 0$ to obtain $\mathbb{F}_p[i]$. The construction takes three steps. We pass from $\mathbb{Z}$ to the polynomial ring $\mathbb{Z}[x]$, then add two rewrite rules: first, $x^2 + 1 = 0$; second, $p = 0$.

Suppose we reverse the order of the two rewrite rules. After passing from $\mathbb{Z}$ to $\mathbb{Z}[x]$, apply the rewrite rules in the opposite order: first, $p = 0$; second, $x^2 + 1 = 0$. Applying $p = 0$ takes us from $\mathbb{Z}[x]$ to $\mathbb{Z}_p[x]$, or $\mathbb{F}_p[x]$. Applying $x^2 + 1 = 0$ takes us from $\mathbb{F}_p[x]$ to $\mathbb{F}_p[x]_{x^2+1}$. It is natural to expect that the ring we produce by applying the two rewrite rules to $\mathbb{Z}[x]$ is the same regardless of the order in which the rewrite rules are introduced. Thus, the rings $\mathbb{F}_p[i]$ and $\mathbb{F}_p[x]_{x^2+1}$ should be the same, and so they are.

Theorem 16.9. *For a prime number p, the rings $\mathbb{F}_p[x]_{x^2+1}$ and $\mathbb{F}_p[i]$ are the same.*

Exercise 16.9. Prove Theorem 16.9 by following the steps below:

1. Review the construction of the polynomial congruence rings in order to observe that the ring $\mathbb{F}_p[x]_{x^2+1}$ consists of elements of the form $c + d\gamma$, where c and d are in $\mathbb{F}_p$, the element γ satisfies the rule $\gamma^2 = -1$, and multiplication is given by the rule

$$(c + d\gamma)(e + f\gamma) = (ce - df) + (cf + de)\gamma.$$

2. Compare this to the defining description of the ring $\mathbb{F}_p[i]$ given above. Notice that the descriptions are the same, except that we use γ in one case and i in the other.
3. Conclude that $\mathbb{F}_p[x]_{x^2+1}$ and $\mathbb{F}_p[i]$ are essentially the same rings; that is, they are identical except for a change in notation.

Theorem 16.9 has the following consequence.

Theorem 16.10. *For a prime number p, the ring $\mathbb{F}_p[i]$ is a field if and only if the ring $\mathbb{F}_p[x]_{x^2+1}$ is a field.*

16.3 Fermat's Theorem

We wish to determine which prime numbers p remain irreducible in $\mathbb{Z}[i]$ and which prime numbers p factor nontrivially in $\mathbb{Z}[i]$. In Theorem 16.8, we saw that the primes p that are irreducible in $\mathbb{Z}[i]$ are those for which the ring $\mathbb{F}_p[i]$ is a field. By Theorem 16.10, the ring $\mathbb{F}_p[i]$ is a field if and only if the ring $\mathbb{F}_p[x]_{x^2+1}$ is a field, and by Theorem 14.11, the ring $\mathbb{F}_p[x]_{x^2+1}$ is a field if and only if $x^2 + 1$ is irreducible in $\mathbb{F}_p[x]$. This leads to the following theorem:

Theorem 16.11. *A prime number p is irreducible in $\mathbb{Z}[i]$ if and only if the polynomial $x^2 + 1$ is irreducible in $\mathbb{F}_p[x]$.*

Exercise 16.10. Prove Theorem 16.11.

Theorem 16.11 may come as somewhat of a surprise, since it relates the irreducibility of the prime number p in $\mathbb{Z}[i]$ to the irreducibility of the polynomial $x^2 + 1$ in $\mathbb{F}_p[x]$. However, its truth is evident once we realize that the two rings $\mathbb{F}_p[i]$ and $\mathbb{F}_p[x]_{x^2+1}$ are the same.

We can rephrase the condition that $x^2 + 1$ is irreducible in $\mathbb{F}_p[x]$. After all, $x^2 + 1$ factors nontrivially in $\mathbb{F}_p[x]$ if and only if $x^2 + 1$ has a root in $\mathbb{F}_p$, and $x^2 + 1$ has a root in $\mathbb{F}_p$ if and only if the equation $x^2 = -1$ is solvable in $\mathbb{F}_p$, that is, if -1 is a square in $\mathbb{F}_p$. Thus Theorem 16.11 can be restated as follows:

Theorem 16.12. *A prime number p is irreducible in $\mathbb{Z}[i]$ if and only if -1 is not a square in $\mathbb{F}_p$.*

We concluded Section 16.1 with Theorem 16.6, which states that the primes p that are irreducible in $\mathbb{Z}[i]$ are those for which the equation $x^2 + y^2 = p$ has no integer solutions, while the primes p that factor nontrivially in $\mathbb{Z}[i]$ are those for which $x^2 + y^2 = p$ has integer solutions. Combining this with Theorems 16.11 and 16.12, we obtain the following result:

Theorem 16.13. *For a prime number p, either*

1. *the equation $x^2 + y^2 = p$ has no integer solutions;*
2. *p is irreducible in $\mathbb{Z}[i]$;*
3. *$x^2 + 1$ is irreducible in $\mathbb{F}_p[x]$; and*
4. *-1 is not a square in $\mathbb{F}_p$;*

or

1. *the equation $x^2 + y^2 = p$ has integer solutions;*
2. *p factors nontrivially in $\mathbb{Z}[i]$;*
3. *$x^2 + 1$ factors nontrivially in $\mathbb{F}_p[x]$; and*
4. *-1 is a square in $\mathbb{F}_p$.*

Exercise 16.11. To prove Theorem 16.13, all one needs to do is combine various theorems that preceded it. Make sure that you understand this, and then write out a proof of Theorem 16.13 using the appropriate earlier results.

It remains to identify the primes p for which the first set of conditions of Theorem 16.13 holds and the primes p for which the second set of conditions holds. But we already know this. The simplest prime to handle is 2, for which the second set of conditions holds: $1^2 + 1^2 = 2$, the prime 2 factors in $\mathbb{Z}[i]$ as $(1+i)(1-i)$, the polynomial $x^2 + 1$ factors in $\mathbb{F}_2[x]$ as $(x+1)^2$, and -1 equals the square 1. The odd primes can be divided into two classes, those congruent to 1 modulo 4 and those congruent to 3 modulo 4. For these classes, we can use Theorem 13.6. Let us repeat its statement for convenience.

Theorem 16.14 (Theorem 13.6). *Let p be an odd prime number.*

1. *If $p \equiv 1 \pmod 4$, then -1 is a square in $\mathbb{F}_p$.*
2. *If $p \equiv 3 \pmod 4$, then -1 is not a square in $\mathbb{F}_p$.*

The first part of Theorem 13.6 was proved by an elementary argument based on the congruence known as Wilson's theorem. Combining this first part with Theorem 16.13, we obtain Theorem 16.15 below.

Theorem 16.15. *Suppose p is a prime number satisfying $p \equiv 1 \pmod 4$.*

1. *The equation $x^2 + y^2 = p$ has integer solutions, p factors nontrivially in $\mathbb{Z}[i]$, the polynomial $x^2 + 1$ factors nontrivially in $\mathbb{F}_p[x]$, and -1 is a square in $\mathbb{F}_p$.*

2. *There are eight solutions to the equation $x^2 + y^2 = p$. Each solution (a, b) corresponds to a pair of irreducible Gaussian integers $a + bi$ and $a - bi$ such that $p = (a + bi)(a - bi)$.*

Exercise 16.12. Prove Theorem 16.15.

The second part of Theorem 13.6 was proved by an appeal to the congruence theorem of Fermat, Theorem 7.7. Combining this with Theorem 16.13, we obtain Theorem 16.16 below.

Theorem 16.16. *Suppose p is a prime number satisfying $p \equiv 3 \pmod 4$. Then the equation $x^2 + y^2 = p$ has no integer solutions, p is irreducible in $\mathbb{Z}[i]$, the polynomial $x^2 + 1$ is irreducible in $\mathbb{F}_p[x]$, and -1 is not a square in $\mathbb{F}_p$.*

Exercise 16.13. Prove Theorem 16.16.

Theorems 16.15 and 16.16 combine to complete our description of the primes p that are irreducible in $\mathbb{Z}[i]$.

To handle the prime numbers p that are congruent to 3 modulo 4, we could have proceeded differently. After all, we proved in Exercise 6.11, by the most elementary of congruence considerations, that $x^2 + y^2 = p$ has no integer solutions if p is congruent to 3 modulo 4. Let us review both the theorem and the proof.

Theorem 16.17. *Suppose p is a prime number satisfying $p \equiv 3 \pmod 4$. Then p is not the sum of two squares; that is, the equation $x^2 + y^2 = p$ has no integer solutions.*

Proof. The square of an odd integer is odd, and the square of an even integer is even. Also, the sum of two odd integers is even, as is the sum of two even integers. Therefore, since p is odd, if $a^2 + b^2 = p$, then one of a or b must be odd, and the other must be even. Say a is even and b is odd. Then there exist integers m and n such that $a = 2m$ and $b = 2n + 1$, yielding $p = 4m^2 + 4n^2 + 4n + 1$. From this we see that $p - 1 = 4\left(m^2 + n^2 + n\right)$, so that 4 divides $p - 1$. In other words, the existence of a solution to the equation $x^2 + y^2 = p$ implies that $p \equiv 1 \pmod 4$. We have proved the contrapositive of the assertion of the theorem, and therefore we have proved the assertion of the theorem itself, namely, that if $p \equiv 3 \pmod 4$, then $x^2 + y^2 = p$ has no integer solution.

Combining Theorem 16.17 with Theorem 16.13, we obtain Theorem 16.16 without using the second part of Theorem 13.6. Indeed, we obtain in this way another proof of the second part of Theorem 13.6 itself.

Let us review the approach we have taken in proving Theorems 16.15 and 16.16. The question of whether an odd prime number p is irreducible in $\mathbb{Z}[i]$ was shown to be equivalent to three other questions, the unsolvability of

$x^2 + y^2 = p$ in $\mathbb{Z}$, the irreducibility of $x^2 + 1$ in $\mathbb{F}_p[x]$, and the nonexistence of a square root of -1 in $\mathbb{F}_p$. For a particular choice of p, if we decide any one of these four questions, we get an answer to the other three. For $p \equiv 3 \pmod 4$, we can easily attack the question of the solvability of $x^2 + y^2 = p$. Simple congruence considerations show that it is unsolvable, and this allows us to deduce that p is irreducible in $\mathbb{Z}[i]$, that $x^2 + 1$ is irreducible in $\mathbb{F}_p[x]$, and that -1 is not a square in $\mathbb{F}_p$. For $p \equiv 1 \pmod 4$, we can attack the question of -1 being a square in $\mathbb{F}_p$. By showing, again by congruence considerations, that -1 is a square, we are able to deduce that $x^2 + 1$ factors nontrivially in $\mathbb{F}_p[x]$, that p factors nontrivially in $\mathbb{Z}[i]$, and that the equation $x^2 + y^2 = p$ is solvable in $\mathbb{Z}$. Our ability to interrelate these conditions gives us the flexibility to choose different strategies for determining which conditions hold for a given prime p.

Theorem 16.15 contains within itself a powerful existence theorem. Let us state this theorem separately for emphasis.

Theorem 16.18. *Suppose p is a prime number satisfying $p \equiv 1 \pmod 4$. Then the equation $x^2 + y^2 = p$ has integer solutions. In fact, there are exactly eight pairs of solutions, of the form $(\pm a, \pm b)$ and $(\pm b, \pm a)$ for a pair of distinct integers a and b.*

Especially worthy of note is the role played by the unique factorization theorem for $\mathbb{Z}[i]$ in proving the second statement of Theorem 16.18, that there are exactly eight pairs of solutions.

The existence of solutions is half of a celebrated theorem of Fermat. We stated Fermat's theorem as Theorem 6.1 but were not in a position to prove it. Now we have succeeded. Recall the full statement:

Theorem 16.19. *An odd prime number p is the sum of the squares of two integers if and only if p is congruent to 1 modulo 4.*

The meaning of Fermat's theorem can be understood by a middle-school student if the congruence condition is replaced by the condition that the remainder when p is divided by 4 is 1. Yet despite the elementary nature of the theorem's statement, a proof understandable by a middle-school student is not easily given. The proof we have given depends on ideas and theorems from throughout this book, illustrating the interconnectedness of a wide range of ideas from number theory and algebra. This high point is a good place at which to conclude our treatment of integers, polynomials, and rings.

Just as a novelist does not usually bring the narrative to an end at the dramatic climax, but continues with a dénouement in which loose odds and ends are tidied up, we shall conclude our book with a final chapter in which we discuss some ideas that we were somehow unable to fit comfortably into the main narrative.

17

Finite Fields

17.1 Primitive Roots

In Section 6.4 we defined the rings $\mathbb{Z}_m$ and showed that when m is prime, the ring $\mathbb{Z}_m$ is a field. To emphasize that these are Finite Fields with a *prime* number of elements, we introduced the notation $\mathbb{F}_p$ for the finite field with p elements. Then in Section 13.3 we stated in Theorem 13.9 the remarkable fact that every field $\mathbb{F}_p$ has a primitive root, that is, an element a that *generates* the nonzero elements of $\mathbb{F}_p$ in the sense that $\{\, a, a^2, a^3, \ldots, a^{p-1} \,\}$ is a complete set of the $p-1$ nonzero elements of $\mathbb{F}_p$. Equivalently, a is an element of $\mathbb{F}_p$ of order $p-1$; that is, $a^{p-1} = 1$, and $a^j \neq 1$ for $j = 1, 2, \ldots, p-2$. We did not prove Theorem 13.9 at that time; we will do so now.

The idea of the proof of Theorem 13.9 is actually quite simple. Every element of $\mathbb{F}_p$ has *some* order, which has to be a number in the set $\{\, 1, 2, \ldots, p-1 \,\}$ (why?), and in fact, we shall see that the order has to be a divisor of $p-1$. We are going to count the elements of various orders and show that there are simply not enough elements of orders less than $p-1$ to account for all the $p-1$ elements of $\mathbb{F}_p$. We will have to conclude, then, that the field $\mathbb{F}_p$ has at least one element of order $p-1$, that is, a primitive root.

Exercise 17.1. Explain why the order of an element of $\mathbb{F}_p$ is one of the numbers $1, 2, \ldots, p-1$. (Hint: This exercise is an easy application of Fermat's theorem (Theorem 7.7). Alternatively, you can argue that if the order of an element of $\mathbb{F}_p$ is greater than $p-1$, then $\mathbb{F}_p$ will have more than p elements. The order of an element of $\mathbb{F}_p$ is a positive integer by definition. Your task is to explain why it must be less than p.)

Fix a prime number p and let 1 denote the multiplicative identity of $\mathbb{F}_p$. By Fermat's theorem, every nonzero a in $\mathbb{F}_p$ satisfies $a^{p-1} = 1$. The following proposition shows that the order of a is a divisor of $p-1$.

Proposition 17.1 *Let a be a nonzero element of $\mathbb{F}_p$ of order e; that is, $a^e = 1$ but $a^d \neq 1$ for any positive integer $d < e$. Then e divides $p-1$.*

Exercise 17.2. Prove Proposition 17.1. (Hint: Fermat's theorem yields that $a^{p-1} = 1$. Now use the division theorem to write $p-1$ as $eq + r$ with $0 \leq r < e$ and show that we must have $a^r = 1$. Deduce from the definition of order that r must be 0, so that e divides $p - 1$.)

We are now ready to start counting how many elements of each possible order there are in $\mathbb{F}_p$. Proposition 17.1 tells us that the possible orders are the divisors of $p - 1$. Let us write $N_p(e)$ for the number of elements of $\mathbb{F}_p$ of order e. We are going to obtain a bound on the number of elements of each order, and then show that there must be at least one element of order $p - 1$ to account for all $p - 1$ nonzero elements of $\mathbb{F}_p$.

Recall the definition of the Euler ϕ-function, which we gave in Section 7.3: For a positive integer m, $\phi(m)$ is equal to the number of integers in the set $\{1, 2, \ldots, m - 1\}$ that are relatively prime to m.

Proposition 17.2 *Let p be a prime number, let e be a divisor of $p - 1$, and let $N_p(e)$ be the number of elements of $\mathbb{F}_p$ of order e. Then*

$$N_p(e) \leq \phi(e),$$

where ϕ is the Euler ϕ-function.

Proposition 17.2 tells us that we can obtain a bound on the number of elements of order less than $p - 1$ by adding up the values of $\phi(e)$ as e runs over the positive divisors of $p - 1$ that are less than $p - 1$. The field $\mathbb{F}_p$ has $p - 1$ nonzero elements, and if we can show that there are fewer than $p - 1$ elements whose orders are less than $p - 1$, then any remaining elements must have order $p - 1$. These elements are the primitive roots.

Before we prove Proposition 17.2, let us see how it can help us by looking at a concrete example. Let us consider the field $\mathbb{F}_{11}$. Propsition 17.1 tells us that the order of every nonzero element of $\mathbb{F}_{11}$ is a divisor of 10. Thus the only possible orders of elements in $\mathbb{F}_{11}$ are 1, 2, 5, and 10. Proposition 17.2 tells us that in $\mathbb{F}_{11}$ there are at most $\phi(1)$ elements of order 1, at most $\phi(2)$ elements of order 2, and at most $\phi(5)$ elements of order 5. Thus there are at most $1 + 1 + 4 = 6$ elements of order less than 10. But there are 10 nonzero elements in $\mathbb{F}_{11}$ altogether. Therefore, there must be at least four elements of order 10 in $\mathbb{F}_{11}$. That is, $\mathbb{F}_{11}$ has a primitive root, in fact, at least four of them.

Exercise 17.3.   Calculate the order e of each nonzero element of $\mathbb{F}_{11}$. How many elements are there of each order? How do these values compare with the corresponding values of $\phi(e)$?

Now do the same for $\mathbb{F}_{13}$.

Exercise 17.4. Prove Proposition 17.2 as follows:

1. If there are no elements of order e, then $N_p(e) = 0$, and the inequality holds. In this case we are done. Therefore, we may assume that $N_p(e)$ is positive. Under this assumption, let b be an element of $\mathbb{F}_p$ of order e.

2. Observe that the elements $b, b^2, b^3, \ldots, b^e$ all satisfy the equation $x^e = 1$. In other words, they are all roots in $\mathbb{F}_p$ of the polynomial $x^e - 1$. Observe also that they are all distinct. Thus we have found e distinct roots in $\mathbb{F}_p$ of a degree-e polynomial with coefficients in $\mathbb{F}_p$. Using Theorem 9.10, deduce that these e elements are *all* the roots of $x^e - 1$ in $\mathbb{F}_p$. In particular, observe that every element of order e in $\mathbb{F}_p$ must be on this list of elements.

3. If m is an integer in the range from 1 to e, show that

$$(b^m)^{(e/(m,e))} = 1.$$

(Hint: Use the fact that m is divisible by (m, e).) In particular, observe that if $(m, e) \neq 1$, then b^m has order less than e.

4. Deduce that at most $\phi(e)$ of the numbers in the list $b, b^2, \ldots, b^e$ have order e, so that $N_p(e) \leq \phi(e)$.

As we have seen in our examples, the bound given in Proposition 17.2 can help in establishing the existence of primitive roots, and therefore, we make a slight detour to study the Euler ϕ-function.

Before stating our first result, let us consider an example. We can categorize each of the integers from 1 to 36 according to its greatest common divisor with 36. All the integers a relatively prime to 36, that is, with $(a, 36) = 1$, are placed in one pile; call it P_1. Thus $P_1 = \{1, 5, 7, 11, 13, 17, 19, 23, 25, 29, 31, 35\}$. All the integers m with $(36, m) = 2$ get placed in a second pile, P_2. Thus $P_2 = \{2, 10, 14, 22, 26, 34\}$. Similarly, we obtain

$$P_3 = \{3, 15, 21, 33\},$$
$$P_4 = \{4, 8, 16, 20, 28, 32\},$$
$$P_6 = \{6, 30\},$$
$$P_9 = \{9, 27\},$$
$$P_{12} = \{12, 24\},$$
$$P_{18} = \{18\},$$
$$P_{36} = \{36\}.$$

Notice that the list of greatest common divisors of various integers and 36 is exactly the list of divisors of 36, and that for each such divisor d there is at least one element in P_d, namely, d itself. How many integers are in P_d? If you count, you will find that for each divisor d of 36, there are $\phi\left(\frac{36}{d}\right)$ elements of P_d. This is reasonable, since an integer a is relatively prime to $\frac{36}{d}$ precisely if a has no nontrivial factors with $\frac{36}{d}$, which is the case if and only if d is the only factor that a has in common with 36. In particular, the pile P_1, which consists of the integers relatively prime to 36, contains $\phi(36)$ integers, as we would expect. Since every number from 1 to 36 appears in precisely one of the piles, we have the equality

$$\phi\left(\frac{36}{1}\right) + \phi\left(\frac{36}{2}\right) + \phi\left(\frac{36}{3}\right) + \phi\left(\frac{36}{4}\right) + \phi\left(\frac{36}{6}\right) + \phi\left(\frac{36}{9}\right) + \phi\left(\frac{36}{12}\right)$$
$$+ \phi\left(\frac{36}{18}\right) + \phi\left(\frac{36}{36}\right) = 36.$$

We can write this more compactly as

$$\sum_{d|36} \phi\left(\frac{36}{d}\right) = 36,$$

where the uppercase Greek sigma indicates summation, as usual, and the notation $d \mid 36$ means that the summation is to be taken over all divisors d of 36.

We can simplify this expression further by observing that d is a divisor of 36 if and only if $\frac{36}{d}$ is a divisor of 36, and therefore,

$$\sum_{d|36} \phi\left(\frac{36}{d}\right) = \sum_{d|36} \phi(d),$$

since all we are doing is adding up the same numbers in a different order. We end up with the following equality:

$$\sum_{d|36} \phi(d) = 36.$$

There was nothing special about the integer 36 in our discussion, and so we may replace it by an arbitrary positive integer n to obtain the following theorem:

Theorem 17.3. *Let n be a positive integer and let d be a divisor of n. The number of integers m in the range from 1 to n such that $(m,n) = d$ is $\phi\left(\frac{n}{d}\right)$. Hence,*

$$\sum_{d|n} \phi(d) = n.$$

Exercise 17.5. Prove Theorem 17.3. Proceed as follows.

1. First observe that for a divisor d of n, the integers m between 1 and n such that $(m,n) = d$ must be multiples of d.
2. Now show that if k is a positive integer and d is a divisor of n, then $(kd,n) = d$ if and only if $\left(k,\frac{n}{d}\right) = 1$. Conclude that the number of multiples of d in the range from 1 to n whose greatest common divisor with n is d is equal to the number of integers in the range from 1 to $\frac{n}{d}$ that are relatively prime to $\frac{n}{d}$. This yields the first statement of the theorem.
3. Conclude that
$$\sum_{d|n} \phi\left(\frac{n}{d}\right) = n.$$

Then deduce the second part.

We can now prove that $\mathbb{F}_p$ has a primitive root! We will actually prove more, namely, that $\mathbb{F}_p$ has exactly $\phi(p-1)$ primitive roots. Note that this really is more, since $\phi(n) \geq 1$ for all positive integers n (why?).

Theorem 17.4. *For every prime number p, the field $\mathbb{F}_p$ contains a primitive root. In fact, $\mathbb{F}_p$ contains exactly $\phi(p-1)$ primitive roots.*

Exercise 17.6. Prove Theorem 17.4 as follows:

1. By Proposition 17.1 and Fermat's theorem, every nonzero element has some order e dividing $p-1$. Deduce that

$$\sum_{e|(p-1)} N_p(e) = p - 1.$$

2. Recall from Theorem 17.3 that

$$\sum_{e|(p-1)} \phi(e) = p - 1.$$

3. By Proposition 17.2 you know that $N_p(e) \leq \phi(e)$ for each divisor e of $p-1$. Deduce from this and the two summation formulas above that the equalities $N_p(e) = \phi(e)$ must hold.
4. Choose e to be $p-1$ and obtain the formula $N_p(p-1) = \phi(p-1)$.
5. Conclude that there are $\phi(p-1)$ primitive roots in $\mathbb{F}_p$, and in particular, that primitive roots exist in $\mathbb{F}_p$.

We conclude this section with a curious observation. The equality $N_p(e) = \phi(e)$ exhibits an asymmetry in that the left-hand side depends on the choice of prime number p, while the right-hand side does not: $N_p(e)$ is the number of elements of order e in the field $\mathbb{F}_p$, while $\phi(e)$, the number of positive integers less than e and relatively prime to e, has nothing to do with the choice of p at all. Therefore, if p_1 and p_2 are prime numbers and e is divisor of both $p_1 - 1$ and $p_2 - 1$, then the fields $\mathbb{F}_{p_1}$ and $\mathbb{F}_{p_2}$ have the same number of elements of order e. For example, consider the fields $\mathbb{F}_{37}$ and $\mathbb{F}_{31}$. We can choose $e = 6$, a divisor of both 36 and 30, and observe that each of $\mathbb{F}_{37}$ and $\mathbb{F}_{31}$ has $\phi(6) = 2$ elements of order 6.

17.2 Quadratic Reciprocity

We saw in Section 13.3 that half the nonzero elements of $\mathbb{F}_q$, for an odd prime number q, are squares, and half are not. We discussed how to determine whether -1 is a square, but we did not pursue the issue further. We did, however, point out that knowing that $\mathbb{F}_q$ contains a primitive root would lighten the burden of proof, and now that we have established that each of the fields $\mathbb{F}_q$ indeed contains a primitive root, we can offer new proofs for old theorems: Let a be a primitive root of $\mathbb{F}_q$. Then the elements of $\mathbb{F}_q$ can be listed as $a, a^2, a^3, \ldots, a^{q-1}$. Observe that a cannot be a square in $\mathbb{F}_q$ (why?), and therefore, the squares in $\mathbb{F}_q$ are precisely the $\frac{q-1}{2}$ even powers of a, namely, $a^2, a^4, \ldots, a^{q-1}$. Furthermore, since $a^{(q-1)/2}$ is the element -1 in $\mathbb{F}_q$ (why?), it follows that -1 is a square if and only if $\frac{q-1}{2}$ is even (why?), that is, if and only if q is congruent to 1 modulo 4.

Therefore, in theory, we can determine which elements are squares in $\mathbb{F}_q$ by finding a primitive root and writing out the even powers of that root. However, there is nothing in our proof of the *existence* of a primitive root that tells us how to find that primitive root, and if we wanted to know, say, whether [7] is a square in $\mathbb{F}_{101}$, we would be reduced to trial and (perhaps plenty of) error in finding a primitive root in $\mathbb{F}_{101}$ and then calculating that root's even powers to see whether [7] is in the list. (Some calculation shows that [3] is a primitive root in $\mathbb{F}_{101}$ and that $[7] = [3]^{61}$. Since 61 is odd, [7] is not a square. In fact, the following exercise implies that [7] is itself a primitive root of $\mathbb{F}_{101}$.)

Exercise 17.7. Show that if a is a primitive root of $\mathbb{F}_q$, then a^x is also a primitive root of $\mathbb{F}_q$ if and only if $(x, q-1) = 1$. Observe the corollary that $\mathbb{F}_q$ has $\phi(q-1)$ primitive roots.

The great Gauss considered the problem of squares in $\mathbb{F}_q$ and discovered a marvelous relationship among squares in various fields $\mathbb{F}_q$ for different primes q. This relationship is called the *law of quadratic reciprocity*. A bit of Latin will help to explain this terminology: *reciprocus* means "alternating," and *quadrum* means "square." The law of quadratic reciprocity provides an *alternative* expression for a *square* in a field $\mathbb{F}_q$ in terms of a square in a different field $\mathbb{F}_{q'}$. The idea of alternation is stronger even than a mere alternative would suggest: The law of quadratic reciprocity tells us whether a prime p is a square in $\mathbb{F}_q$ in terms whether q is a square in $\mathbb{F}_p$. For example, the question we asked above, whether [7] is a square in $\mathbb{F}_{101}$, can be reduced to the question whether [101] is a square in $\mathbb{F}_7$. But $[101] = [3]$ in $\mathbb{F}_7$, and listing the squares in $\mathbb{F}_7$ is easier than doing the same in $\mathbb{F}_{101}$. (A bit of calculation shows that [3] is a primitive root in $\mathbb{F}_7$ and is therefore not a square.)

Let us switch our point of view for the moment, considering not the question of squares in the field $\mathbb{F}_q$ but instead the equivalent congruence question in $\mathbb{Z}$: An element a is the square of an element b in $\mathbb{F}_q$ precisely when $a \equiv b^2 \pmod{q}$. Our new problem is therefore to discover which integers n are congruent to squares modulo a particular prime q and which are not. In other words, for each integer n, is there an integer m such that $n \equiv m^2 \pmod{q}$, or does no such m exist? Whatever the answer is for n, we will obtain the same answer for every integer congruent to n modulo q. Thus it suffices to ask the question for the integers n ranging from 0 to $q-1$, though we may at times wish to work with other integers.

The integers that are congruent to squares modulo q are called *quadratic residues modulo q*, and the integers that are not congruent to squares modulo

q are called *quadratic nonresidues modulo* q. For instance, every integer is a quadratic residue modulo 2. We have proved that for an odd prime q, half the integers from 1 to $q-1$ are quadratic residues modulo q and half are not. This is simply a restatement of the theorem that half the nonzero elements in the field $\mathbb{F}_q$ are squares and half are not. We have also proved that -1 is a quadratic residue modulo an odd prime q if and only if q is congruent to 1 modulo 4.

Exercise 17.8. Fix a prime number q.

1. Suppose that two integers a and b are both quadratic residues modulo q or that neither is a quadratic residue modulo q. Prove that the product ab is a quadratic residue modulo q. (Hint: Translate this into a statement in $\mathbb{F}_q$ and observe that you have proved it already.)
2. In contrast, show that if one of a and b is a quadratic residue modulo q and the other is not, then the product ab is not a quadratic residue modulo q.
3. Let n be a positive integer with prime factorization $p_1 \cdots p_t$, where we do not insist that all the prime factors p_i be distinct. Show that n is a quadratic residue modulo q if and only if an even number of the prime factors p_i are quadratic nonresidues modulo q.
4. Continue with the same n, so that $-n$ factors as $(-1)p_1 \cdots p_t$. If $q \equiv 1 \pmod 4$, what condition on the p_i's determines whether $-n$ is a quadratic residue modulo q? What if $q \equiv 3 \pmod 4$?

The exercise above shows that in trying to determine whether an *integer* n is a quadratic residue modulo a prime q, we can instead solve the problem of deciding whether the *prime number* p_i is a quadratic residue modulo q for each of the prime factors p_i of n.

In mathematics, good notation is a powerful tool (just imagine trying to do long division with Roman numerals!), and a handy notation has been devised for working with quadratic residues: Let n be an integer and q a prime number. The *Legendre symbol*

$$\left(\frac{n}{q} \right)$$

is set equal to 1 if n is a quadratic residue modulo q, and to -1 if n is a quadratic nonresidue modulo q. The symbol is named for the French mathematician Adrien Marie Legendre (1752–1833). The beauty of this notation is that if n factors as ab, then the first two parts of the previous exercise can be interpreted as saying that

$$\left(\frac{n}{q} \right) = \left(\frac{a}{q} \right) \left(\frac{b}{q} \right),$$

and the second two parts say that if $n = p_1 \cdots p_t$, then

$$\left(\frac{n}{q}\right) = \left(\frac{p_1}{q}\right) \cdots \left(\frac{p_t}{q}\right)$$

and

$$\left(\frac{-n}{q}\right) = \left(\frac{-1}{q}\right)\left(\frac{p_1}{q}\right) \cdots \left(\frac{p_t}{q}\right).$$

These formulas reduce the problem of determining whether an *integer* is a quadratic residue modulo an odd prime number q to the problem of determining whether certain *prime numbers* are quadratic residues modulo q.

As we remarked near the beginning of this section, we are going to state a "reciprocity" law that relates quadratic residues for primes p and q, that is, a law that relates $\left(\frac{p}{q}\right)$ to $\left(\frac{q}{p}\right)$. The prime $p = 2$, as the sole even prime number, is a special case, and therefore, before proceeding to the general case of p an odd prime, we are going to state a theorem that tells us for which odd primes q the prime 2 is a quadratic residue modulo q and for which odd primes q the prime 2 is a quadratic nonresidue. Our result will be analogous to that of Theorem 13.6, which provides the same service for -1. Before stating the theorem, let us look at some data, as we did for -1, and try to guess what the theorem should say. As before, let us examine all the odd primes q up to $q = 101$, listing those for which 2 is a quadratic residue and those for which 2 is a quadratic nonresidue. If you do this, you will find that 2 is a quadratic residue modulo the primes

$$7, \ 17, \ 23, \ 31, \ 41, \ 47, \ 71, \ 73, \ 79, \ 89, \ 97,$$

and that 2 is a quadratic nonresidue modulo the primes

$$3, \ 5, \ 11, \ 13, \ 19, \ 29, \ 37, \ 43, \ 53, \ 59, \ 61, \ 67, \ 83, \ 101.$$

For instance, to see that 2 is a quadratic residue modulo 23, notice that 2 is congruent modulo 23 to 25, which is 5^2. Or to see that 2 is a quadratic residue modulo 79, notice that 2 is congruent modulo 79 to 81, which is 9^2. It takes more work to verify that 2 is a quadratic nonresidue modulo a given prime q, for in that case we must square each of the numbers from 0 to $q - 1$ and check that none of the resulting squares is congruent to 2.

You should check, at least for $q = 3, 5, 11$, and 13, that 2 is a quadratic nonresidue modulo q.

Is there a pattern to our lists? In contrast to the question whether -1 is a quadratic residue, it is not the case that the congruence class of q modulo 4 makes a difference. We see that 2 is a residue for primes congruent to 1 modulo 4 as well as for primes congruent to 3 modulo 4. And the same situation obtains for 2 a nonresidue. Nonetheless, there is a pattern. What matters is the congruence class of q modulo 8. Every integer is congruent modulo 8 to exactly one of $0, 1, 2, 3, 4, 5, 6, 7$. An odd integer is congruent modulo 8 to exactly one of $1, 3, 5, 7$. With this in mind, look back at the two lists. Every prime number on the first list is congruent to 1 or 7 modulo 8; every prime number on the second list is congruent to 3 or 5 modulo 8. This suggests the following theorem, which we state without proof.

Theorem 17.5. *Let q be an odd prime number. The integer 2 is a quadratic residue modulo q if q is congruent to 1 or 7 modulo 8, and is a quadratic nonresidue modulo q if q is congruent to 3 or 5 modulo 8.*

Theorem 17.5 is a wonderful result, providing a simply stated criterion for whether 2 is a quadratic residue modulo a prime. For instance, is 2 a quadratic residue modulo the prime number 331 577? In other words, is there an integer m with the property that $m^2 - 2$ is divisible by 331 577? Theorem 17.5 tells us the answer. (What is it?)

Now that we have taken care of the special case of the prime 2, we are ready to face the question of which odd primes p are quadratic residues modulo an odd prime q, and which are nonresidues. Here, the story gets more interesting. Since we are dealing with all odd primes together, we are not going to obtain, as we did for -1 and for 2, a simple rule. What we obtain instead is a beautiful and famous theorem, the *law of quadratic reciprocity*, that stands the question on its head. Rather than giving us a straight answer, the law relates the question whether p is a quadratic residue modulo q to the question whether q is a quadratic residue modulo p. There is no reason to think a priori that the answers to these two questions should be related.

Since the two questions are symmetric, the only difference being the order of p and q, it might seem that a relationship between $\left(\frac{p}{q}\right)$ and $\left(\frac{q}{p}\right)$ would simply lead us in circles, but in fact, it does not. The reason is that if p is smaller than q, then the question whether q is a quadratic residue modulo p is easier than that whether p is a quadratic residue modulo q. For instance, if we want to know whether 11 is a quadratic residue modulo 83, the law of quadratic reciprocity will tell us that we can answer this question if we can answer the question whether 83 is a quadratic residue modulo 11. But this is a much easier question, for we can replace 83 by 6, since 83 is congruent to 6 modulo 11, and so 83 is a quadratic residue modulo 11 if and only if 6 is. To determine whether 6 is a quadratic residue modulo 11, we must answer whether 2 and 3 are quadratic residues modulo 11. We can check directly that 2 is not, or use Theorem 17.5. We can check directly that 3 is a quadratic residue modulo 11, or we can use the law of quadratic reciprocity again to flip the problem over and ask whether 11 is a quadratic residue modulo 3, which is equivalent to asking whether 2 is a quadratic residue modulo 3 (is it?).

In general, as in this example, if p and q are odd primes with $p < q$, and we are faced with the problem of determining whether p is a quadratic residue modulo q, the law of quadratic reciprocity allows us to switch to the question whether q is a quadratic residue modulo p, allowing us to work with a smaller modulus. We can reduce q modulo p to get a number smaller than p, and then we can factor this as a product of primes, and we can now work with primes r that are smaller than p. In this way, we can keep reducing the size of the problem until we are left with such small numbers that we can decide our question easily.

We have yet to describe exactly how the law of quadratic reciprocity relates the question whether p is a quadratic residue modulo q to the question whether q is a quadratic residue modulo p. It turns out that the relation is simple, but subtle. It is instructive to try to guess what this relationship must be by calculating $\left(\frac{p}{q}\right)$ for all odd primes p and q less than some number, say 101, and then studying the results, comparing $\left(\frac{p}{q}\right)$ to $\left(\frac{q}{p}\right)$. Euler and Legendre must have made such a study, in order to be able to guess the relationship that has come to be known as the law of quadratic reciprocity. They were unable to prove it, however. As noted earlier, Gauss found a proof when he was 19 years old and published it in 1801. Gauss was so fascinated by this result that he found several more proofs, and dozens more have been given since. Many of the most profound developments in number theory in the two hundred years since Gauss's first proof have arisen from research aimed at generalizing this law. Here is its statement.

Theorem 17.6. *Let p and q be two distinct odd prime numbers.*

1. *Suppose one or both of p and q is congruent to 1 modulo 4. Then p is a quadratic residue modulo q if and only if q is a quadratic residue modulo p:*

$$\left(\frac{q}{p}\right) = \left(\frac{p}{q}\right).$$

2. *Suppose instead that p and q are both congruent to 3 modulo 4. If p is a quadratic residue modulo q, then q is a quadratic nonresidue modulo p; if p is a quadratic nonresidue modulo q, then q is a quadratic residue modulo p:*

$$\left(\frac{q}{p}\right) = -\left(\frac{p}{q}\right).$$

Let us use Theorem 17.6 to determine whether 11 is a quadratic residue modulo 83. Both 11 and 83 are congruent to 3 modulo 4. Thus the answer will be the opposite of the answer to the question whether 83 is a quadratic residue modulo 11. This question, in turn, is equivalent, as we noted above, to the question whether 6 is a quadratic residue modulo 11, and we can answer that question by factoring $6 = 2 \times 3$ and determining whether 2 and 3 are quadratic residues modulo 11. We know by Theorem 17.5 that 2 is not a quadratic residue modulo 11. Therefore, 6 is a quadratic residue modulo 11 if 3 is not, and 6 is not a quadratic residue modulo 11 if 3 is. Using quadratic reciprocity again, the answer to the question whether 3 is a quadratic residue modulo 11 is the opposite of the answer to the question whether 11 is a quadratic residue modulo 3, or, equivalently, whether 2 is a quadratic residue modulo 3. Since 2 is a quadratic nonresidue modulo 3, we find that 3 is a quadratic residue modulo 11, so that 6 is not a quadratic residue modulo 11,

and 11 is a quadratic residue modulo 83. This calculation is much simplified
if we use the Legendre symbols:

$$\left(\frac{11}{83}\right) = (-1)\left(\frac{83}{11}\right) = (-1)\left(\frac{6}{11}\right)$$

$$= (-1)\left(\frac{2}{11}\right)\left(\frac{3}{11}\right) = (-1)(-1)\left(\frac{3}{11}\right)$$

$$= (-1)(-1)(-1)\left(\frac{11}{3}\right) = (-1)(-1)(-1)\left(\frac{2}{3}\right)$$

$$= (-1)(-1)(-1)(-1) = 1.$$

Exercise 17.9. Answer the following questions, using the law of quadratic
reciprocity.

1. Is 12 a quadratic residue modulo 29?
2. Is -17 a quadratic residue modulo 37?
3. Is 56 a quadratic residue modulo 37?
4. Is 14 a quadratic residue modulo 137?
5. Is 55 a quadratic residue modulo 179?

Exercise 17.10. One reason we want to know whether an integer is a
quadratic residue modulo an odd prime q is so that we can determine which
elements of $\mathbb{F}_q$ are squares. This will allow us to decide which quadratic equations over $\mathbb{F}_q$ are solvable, or which quadratic polynomials in $\mathbb{F}_q[x]$ are irreducible. We need merely determine whether the discriminant of the polyno-
mial has a square root in the field. Answer the following questions:

1. Is the polynomial $x^2 - [7]x + [1]$ irreducible in $\mathbb{F}_{47}[x]$?
2. Is the polynomial $x^2 + [8]x + [4]$ irreducible in $\mathbb{F}_{37}[x]$?
3. Is the polynomial $x^2 + x + [1]$ irreducible in $\mathbb{F}_{71}[x]$?

17.3 Classification

In Section 14.5 we learned how to construct a field extension K of a "base"
field F by choosing a monic irreducible polynomial $m(x)$ in $F[x]$, letting γ
be root of $m(x)$, and adjoining γ to the field F by implementing the rewrite
rule $m(\gamma) = 0$. When the base field is $\mathbb{Q}$, the field of rational numbers, the
field extensions K obtained in this way can be identified with subfields of the
complex numbers $\mathbb{C}$, and the elements of K are called algebraic numbers. If
we choose as our base field not $\mathbb{Q}$ but a finite field $\mathbb{F}_p$, the construction of
extension fields K is also of interest, because we obtain new finite fields.

Suppose that we start with the finite field $\mathbb{F}_p$, which has p elements, and
select a monic irreducible polynomial $m(x)$ in $\mathbb{F}_p[x]$ of degree n. Our construction of the new field K containing $\mathbb{F}_p$ and containing a root of $m(x)$ involves
creating a new element γ that is to be a root of $m(x)$ and taking K to be the

set of all polynomial expressions in γ of degree less than n with coefficients in $\mathbb{F}_p$, with the arithmetic operations in K subject to the rewrite rule $m(\gamma) = 0$. Thus a typical element of K looks like

$$a_0 + a_1\gamma + a_2\gamma^2 + \cdots + a_{n-1}\gamma^{n-1},$$

where the coefficients a_i are in $\mathbb{F}_p$. It is easy to determine the number of elements in K. There are p possibilities for each coefficient a_i, and n different coefficients. Thus there are

$$\overbrace{p \times p \times \cdots \times p}^{n \text{ times}} = p^n$$

different elements in K. We have built a field of size p^n.

Two natural directions of study now present themselves: applications of finite fields, and the general development of the theory of finite fields. Among applications are the construction of Latin squares in statistical design theory, the construction of finite projective planes in geometry, and the construction of error-correcting codes in coding theory. We will not, alas, pursue these applications, but you can learn about them in many books. In each of the applications mentioned, one requires a finite field with exactly t elements, for some positive integer t. The primary result on existence of finite fields is the following.

Theorem 17.7. *Let p be a prime number and let n be a positive integer. Then there exists a monic irreducible polynomial $m(x)$ in $\mathbb{F}_p[x]$ of degree n. Hence there is a field $\mathbb{F}_p[x]_{m(x)}$ of size p^n.*

Theorem 17.7 ensures the existence of finite fields of any prime-power size. What about other sizes? Surprisingly, no other sizes are possible. Or perhaps it is not so surprising, since our experience with congruence rings with a composite (nonprime) number of elements showed the possibility of zero-divisors lurking about. This is precisely the problem with finite fields: If the number of elements of a finite ring is not a prime power, then the ring has zero-divisors.

It turns out to be fairly easy to prove that every finite field has prime-power size, provided that one has sufficient background in linear algebra. If K is a finite field, one first shows that there is a prime number p such that inside K is a copy of $\mathbb{F}_p$. To do so, consider in K the sums of the additive identity 1. In other words, consider $1, 1+1, 1+1+1$, etc. Let us write $[t]$ for the element in K that one gets by adding 1 to itself t times (so $1+1+1 = [3]$). Thus, $[t]$ is not itself an integer, but an element of our field K. Since K is finite, the elements $[1], [2], [3], \ldots$ cannot all be different elements of K. There must, then, be two distinct positive integers r and s such that $[r] = [s]$. It follows from this that there is a positive integer m such that the element $[m]$ in K is zero ($0 = [r] - [s] = [r - s]$). There must be a smallest such positive integer, call it m_0, with the property that $[m_0]$ in K is zero. We now argue that if

m_0 is not prime, but factors nontrivially as ab, then $[a][b] = [ab] = [m_0] = [0]$ in K, so that $[a]$ and $[b]$ are zero-divisors. Since K is a field, this cannot happen, and we conclude that m_0 is a prime number p. The resulting elements $[0], [1], \ldots, [p-1]$ in K form a copy of the field $\mathbb{F}_p$.

Here is where the linear algebra comes in. First we observe that linear algebra can be developed over any field, not just $\mathbb{R}$, so that we can apply the theorems of linear algebra to vector spaces over the field $\mathbb{F}_p$. Then we observe that the field K can be thought of as a vector space over $\mathbb{F}_p$. As a vector space, K has a basis, of some size n. Therefore, the elements of K can be written as linear combinations of the n basis elements, with the p elements of $\mathbb{F}_p$ as coefficients. It follows immediately that K has size p^n. The result of these considerations is the following theorem.

Theorem 17.8. *If K is a finite field, then there is a prime number p such that K contains a copy of the field $\mathbb{F}_p$. Moreover, there is a positive integer n such that K has size p^n. Thus the only possible sizes of finite fields are powers of prime numbers.*

An alternative proof of Theorem 17.8 can be given by employing a generalization of the theorem that $\mathbb{F}_p$ has a primitive root, namely, that every finite field K contains a primitive root. In other words, if K has r elements, then there is an element a in K such that $a, a^2, \ldots, a^{r-1}$ constitutes a complete collection of the nonzero elements of K. A proof of this for the fields $\mathbb{F}_p$ was laid out in Section 17.1. The more general proof requires some ideas that we have not discussed.

Our last major result on the nature of finite fields is the following:

Theorem 17.9. *Suppose K is a finite field of size p^n, for some prime number p and some positive integer n.*

1. *There is a monic irreducible polynomial $m(x)$ of degree n in $\mathbb{F}_p[x]$ such that K contains a root γ of $m(x)$ and such that every element of K is a polynomial expression of degree less than n in γ with coefficients in $\mathbb{F}_p$. The field K can be identified with $\mathbb{F}_p[x]_{m(x)}$.*
2. *All fields of size p^n are the same, in the sense that they all can be identified with $\mathbb{F}_p[x]_{m(x)}$.*

Theorem 17.9 shows that our construction in Theorem 17.7 of new finite fields of the form $\mathbb{F}_p[x]_{m(x)}$ yields all possible finite fields. Every finite field has size p^n, for some prime number p and some positive integer n, and for each such prime power p^n there is essentially only one finite field of that size.

Index

Undergraduate Texts in Mathematics

(continued from page ii)

Franklin: Methods of Mathematical Economics.

Frazier: An Introduction to Wavelets Through Linear Algebra

Gamelin: Complex Analysis.

Gordon: Discrete Probability.

Hairer/Wanner: Analysis by Its History. *Readings in Mathematics.*

Halmos: Finite-Dimensional Vector Spaces. Second edition.

Halmos: Naive Set Theory.

Hämmerlin/Hoffmann: Numerical Mathematics. *Readings in Mathematics.*

Harris/Hirst/Mossinghoff: Combinatorics and Graph Theory.

Hartshorne: Geometry: Euclid and Beyond.

Hijab: Introduction to Calculus and Classical Analysis.

Hilton/Holton/Pedersen: Mathematical Reflections: In a Room with Many Mirrors.

Hilton/Holton/Pedersen: Mathematical Vistas: From a Room with Many Windows.

Iooss/Joseph: Elementary Stability and Bifurcation Theory. Second edition.

Irving: Integers, Polynomials, and Rings: A Course in Algebra

Isaac: The Pleasures of Probability. *Readings in Mathematics.*

James: Topological and Uniform Spaces.

Jänich: Linear Algebra.

Jänich: Topology.

Jänich: Vector Analysis.

Kemeny/Snell: Finite Markov Chains.

Kinsey: Topology of Surfaces.

Klambauer: Aspects of Calculus.

Lang: A First Course in Calculus. Fifth edition.

Lang: Calculus of Several Variables. Third edition.

Lang: Introduction to Linear Algebra. Second edition.

Lang: Linear Algebra. Third edition.

Lang: Short Calculus: The Original Edition of "A First Course in Calculus."

Lang: Undergraduate Algebra. Second edition.

Lang: Undergraduate Analysis.

Laubenbacher/Pengelley: Mathematical Expeditions.

Lax/Burstein/Lax: Calculus with Applications and Computing. Volume 1.

LeCuyer: College Mathematics with APL.

Lidl/Pilz: Applied Abstract Algebra. Second edition.

Logan: Applied Partial Differential Equations.

Lovász/Pelikán/Vesztergombi: Discrete Mathematics.

Macki-Strauss: Introduction to Optimal Control Theory.

Malitz: Introduction to Mathematical Logic.

Marsden/Weinstein: Calculus I, II, III. Second edition.

Martin: Counting: The Art of Enumerative Combinatorics.

Martin: The Foundations of Geometry and the Non-Euclidean Plane.

Martin: Geometric Constructions.

Martin: Transformation Geometry: An Introduction to Symmetry.

Millman/Parker: Geometry: A Metric Approach with Models. Second edition.

Moschovakis: Notes on Set Theory.

Owen: A First Course in the Mathematical Foundations of Thermodynamics.

Palka: An Introduction to Complex Function Theory.

Pedrick: A First Course in Analysis.

Peressini/Sullivan/Uhl: The Mathematics of Nonlinear Programming.

Undergraduate Texts in Mathematics